Visual Mechanic Knowledge

The Workshop Drawings of Isaac Ebenezer Markham (1795–1825), New England Textile Mechanic

Portrait of Isaac Ebenezer Markham, artist and date unnknown ($24^{1}/_{2} \times 20^{1}/_{2}$ inches). Collection of Sheldon Museum, Middlebury, Vermont. Gift of Thomas A. McLeod, 1889.

Visual Mechanic Knowledge

The Workshop Drawings of Isaac Ebenezer Markham (1795–1825), New England Textile Mechanic

David J. Jeremy and Polly C. Darnell

American Philosophical Society
Philadelphia • 2010

Memoirs of the
American Philosophical Society
Held at Philadelphia
For Promoting Useful Knowledge
Volume 263

ISBN: 978-0-87169-263-4
US ISSN: 0065-9738

Library of Congress Cataloging-in-Publication Data

Jeremy, David J.
Visual mechanic knowledge : the workshop drawings of Isaac Ebenezer Markham (1795-1825), New England textile mechanic / David J. Jeremy and Polly C. Darnell.
p. cm.
Includes bibliographical references and index.
ISBN 978-0-87169-263-4
1. Cotton machinery—Design and construction—History—19th century. 2. Textile machinery—Design and construction—History—19th century. 3. Markham, Isaac, 1795-1825. 4. Drafters—New England—Biography. I. Darnell, Polly C. II. Title.
TS1583.J47 2010
677'.0285097309034—dc22

2010046580

To Jean and Willie,
our spouses

Contents

☆ ☆ ☆ ☆ ☆ ☆ ☆

Acknowledgments

PUBLICATION OF THESE DRAWINGS HAS BEEN A LONG TIME COMING. In 1981, Bruce Moseley, a former member of the Sheldon Museum staff, identified Markham's sketch of the Boston Manufacturing Co. mills at Waltham, MA, and word was sent out on the industrial archaeology grapevine that related textile machine drawings may also exist. When the late Michael Brewster Folsom first identified the drawings, he circulated them to a number of scholars (Richard Candee, Larry Gross, Steven Lubar, Pat Malone, Gail Fowler Mohanty, Ted Penn, Merritt Roe Smith, Robert Vogel, and David Jeremy). These individuals provided initial assessments of the drawings' significance. So too did Helena Wright, Curator of Graphic Arts at the Smithsonian Institution, who has consistently and generously provided advice during each phase of the project.

Over the years the staff of the Sheldon Museum Research Center, which holds the Markham papers, along with an outstanding collection of manuscripts and printed material documenting the local area, has also been remarkably generous. Scott Morrison made note of every textile reference he found as he identified and organized the manuscript collection and shared his extensive knowledge of Middlebury during Markham's lifetime; Wiz Dow compiled all those references into a database; Jane Ploughman did the initial measurements and descriptions of the drawings; and Orson Kingsley, while still new to the institution, responded immediately to all requests, as did Collections Manager Mary Towle-Hilt.

Middlebury College's Special Collections & Archives, headed by Andrew Wentink, scanned the Markham drawings. Danielle Rougeau, Assistant Curator, and Amanda Gustin were unfailingly helpful. During the last stage of the project we also appreciated the encouragement of current Sheldon Museum director, Dr. Jan Albers, and the map-making abilities of Middlebury College student Ted Kwasnik.

Martha Hill, of Orwell, Vermont, and Charlottesville, Virginia, and the late Deborah Clifford improved the biography of Markham through their comments on the manuscript. Martha also cheerfully provided practical assistance with the final dexcriptions of the drawings. Willie Jansen provided considerable personal support and encouragement throughout.

The work in manuscript form has been read by three referees to whom the authors are indebted. One anonymous referee called for an expansion of the work into "a study of technological knowledge, its acquisition and transmission." This prompted the primary author to research and write the introductory section on visual knowledge. The other referees, Dr. Carolyn C. Cooper of Yale University and Professor Laurence Gross of the University of Massachusetts, Lowell, made suggestions that helped to reshape the overall structure of the work. In addition, parts of the book have been read by Helena Wright, Dr. Ben Thomas of the University of Kent, and by Professor Geoffrey Tweedale of Manchester Metropolitan University Business School, and Dr. Francis Goodall. Robert M. Vogel, formerly of the Smithsonian Institution, gave encouragement in the early stages of the project. When the manuscript was in the final draft, Merritt Roe Smith, Everett and William Cutten Professor of the History of Technology at the Massachusetts Institute of Technology, kindly went through Part I, making very useful observations on structure and sources. The positive comments of all these readers are much appreciated.

The primary author could not have developed the knowledge context to Markham's drawings without assistance from many directions. The staff and library resources of the Manchester Central Reference Library, the John Rylands University Library, and the Manchester Metropolitan University Library enabled him to read most of the scarce works cited, many of them in electronic form. The British Library in London, the Brotherton Library at the University of Leeds, Chetham's Library in Manchester, the Sydney Jones Library at the University of Liverpool, and Cambridge University Library provided the rest.

Professor Robert D. Arner of the University of Cincinnati, in generously providing details of the publishing history of Rees's *Cyclopaedia*, has helped to establish with hitherto unrealized precision the remarkable speed of the transatlantic transfer of published information about Britain's new textile technology. Tony Woolrich advised about some of the mechanic literature of the early industrial period. Dr. Ben Thomas gave valuable pointers to the literature on the history of engravings. Lucy Salt and Matt Edwards of the City of Derby Museum and Art Gallery were helpful in providing information and images of the Joseph Wright portrait of Sir Richard Arkwright. Hugh Alexander of the Image Library at the National Archives in London has facilitated the filming and reproduction of the British patent drawings presented here for the first time in their original form; whereas Jessamy Sykes, in the Advice and Records Knowledge section of the National Archives, located a miscatalogued patent drawing. Marie E. Lamoureux of the American Antiquarian Society and Jean McDonald of Paisley Central Library supplied professional help. Mark H. Jones, Connecticut State Archivist, measured the Scholfield billy drawings. Clare Sheridan, Librarian of the American Textile History Museum at Lowell, Massachusetts, aided in securing images of early textile machinery and in unfailingly answering queries from the unique resources of the ATHM.

On the drawings themselves, Dr. R. Hugh Gong, Reader in Textile Technology, School of Materials, at the University of Manchester clarified roving-frame mechanisms. Ian Birtles, loom tackler of Quarrybank Mill Museum at Styal, Cheshire, explained the finer technicalities of power-loom weaving. Dr. Richard Potton, Physics Discipline, School of Computing, Science and Engineering, University of Salford, analyzed Markham's batten motion. Three practicing engineers, with mechanical training and experience, gave technical and computing assistance: Peter Farrant, MBCS CITP; Michael Middleton, CEng; and Peter Nicoll, CEng. Richard Rigby, BSc, also supplied useful comments. Professor Mark Mon-Williams of Leeds University offered advice on cognitive psychology.

The primary author, in retirement, is very grateful for much personal support. Breaks with his daughters, Joanna and Rebecca, and their families, gave time (usually when walking over hills in the English Lakes or the Yorkshire Dales) to reflect on Markham from a distance; Helen and Paul Evans kindly provided hospitality in London. So, in Cambridge, did Margaret and Richard Potton. Jean Jeremy has shown more than a passing interest in the completion of the work and was first to spot what happened behind Arkwright's left arm.

At the American Philosophical Society, the editor, Mary McDonald, has overseen the production of the book. To her and to her editorial committee we are most grateful for their faith in the Markham project in what turned out to be an extremely adverse economic climate. And to Pamela Lankas and her colleagues at IGS, we are grateful for their superb skills in book production.

We are very grateful to all the above for their good will and encouragement, and most of all to our spouses for their tolerance, patience, and love. What remains is of course our own responsibility.

David J. Jeremy
Whaley Bridge,
Derbyshire, UK

Polly Darnell
New Haven,
Vermont, USA

A Guide to the Book

☆ ☆ ☆ ☆ ☆ ☆ ☆

ISAAC EBENEZER MARKHAM'S SIXTY OR SO DRAWINGS ARE THE earliest-known set of textile machine maker's workshop drawings produced in the United States. Nothing similar has apparently survived from the antebellum decades. Prepared between 1814 and 1825, when their author died, they form an important piece in the jigsaw puzzle of American industrial development.

Made at the cusp of change between preindustrial and industrial forms of technological knowledge and activity, between handicraft and factory production, Markham's drawings are particularly significant. As comparisons with contemporary European machine drawings reveal, they exemplified a fundamental shift in graphical knowledge of two-dimensional machine representation, from the perspective to the projective—a shift that was part of a much wider knowledge revolution in the United States. They also testify to the ubiquity of inventiveness even in the remotest of early machine shops, at the extremity of the well-known American and transatlantic mechanic networks. Local community networks locked into mechanic networks could assist the process of invention, as Markham's biography shows. In the hands of an ingenious, well-informed machinist, modest technical changes could be introduced; with knowledge of a major technological breakthrough, radical developments like the power loom could be transplanted from abroad, as Markham's power-loom drawings illustrate. It is hard to avoid the conclusion that the role of drawings and drawing skills, in creating and diffusing knowledge and in the formation of machinists' networks, has been underestimated.

THE SHAPE OF THE BOOK

A collection of such significance requires an examination of its provenance, a biography of the draftsman, and a portrayal of the historical and technical contexts shaping both draftsman and drawings. Part I is intended to recreate the industrial and social and intellectual worlds inhabited by Isaac Markham, the textile machine builder who made the drawings. It begins in chapter 1 with a biography of Markham showing how he grew up in Middlebury, a remote Vermont community, when the American Republic was young; how he was given opportunities to learn about new technology; and what he bequeathed after his life was tragically cut short.

From the world of the rural community Markham moved into a world of rapid industrial change. Roads and canals were beginning to link East Coast cities to westward communities when Markham was in his twenties. Immigrants were bringing ideas and techniques from Europe, which introduced new ways of production that took manufacturing out of the home and into the workshop and then the factory. Cut off from European markets by the Napoleonic wars and then the War of 1812, American merchants invested in spinning factories to which power looms were soon added. Bright young men found job opportunities in the clusters of machine shops and mills that sprang up close to or within Philadelphia; New York; Providence, Rhode Island; and Boston. Among them was Isaac Markham. This world of industrial change is explored further in chapter 2.

Through his apprenticeship in a Middlebury cotton mill and later work at the Boston Manufacturing Co.'s mills at Waltham and the Dover Manufacturing Co.'s factories at Dover, New Hampshire, Markham imbibed the intellectual milieu of the transatlantic fraternity of mechanicians. He would have soon realized that two kinds of knowledge about the new technology were crossing the Atlantic. One was about the designs of new kinds of machines that were transforming the textile industries of Britain, soon followed by continental Europe. The other was about the new way of depicting those designs. Chapter 3 focuses on this revolution in making two-dimensional records of three-dimensional machines. In particular it compares British and American drawing techniques showing how these drastically changed in Markham's lifetime. Markham was clearly heavily influenced by the new ways of recording visual knowledge, as his drawings demonstrate. Chapter 4 summarizes why Markham and his drawings are so important, whereas chapter 5 sets out how the drawings have been edited.

In Part II, Markham's drawings and papers are presented, arranged in nine sections. Though not complete, the notebooks, loose sheets, and sketchbook supplement the drawings. These are arranged according to textile-processing stage and each is prefixed by a short general explanation of the technology used at that stage. These explanations are derived from sources familiar to Markham. Particularly interesting features are noticed in the commentary on individual drawings.

Since North Americans did not have the relatively abundant supplies of labor available in Britain, not least because immigrant populations moved westward, American mechanics had to find ways of adapting imported designs to compensate. Markham's drawings show evidence of this particularly at the warping and weaving stages.

Part III is a collection of specialized studies that investigate in detail particular aspects of Markham's drawings. Testing the genuineness of the papers themselves, investigating what published sources about textile technology were available to Markham, estimating how much of Markham's drawings evidenced inventiveness, testing his own pursuit of the new drawing techniques, and gauging how complete might be this set of workshop records: these are the topics of the micro-studies in the Appendices.

NOTE ON THE ILLUSTRATIONS

Throughout the volume there are a large number of illustrations. The primary (master) versions of the illustrations are mostly found in Part II and Part III, Appendices 3 and 4. A few primary versions appear in Part I, because they predate or are outside the range of machines recorded in Markham's drawings. Derived copies are smaller copies of primary versions and are confined to Part I. These derived copies are designed to help the reader to see immediately what is being discussed and, like all illustrations in Part I, are intended to augment or expand the narrative and argument.

Primary illustrations throughout the book are accompanied by a legend that usually provides artist, size, location, or other relevant information, and in the case of the Markham drawings themselves, a fuller analysis of physical condition and explanation of the technology.

Derived drawings are accompanied by a short identification and a cross-reference to the primary drawing in Parts II or III.

Enlargements of some machine parts have been made when a particular element is of special interest and could not otherwise be seen clearly.

CROSS-REFERENCES

Cross-references are provided to help the reader move forward through the volume.

In Part I allusions in the text to a derived drawing are cross-referenced to two places, first to a derived drawing adjacent or close to the point at which it is mentioned; second to the primary version of that drawing found later in Parts II or III. Other cross-references in Part I will be to matter in the general introductions in Part II, chapter 6, or to one of the micro-studies in the Appendices.

MEASUREMENT OF THE DRAWINGS

1. The measurements of all drawings and paintings in this volume are presented as height x width unless otherwise stated.
2. Measurements have been provided only for original drawings or paintings. Thus none are given for the reconstructed United States patent drawings, the originals having been destroyed by the Patent Office fire of 1836.

NOTE ON ENGLISH PATENT DRAWINGS

For these inventions, the master patent drawings (originally colored) are reproduced here for the first time. There is only one reason for preferring them, rather than their mid-Victorian black-and-white line copies: they are the primary historical evidence, the documents that patentees held and that, sometimes, their rivals wanted to see. The Victorian copies are almost as good unless color and tonal values are deemed important. At the least, color might indicate the sort of machine drawing conventions coming into use. Thus the Hargreaves jenny patent (1770) shown in Figure 3.3 appears to be almost black-and-white: paying little attention to any color code. Burnt sienna in Garlick's double speeder (1797) Figure A3.12, and in Raynor's double speeder (1813) Figure A3.13, indicates a wooden machine frame. A more yellowish tint in Johnson's dresser patent (1804), Figure A3.15, also conveys a hint of timber, though a deep yellow in the Johnson dresser patent of 1803, Figure A3.14, has been allocated to a large gear wheel on the warping frame: suggesting brass? A clear difference between wood and metal is seen in Horrocks's power-loom patent of 1813, Figure A3.17. Shades of burnt sienna indicate wooden parts; shades of gray, iron framing, cast-iron wheels, and gear wheels. Black in Figure 6 of the Horrocks patent, Figure A3.18, marks a crank arm AB that turns on the crankshaft A: a crucial part of the compound lever by which Horrocks achieved his variable batten speed motion.

Normally only two or three original patent drawings would have been made. One, the legal master, written and drawn on vellum made from animal skin (pre-1820 drawings were sometimes made on a textile, like linen), was sewn into the most recent patent roll, stored in any one of three central government offices at opposite ends of Chancery Lane, London: the Petty Bag Office, the Rolls Chapel Office, and the Enrolment Office. Another would presumably have been retained by the patentee.

This chaotic system was reformed under the Patent Law Amendment Act of 1852. All original patents were centralized, the patent indexing system was unified, and all future patents of invention were required to be printed and published. Under the initiative and supervision of the first Superintendent of the new Patent Office, Bennet Woodcroft (1803–1876), all the 14,359 patents (specifications, drawings, and enrolments) issued in the period 1617–1852, were also printed and published. This was a mammoth task, producing hundreds of volumes, but it was completed in five years. Whole sets were given to major public libraries around the country. Woodcroft also produced a trio of catalogs providing subject, name, and chronological indexes for patents in this early 1617–1852 era. In the course of reorganizing, publishing, and indexing the first two-and-a-half centuries of patents of invention, each patent was given a unique sequence number, beginning with the first in 1617. Today scholars accessing English patents normally use the published Victorian copies and refer to them by the identity numbers pioneered by Woodcroft. Here these numbers are provided as "GB Patent no..." (Jamieson, "Introduction").

EXCHANGE RATES

In Markham's lifetime £1 = $4.80 approximately.

INDEX

The index here is not a concordance, listing the name of every person or place mentioned in the book. Instead it focuses on those persons, places, and topics that are central to the big themes of the book—Isaac Ebenezer Markham's life, the textile industry and its technology, and visual knowledge.

List of Figures

☆ ☆ ☆ ☆ ☆ ☆ ☆

FIGURES FOR CHAPTER 1

FIGURES FOR CHAPTER 2

FIGURES FOR CHAPTER 3

FIGURES FOR CHAPTER 6

FIGURES FOR APPENDICES

List of Tables

☆ ☆ ☆ ☆ ☆ ☆ ☆

TABLES FOR CHAPTER 3

TABLES IN APPENDICES

☆ ☆ ☆ ☆ ☆ ☆ ☆

DR. DAVID J. JEREMY HAS INVESTIGATED VARIOUS ASPECTS OF TEXTILE, business, and church history for more than forty years. He has held research appointments at Princeton University, the Smithsonian Institution, and the London School of Economics. In the early 1970s he was Curator of the Merrimack Valley Textile Museum at North Andover, Massachusetts (now the American Textile History Museum at Lowell, Massachusetts). His *Transatlantic Industrial Revolution: The Diffusion of Textile Technologies between Britain and America, 1790–1830s* (MIT Press, 1981), based on his PhD (LSE), was awarded the Dexter Prize of the Society for the History of Technology and the John H. Dunning Prize of the American Historical Association. Retiring as Professor of Business History at Manchester Metropolitan University Business School, he was made an emeritus professor in 2004. In retirement he works and writes from his home in the Peak District of Derbyshire, England.

Polly Darnell is currently the Archivist and Librarian at the Shelburne Museum in Shelburne, Vermont. Between 1980 and 1995, she ran the Research Center of the Sheldon Museum in Middlebury, Vermont. She has taught and written about archives and community history and been active in regional and national archival organizations. She has a B.A. in American History from Goddard College and an M.L.S. from the State University of New York at Albany.

☆ ☆ ☆ ☆ ☆ ☆ ☆

Part I

INTRODUCTION: CONTEXTS FOR ISAAC MARKHAM, HIS PAPERS, AND HIS DRAWINGS

☆ ☆ ☆ ☆ ☆ ☆ ☆

Chapter 1

Isaac Ebenezer Markham (1795–1825)

☆ ☆ ☆ ☆ ☆ ☆ ☆

ISAAC MARKHAM GREW UP IN THE TOWN of Middlebury in western Vermont in a place and era that encouraged developing industries and embraced new technologies. He used those advantages to develop fully his own mechanical abilities, learning from both local and itinerant mechanics, after which he traveled to work in other machine shops, eventually returning to Middlebury to superintend the cotton factory where his career began. When he died of typhus in 1825, at the age of thirty, he was mourned as a major loss to the community.

MIDDLEBURY IN THE LATE EIGHTEENTH CENTURY

At the time of Isaac Markham's birth in 1795, Middlebury was a relatively new settlement, scarcely thirty years old, founded by settlers primarily from Litchfield County, Connecticut, in search of more fertile farmland. The settlement lay on the eighty-mile-long littoral of Lake Champlain, ten miles away to the west. Less than half a dozen miles to the east, rose the densely wooded Green Mountains, in places nearly 4,000 feet high and running for a hundred miles and more north–south, parallel to the Lake and, further south, the upper Hudson valley (Fig. 1.1). Transportation routes heavily depended on waterways. Toward New York City the route was south along Lake Champlain, across the isthmus at the old fortress town of Ticonderoga and into Lake George, overland for five miles to Glens Falls and the Hudson Falls, south down the Hudson River to Albany, and again down the Hudson to New York City, 250 miles away from Middlebury as the crow flies. Physically it was easier to go north to Canada: down the Richelieu River and overland to the St. Lawrence at Montreal. However, the northern route posed political obstacles. Until 1763 the French, and after 1783 the British, controlled Canada.

Settlement by southern New Englanders didn't begin until after the French and Indian War ended in 1763, then was slowed by competing claims to the territory by New York and New Hampshire; governors of both territories were issuing grants to the same land. Before the land grants were resolved, the Revolution again disrupted the flow of settlers, like Isaac Markham's grandparents, who in 1770 had settled in Addison along Lake Champlain, a thoroughfare for British troops coming from Canada.

MARKHAM'S FAMILY BACKGROUND IN VERMONT

Isaac Markham's mother, Sarah (Sally) Gould Kellogg (1767–1850), was a child of three when her family moved to the New Hampshire Grants (Fig. 1.2). During the war the family retreated from Addison to

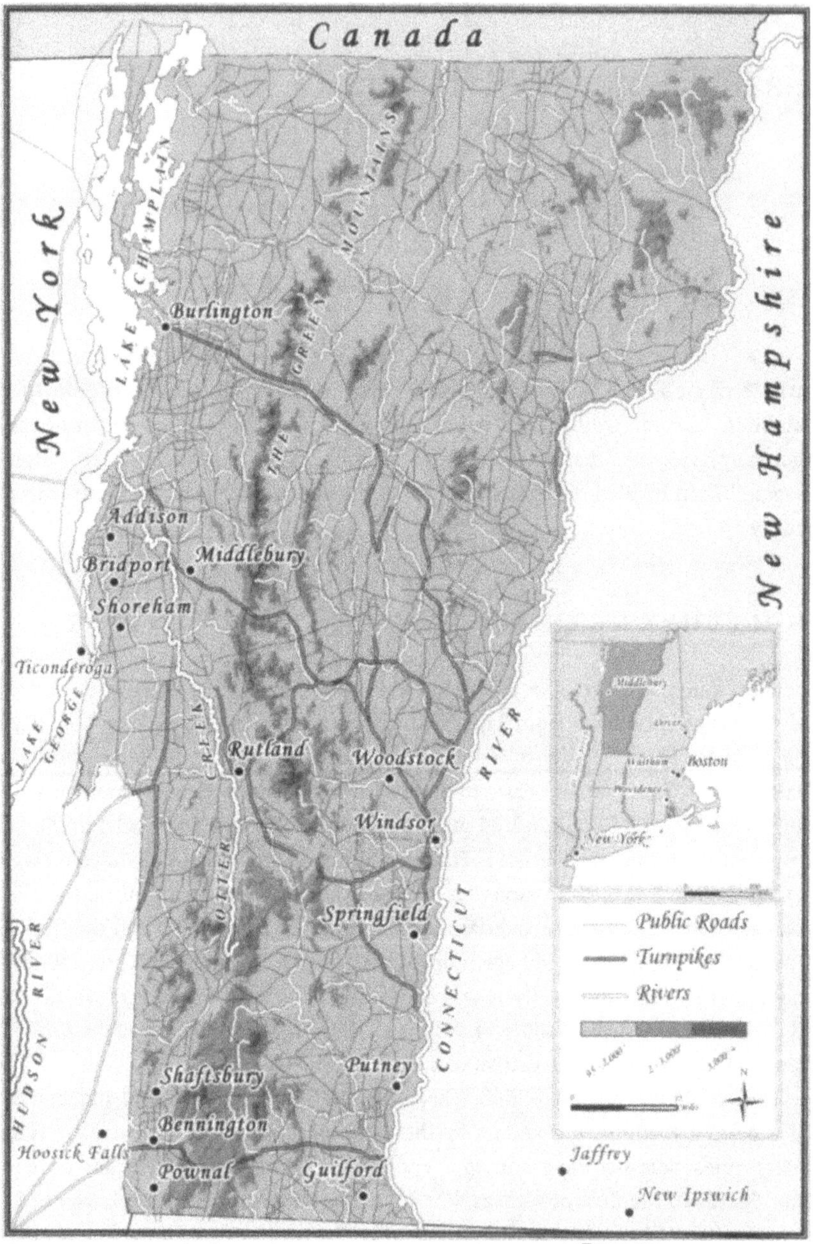

FIGURE 1.1 Map of Vermont, ca. 1820, compiled by Ted Kwasnik, 2008, showing rivers, roads, mountain ranges, major Vermont tow by population from 1790 to 1820, and locations in Vermont and adjoining states mentioned in the biography of Markham. The in relates Vermont to New York, Boston, Providence, and sites where Markham worked: Middlebury, Waltham, and Dover.

Rev. Isaac Markham (ca. 1711–1759)
m.
Jemima Pease (born 1712)

Benjamin Kellogg (died 1777)
of Canaan, CT, and Addison, VT
m. ? (wife's name unknown; died 1780).
They had one son and three daughters, including:

Ebenezer Markham (1749–1813)
Born Enfield, Connecticut
Married 1st 1775, Catherine Lydius of Schenectady, NY (died 1789)
None of their children, born 1776–1786, survived to adulthood.

married 2nd **Sarah (Sally) Gould Kellogg** (1767–1850)

Catherine Lydius Markham
(1790–1885) m. McLeod

David Kellogg Markham
(1793–1833)

Isaac Ebenezer Markham
(1795–1825)

Thomas Henry McLeod

URE. 1.2 Genealogy of Isaac Markham's family, including his nephew, Thomas Henry McLeod, who gave Markham's drawings to Sheldon Museum.

ces: Family Bible of Ebenezer and Catherine Markham, brought 1779. Privately owned in 1988 by Ellis McLeod, Urbana, Illinois.
ry L. Sheldon papers, Scrapbook 156 Pictures of 1800 Middlebury residents, Sheldon Museum, Middlebury, Vermont.
McLeod conversation with Henry Sheldon, ca. 1901 (catalog # 901213.2). Family trees on Ancestry.com

southern Vermont in the wake of the American fleet that delayed the advance of the British from Canada. They returned after the Battle of Bennington in 1777, thinking they would be safe, but Sarah's father and brother were captured by the British. Months later the rest of the family were dispossessed in an Indian raid, their house burned, livestock killed, provisions and other belongings carried off. In company with other refugee women and children they made their way to southern Vermont (Fig. 1.3). Sarah later wrote:

> Our family was completely broken up. My Father had died in prison at St. John's, my only brother [was] in Quebec jail. My mother and four daughters had to seek a shelter where they could find one. We never after that time kept house; some of the time we fell into humans [or humane] hands. At other times we had to suffer all the abuse and censure of a Tory faction.[1]

Isaac's father, too, was affected by the war. Ebenezer Markham (1749–1813), the son of a Connecticut minister, had gone to sea at the age of ten, was a ship's captain by twenty, and a Montreal merchant by the age of twenty-six. For assisting Americans during the war, he was imprisoned for three years and his property, worth $15,000, was confiscated.

[1]Philip Battell, *Minute Book of Conversations*, 1848, Swift Family Papers, Sheldon Museum, Middlebury, Vermont.

FIGURE 1.3 View of Middlebury, by George Heriot (1766–1844), ca. 1810, watercolor ($6^1/2$ x 9 inches, or 16.5 x 23 cm). Painted by a Canadian artist, this painting shows the buildings around the falls during Markham's youth. The large building on the right with four chimneys may be the Markham family's home and tavern.

Courtesy of the Sheldon Museum.

Starting over after the war, Ebenezer moved first to Ticonderoga, New York, where he was given command of a fleet of boats on Lake George and Lake Champlain. In 1789, the year he married Sally Kellogg, he resigned his command, and set up as a merchant in Bridport, Vermont, on Lake Champlain. Within a couple of years he had moved on to the neighboring town of Shoreham, where he set up a forge, trip-hammer, and nail-making machine.[2]

In moving to Vermont, the Markhams joined a steady stream of settlers following waterways and trails to spread over Vermont's most fertile land to convert the forests to farms. Between 1790 and 1820 the state's population increased by 8 percent a year. Most of the new settlers planned to farm, but some

[2]Henry L. Sheldon, Scrapbook 156, "Twentieth Century Memorial of Middlebury, Vt. Copies of All Known Pictures of Persons Living in Town at the Beginning of the 19th Century" Henry L. Sheldon Papers; Goodhue, Shoreham, 37.

were induced to build sawmills and gristmills, supplementing their farms and providing an essential service when the lack of roads limited the transport of grain and lumber. As more forest was converted to farmland, more specialized mills and craftsmen could be supported. They were risk-takers trying to improve their prospects; land and opportunities often changed hands after a few years as some succeeded, others failed, and many moved on.[3]

None of Capt. Markham's ventures recouped his fortunes. In 1795, four years after setting up in Shoreham, he was "committed to the liberties of the jailyard" in Middlebury for debt. Not confined to the jail, he could work in the area to pay off his debts. He tended bar at the local tavern during 1795, then in 1796 set up his nail factory in a room at the end of one of the three sawmills around the falls, and little by little built up a home and new businesses.[4]

Middlebury was becoming the central town of the region. It had abundant waterpower at two falls on Otter Creek, Vermont's longest waterway. Grist and sawmills had been built on both sides of the upper falls in the mid-1780s shortly after settlers returned to the area following the Revolution. A ford above the falls, and by 1787 a bridge, ensured that converging roads would bring business to the village. The town's first store, built in 1790 at one end of the bridge, facilitated trade among the 395 residents who lived in Middlebury in 1791, when Vermont became the 14th state. Its designation as the county seat, also in 1791, brought the courts, lawyers, jail, and additional commerce. The owner of 100 acres on the east side of the falls, Gamaliel Painter, further developed the village by encouraging a variety of mills and artisans.

Capt. Markham's daughter later described the village in 1796, when the whole family moved to Middlebury: The land was covered with forest, with small individual clearings near the falls for about thirty-nine small houses, two grist mills, two sawmills, a forge and gun factory, two taverns, and a fuller.[5] Year by year the enterprises and inhabitants of Middlebury were joined by others—a potashery, a distillery, tanners, carpenters, blacksmiths, a cabinetmaker, more merchants and taverns, another forge and a furnace, a fuller, a hatter, a clothier, a carding mill, a saddler, carriage-maker, a marble quarry and mill —but many businesses were fleeting.[6] Within two decades of the first mills being set up, the first mill owners had sold out and moved on; the crude early buildings were adapted, replaced, destroyed, or left to rot; a much wider variety of skills appeared; and new and larger industries were being established. Isaac Markham grew up in the midst of them.

When they first moved to Middlebury in 1796, the Markhams lived in the nail factory while they built a house, which, by 1800, they enlarged to include a public house. The building was then substantial enough to serve as a meeting place for the Episcopal Church, and to serve as the center of Capt. Markham's various businesses, which included a tavern, forge, hay weighing, and hats.[7] The hats were evidently made on shares with a couple of hatters, with Markham supplying the wool, which he received in trade. His account book also includes steps in the production of cloth: sending wool to a carding machine; paying women to spin cotton and tow, employing both a man and several women to weave; and having cloth (fustian, "wooman's ware" and gray coating) dressed at the local fulling mill. In addition, Capt. Markham was appointed the keeper of the jail in 1801.

---

[3]Sherman, *Freedom and Unity*, 131; Potash, *Vermont's Burned Over District, passim.*

[4]Swift, *Middlebury*, 289.

[5]Swift, *Middlebury*, 298–299.

[6]Swift, *Middlebury*, 241, 298; Henry L. Sheldon, Scrapbook 159, "List of Town Officers, Merchants and Professional Men in Middlebury, 1762–1906," ca. 1880–1906.

[7]Ebenezer Markham, Daybook, 1791, 1801–1806, Markham Family Papers. The daybook portion from 1801 to 1806 shows transactions for the tavern (which did a steady business in "nips of sling" and other drinks), hats, meals, lodging, weighing hay, nails, and makes reference to three other account books that don't survive: "Forge Book," "Barr Book," and "mem'dum of weying hay."

None of this meant that all was well with the Markhams. The property belonged to Sally Markham's brother, Isaac Kellogg; Capt. Markham's Montreal debts caught up with him; and his health declined. In 1807 the state legislature passed an Act for the relief of Ebenezer Markham:

> Whereas Ebenezer Markham, a prisoner in the common gaol in Middlebury...has had a paralytic fit, almost wholly lost the use of his limbs, and is advised by his physicians that change of air and the use of medical springs are the most profitable means of restoring him his health, and to the use of his limbs: therefore, it is hereby enacted...that the said Ebenezer Markham be...released and discharged from his said confinement...for and during...one year from the passing of this act.[8]

In 1808 the relief was extended for two years; in 1810 for five years. He died in 1813.

ISAAC MARKHAM: EARLY YEARS

In spite of financial difficulties, the Markham children—Catherine (1790–1885), David (1793–1833), and Isaac (1795–1825)—were educated by private tutors for several years. They also attended the local schools.[9] According to his grandson, Thomas McLeod, "Cap. Markham took great pains in the education of his children. They were instructed in horsemanship, dancing and every thing but music."[10] Capt. Markham in 1800 became a member of the private Center Library, from which the family could borrow books on history, philosophy, and biography.[11]

The village itself would have provided an education, between news being relayed in the tavern and the activity the children saw every day (Fig. 1.3). David Markham later described the footpaths he took to school in the winter of 1804–1805. Landmarks he passed daily included "the Old Factory & coal house," "the coal bed," "an old house commonly called the filing shop," and a marble quarry.[12]

Family legend claimed that young Isaac didn't just pass by such interesting sites:

> [T]wo men were engaged secretly in contriving and building a picking machine. No persons were admitted to a sight of the machine, lest the secret should be discovered, before a patent was obtained. But Isaac, being a boy, was admitted without suspicion. When he went home he said he could contrive a better machine, and, with such tools and materials as he had, formed a model, which, it is said, was adopted by the men instead of their own.[13]

Middlebury was as ripe as any town for the adoption and adaptation of new technologies developed elsewhere. Though remote for the transportation of goods, news and ideas traveled easily to Middlebury. Middlebury residents had close ties to southern New England and could keep up with developments there and elsewhere through letters from family and friends, new settlers arriving, people passing through,

---

[8] Vermont, *Acts and Laws*, 1807, 143. See also Vermont, *Acts and Laws*, 1808, 1810.

[9] Ebenezer Markham, Daybook: On November 30, 1801, Chester Wright bought some nails and "commenced schooling David" who was then eight years old. Wright himself was, or had recently been, a student at the Addison County Grammar School where he prepared to attend Middlebury College, which he entered in 1802. On September 29th, 1802, not only David, but also Catherine and Isaac commenced school with him. It's not clear whether they were being tutored or taught with other children at a school. After college, Wright became the principal of the Grammar School, where David learned Latin, prior to studying law.

[10] Thomas Henry McLeod conversation with Henry L. Sheldon, ca. March 13,1901 (catalog # 901213.2), Sheldon Papers: Catherine's son, Thomas McLeod, said that the three children "were under the instruction of a Mr. Black, a Endingburgh [*sic*] University man four years" and that Isaac "was educated under the instruction of a Mr. Ethrington a graduate of the University of Cambridge England." Because Sheldon was deaf, visitors wrote down what they had to say. Sheldon subsequently saved these "conversations." In the same conversation, McLeod said that "Isaac thought of nothing but his machinery. He studied that night and day. He thought everybody was his friend and honest."

[11] Center Library (Middlebury, VT), Volume 4 (1795–1805), Records of the Middlebury Congregational Church, 1790.

[12] David K. Markham Affidavit, January 7, 1817, Judd and Harris Family Papers.

[13] Swift, *Middlebury*, 336.

and the weekly newspaper started in 1801. As elsewhere, there would have been a receptive audience for stories of improvements in manufacturing.

The first carding mill in Middlebury, and perhaps the county, was built in 1801 by Artemas Nixon. It was "considered a great curiosity and nearly everyone went to see it and give it more or less patronage."[14] Within three years, Nixon was advertising that he had "2 new carding machines of the very best kind now in motion, in addition to the two heretofore used by him, all of them newly covered with the best new patent twilled cards."[15]

With industries just getting established, experimentation was under way even in Middlebury. In 1799 or 1800 three men—Josiah Nichols, Daniel Pettibone, and Ezekiel Chapman—working the trip-hammer at the falls discovered and patented a process for welding cast steel.[16] Another man, Eben Judd, who had also been imprisoned for debt in Middlebury, set up an innovative mill to saw marble, which he quarried from the creek bed and its surroundings.[17]

ISAAC MARKHAM: WORKING LIFE

In 1812 Isaac Markham started working for David Page, Jr., who built the first of two cotton mills in Middlebury (Fig. 1.4). Markham worked under several machine builders and was making drawings by 1814, the date on the cover of a portfolio of his drawings of machine parts.

Page moved to Middlebury from Jaffrey, New Hampshire, in 1807, set up as a merchant, and with an architect partner, Lavius Fillmore, bought the mills and waterpower rights on the east side of the falls. They replaced the old mills with a bigger, better, and more permanent gristmill of stone, which had a flume of Fillmore's design cut out of solid rock, the water being exposed only where it flowed into six tub wheels before entering a subterranean outlet.[18]

In 1811 Page built the stone cotton factory (which also had carding machines for wool) next to the gristmill, three stories high at the end fronting on Main Street and six stories high at the other. Here Markham worked under three machine builders—John Houghton, Jesse Holton, and Joseph Gordon. To construct the machinery, Page recruited two practical men involved in the cotton mills built in 1804 and 1808 at New Ipswich, New Hampshire, ten miles from his hometown of Jaffrey.[19]

Houghton had come to Middlebury in 1810 and was working mainly as a cabinetmaker, though also turning his hand to other jobs, including tailoring, making varnish, and haying, according to a surviving fragment of his account book. In March through June of 1813 he worked eighty and a half days for Page and Fillmore, without specifying the jobs he did. He worked for them again in the winter of 1814 for thirty-six days. In 1819 he was paid by the second cotton mill in town for such jobs as "turning a drum," "turning a yarn beam &c," and "drilling gudgen and wedging the same," as well as for 25 days of work not specified.[20]

---

[14] Weeks, *Salisbury*, 297.

[15] *Middlebury Mercury*, 15 August 1804; Pliny Earle's patent of 1803, see p. 225.

[16] Swift, *Middlebury*, 330. Their patent for welding cast steel to iron was dated 22 December 1802. Pettibone (whose first patent, in 1797, was for breaking dough and paste) registered other patents: for welding steel, 22 March 1806 (when he was located at Roxbury, Norfolk, Connecticut); for plane irons and scythes, 6 May 1813 (when he was living in Philadelphia); and for boring gun barrels, 12 February 1814 (when he was located in Philadelphia). Several other armory workers also claimed the invention of the screw auger. M. R. Smith, *Harpers Ferry Armory*, 96–97; United States Patent Office, *List of Patents, 1790–1836*.

[17] Swift, *Middlebury*, 335–337.

[18] Hall, *Statistical Account* quoted in Swift, *Middlebury*, 332.

[19] Bagnall, *Textile Industries I*, 368–372.

[20] Houghton, Account book, 1810–1819.

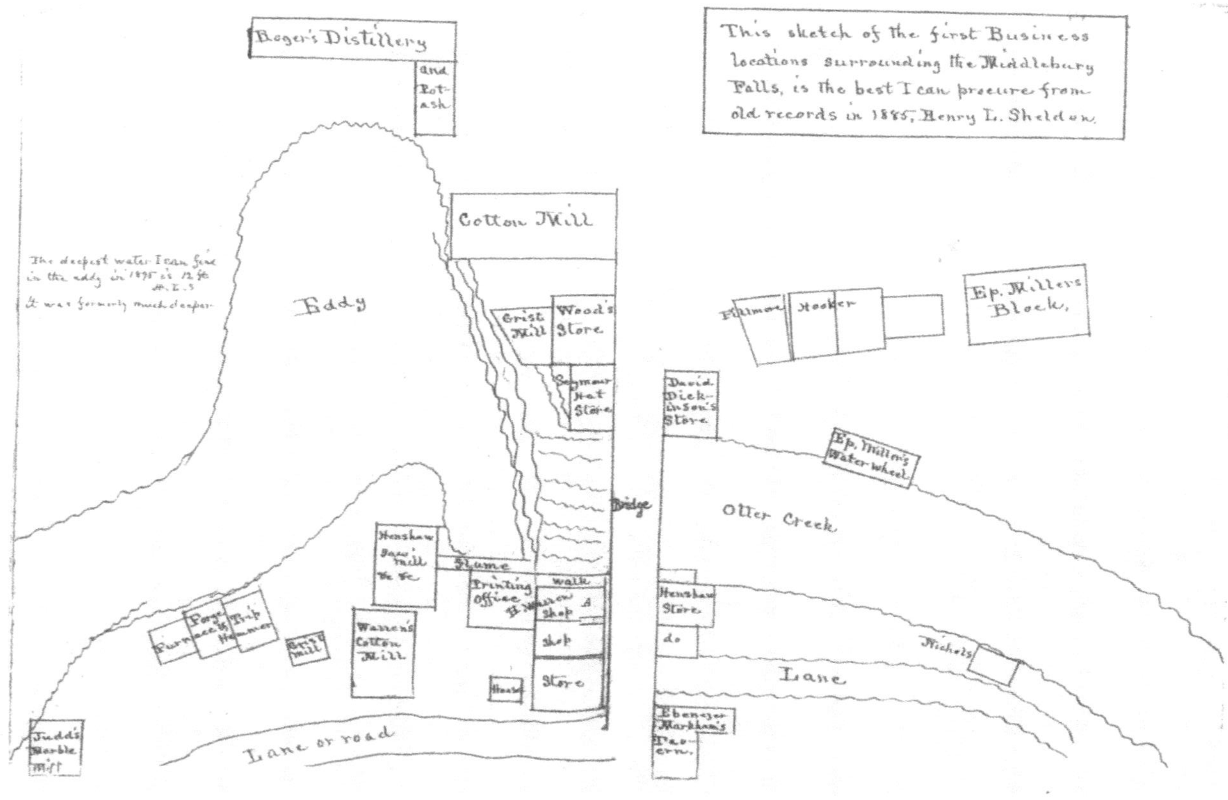

FIGURE 1.4 Map/plan of Middlebury ca. 1815, created in 1885 by Henry Sheldon from research and conversations with residents of the town. This diagrams the layout of the mills and shops around the falls, including the Markham family's tavern. The Masonic Hall was less than a quarter of a mile to the northeast; Middlebury College is barely half a mile to the south of the tavern.

He was evidently followed by Jesse Holton, of New Ipswich, New Hampshire, who testified "that for a number of years I resided in Middlebury Vermont and was Master Builder of the machinery of the cotton factory when it was owned by David Page, Jr."[21] Holton bought property in Middlebury in October of 1814. In an agreement he made in 1819 to run the factory for three years, Holton was already listed as a major creditor, for debts incurred by the factory to him in 1817 and 1818.[22]

Cotton-mill machinery was custom built on site at each factory by builders who had learned what to do by working at other factories. By a long chain of knowledge diffusion, Houghton and Holton evidently brought Arkwright technology to Middlebury. Fifteen years after Samuel Slater reached Rhode Island to set up some of the earliest Arkwright machinery in the United States for the merchants Almy and Brown, one of his employees, Charles Robbins, was recruited by a New Ipswich entrepreneur to build a line of machinery for the first cotton mill (1804) at New Ipswich. The second cotton mill (1808) at New Ipswich was promoted by Daniel Brooks, one of those involved in the earlier mill. Brooks soon after sold his interest to three partners, one of whom was Jesse Holton, earlier a blacksmith of Charlestown near Boston, Massachusetts.[23]

Within months, a new partner, Samuel Batchelder (1784–1879), joined the business and commenced his career as one of the most successful and prominent New England cotton manufacturers. It is likely that Batchelder knew Page because they were both born in Jaffrey, NH.[24] This network of machine builders and entrepreneurs suggests that Markham, when he came to maturity, would be able to tap into a series of connections that would impart knowledge about the infant American cotton industry, its opportunities for aspiring mechanics, and information about its latest technology.

Page's mill was in operation and looking to increase its workforce when it advertised in the local paper in 1814, the same year that Markham dated his portfolio of drawings:

Middlebury Manufacturing Company
Offers by wholesale or retail, at their
Store at the east end of the bridge,
Cotton Yarn
of all numbers from 6 to 40, and all colours, at the
old Providence Peace Prices.
Also Shirtings and Sheetings
from 3-4 to 6-4 wide.
Ginghams, Stripes, Shambrays,
Bedtickings, Vestings, &c. &c.
which they will at all times sell as low
as can be bought in any City in America, of the
same qualities, for ready pay
Also
5 new Woolen
Carding Machines,
Of a superior kind, and
Machine cards for Wool or Cotton,
either in sets, or to repair with—at the Leicester
prices.—They also make Woolen
Billies and Gennies

[21] Jesse Holton, Deposition 13 May 1831, Judd and Harris Family Papers.

[22] Deed, William Page to Jesse Holton, Middlebury Land Records, VIII, 183.

[23] This chain of diffusion of technology illustrated the development of one of the networks explored in Meyer, *Networked Machinists* and in R. Thomson, *Structures of Change*.

[24] Bagnall, *Textile Industries I*, 368–372, 473–483; "Samuel Batchelder," *DAB*.

warranted Superior.
wanted for a year, at the above Factory
30 Females,
from 16 to 30 years of age, who can bring
substantial recommendations of a good character,
as none other will be received. The price is
One Dollar a week.
Also For Sale, a few Pieces of SMUGGLED
Broadcloths & Kerseymeres,
At the above Factory.

Sept. 7, 1814 50 tf.[25]

A greater opportunity for Markham came in 1817, when a Scotsman, Joseph Gordon, "who had been employed in the manufacture of machinery, and had set up several factories in Scotland...immigrated to this country and brought with him drawings of machinery used there. Mr. Gordon built for Mr. Page twenty power looms at that time. These, Mr. Gordon informs us, were the first power looms ever built in the United States, except six in Rhode Island, [*sic*] which were built by a Scotchman, who came to this country the year before. Isaac Markham, an ingenious young mechanic, who had worked under Houghton, was set to work and manufactured the iron part of the machinery."[26] (There were earlier power looms, some not very successful. Francis Cabot Lowell introduced a variation of the so-called Scotch loom to New England, when he commenced textile manufacture in the first mill of the Boston Manufacturing Co. on the Charles River at Waltham, Massachusetts, in 1813–1814. A more sophisticated power loom, incorporating Horrocks's variable-batten speed, was introduced to Rhode Island by William Gilmour, who emigrated from Scotland in 1815.[27])

No information has come to light about Gordon's origins or the connection that brought him to Middlebury. After building the looms at Middlebury, he moved on to New York state, where he eventually built his own cotton factory, The Caledonian, at Hoosick Falls, before returning to Middlebury after a brief stint at the cotton factory in Bennington, Vermont.[28]

These looms made the Middlebury cotton mill the largest in the state. The 1820 census listed eleven manufacturers of cotton yarn and/or cloth in Vermont. Page's was the largest, with 840 spindles and 24 power looms, all in operation producing $14, 890 worth of No. 15 Shirting a year. The other Middlebury mill had 500 spindles and 8 power looms producing No. 14 shirting and yarn. A mill in Bennington had a large capital investment ($26,000 to the Page mill's $25,500), but reported only 204 of its 816 spindles were operating, producing $4,587 worth of yarn and shirting annually. The "General Observation" said "Abandoned by the owners, and rented out for sum hardly sufficient to keep the machinery in repair." Another factory, with 52 looms, 24 wheels, 2 warping machines, and so on was actually the Vermont State Prison, which operated under a contract with the Lippitt Manufacturing Co. of Rhode Island, at an annual cost to the state of three or four thousand dollars.[29]

---

[25] *Vermont Mirror*, September 7, 1814.
[26] Swift, *Middlebury*, 331.
[27] Jeremy, *Transatlantic Industrial Revolution*, 92–103. See also Appendic 4.
[28] In 1823 Gordon started his own cotton factory, The Caledonian, in Hoosick Falls, New York, that employed 154 people running 7,000 spindles and154 looms, producing 30,000 yards of "printing-cloth" a week. Disabled by an accident while building tenement houses, he lost the business (Sylvester, *Rensselaer*, 386). Letters from Joseph Gordon to Ira Stewart dated Bennington, VT, January 3, March 7, and March 19, 1834, Letter Collection, Sheldon Museum.
[29] United States Census Office. Digest of Accounts. Hall's *Statistical Account*, 8, puts the number of looms at 15.

> [T]he building contains at this time (December, 1820) eight hundred and forty spindles for cotton, fifteen power looms (or looms, which are moved by water), with all the warping and dressing apparatus needful, together with two woolen carding machines. The spindles produce a sufficient quantity of yarn daily, for five hundred yards of sheeting. The whole expense of converting the yarn (taking it from the spindle) into cloth, is about four cents a yard. The looms are attended by females. The goods manufactured are exhibited for sale in an apartment of the same building. (Hall, *Statistical Account*, 8)

Having worked on the most advanced machinery in the state, Markham had to leave Vermont if he wanted to improve his skills. Page supported his efforts. In 1817 and 1818 Page had paid for small quantities of drawing paper delivered to Markham, presumably for drawings of machines.[30] In 1819, Page wrote him a letter of recommendation mentioning his drawings.[31]

Middlebury [page torn] 4, 1819

> This may certify that Isaac E. Markum has worked for me a considerable part of his time building Cotton & Woolen Machinery since 1812 and the last year & an half particularly making Power Looms with all its accompanying parts which I have had in operation more than twelve months, my Looms I would not exchange for any other made on any other plan which has come to my knowledge—Mr. Markum has been the superintendant, and draftsman of them—I consider him perfectly capable of doing every part of the iron and brass work (after forged) and that he is honest and entitled to the confidence of the public—the drawings he has is his of his own execution they speake for themselves.
>
> David Page

Page's evaluation of Markham's power looms was, in all likelihood, an informed one. As a merchant, he would have traveled in New England and to Boston, which, with New York City, was one of the main markets for purchasing raw cotton and selling finished cloth. If he was on friendly terms with Samuel Batchelder, he would have visited New Ipswich and seen the cotton mills there. Most likely he was aware of the Waltham power loom. Whether he inspected it and compared it to Markham's designs is unknown. At the very least we may assume that Page evaluated Markham's looms partly on the basis of the cost (and hence profits) and quality of the cloth they made, partly by comparison with some of the 1,300 power looms installed in mills scattered around New England and the state of New York in 1820.[32]

During his years away from Middlebury, Markham worked in at least two cotton mills—in Dover, New Hampshire, and Waltham, Massachusetts.[33] An 1820 inventory of the Dover cotton mill includes "2 warping machines with stop motions with 2 dressing of Adams & Markhams plan," suggesting that he was making innovations.[34]

After Dover, he was on the payroll of the leading cotton manufacturer, the Boston Manufacturing Co., in Waltham, Massachusetts, from September 2, 1820, through December 1, 1821, as a machine designer and builder in the machine shop.[35] The 19th-century town historian cites a family story that when Markham's nephew visited Markham at Waltham and was taken on a tour of one of the Boston Manufacturing Co.'s factories, Isaac Markham pointed out the picking machines and claimed they were of his design.[36]

From Waltham, Isaac Markham wrote a stinging account of management's treatment of the workforce in a letter to his brother.[37]

---

[30] William Slade, Ledger, 1813–1818, Slade Family Papers, 139, 217; Jonathan Hagar Ledger, 1812–1823, Hagar Family Papers, 170.

[31] David Page letter, 1819, Markham Family Papers.

[32] Jeremy, *Transatlantic Industrial Revolution*, 278.

[33] In addition a drawing of "Pasaic" Falls (see Fig. 1.5) suggests he at least visited Paterson, New Jersey, and a fragment of a document, cut to the shape of a machine part (see Fig. 6.83B, p. 315), is signed "Exeter Aug. 17, 1820 Isaac E. Markham."

[34] Agreement between Williams & Wendell and A. & J. Wendell, June 29, 1820, Dover Cotton Factory file, Williams and Wendell, 1815–1831, Case 11, Wendell Collection, Baker Library, as quoted in Jeremy, "Innovation," 66.

[35] Correspondence with late Michael Folsom in Markham Papers Collection File, Sheldon Museum, Middlebury, Vermont, based on Boston Manufacturing Company payroll records, Baker Library. See also Gonzales, "Female Millworkers and the Mechanization of Textile Production" for further information about the labor dispute described in the following letter.

[36] Swift, *Middlebury*, 336–337.

[37] Letter, Isaac Markham to David K. Markham, May 30, 1821, Markham Family Papers.

FIGURE 1.5 "**Pasaic** [*sic*] Falls," by Isaac Ebenezer Markham, ca. 1819, pencil sketch (19.5 x 11 cm). This sketch of Passaic Falls, New Jersey, is the only evidence that Markham visited Passaic.

Courtesy of the Sheldon Museum.

Waltham

May 30, 1821
My Dear Brother
I have sent fifty dollars to Boston & left it with Messrs Goodman & Saville No. 31 Merchants Row to be delivered to your order. I shall shortly let you have as much more it is now let out & I shall collect it in a few days. The reason that I have not procured a draught is because I cannot procure one without making a discount on Current Money or paying a premium or brokerage. They are anxious that I should return to Dover & say they will employ me for five years if I choose. Advise me what is best to do. If I can get employ for the same wages that I used to have had I better not accept. That compy is rather unfair in some treatment towards hands & this compy are perfectly upright in their monied dealings with the workmen tho they have all the lordly & tyrannical feelings that were ever felt by the greatest despots in the world. tho they have ever treated me well but still they are determined their word shall be law & shall be obeyed. A few weeks since they cut down every unmarried man's wages (except mine) that they employ & without giving them the least notice until the day came for payment. The same trick was played off on the girls but they as one revolted & the works stopped 2 days in consequence. I have been busy for some considerable time making some mechanical calculations therefore cannot bring my mind to writing excuse my scrawling affectionately yours I E Markham

As he worked in these different places, Markham made notes and drawings of what he was learning or thinking. Some were done in a small format, some were larger and more finished. A few notes on small paper ruled for keeping accounts are dated, suggesting that he was in Waltham on 11 December 1821 and in Middlebury later in the month. He was again—or still—in Middlebury in September 1822 when he bought an umbrella and geography book, which he didn't pay off completely until May of

1823.[38] Apparently he had settled down in Middlebury by August 1823, when his uncle Isaac Kellogg deeded the family tavern to him.[39]

He eventually became the superintendent and agent of the cotton factory (formerly operated by David Page, Jr. but from ca. 1820 run by J. Hough & Co.) where he had started, a job that required not only technical skills, but also skill in managing the workforce. That he succeeded in the latter was demonstrated by the warmth of the notices of his death in the local newspaper, which gave him far more space than it did the "town father" who had done so much to boost Middlebury.

MARKHAM'S DEATH

Isaac Markham died of typhus, September 16, 1825. His funeral was attended by everyone employed in the cotton factory; several lodges of freemasons; and the members of a Middlebury College literary and debating society, despite the fact that he didn't attend the college. Markham's obituary suggests the high esteem the townspeople of Middlebury had for him:

> We have never witnessed a larger and more respectable assemblage of people upon any similar occasion. It was, however, but a just tribute of respect to the deceased, who, without the advantages of birth or fortune, by the force of talents and virtue alone, had succeeded in the esteem and affections of his fellow citizens.
>
> It is indeed very seldom that the death of an individual in the private walks of life, excites so great a sensation of regret, and so much sincere and heartfelt sorrow as that of Mr. Markham. Distinguished as he was for noble and generous feelings, for independent and manly sentiments, warm and unwavering friendship, undeviating and exact rectitude, he was still more distinguished by a mind of singular felicity of invention, originality and strength of conception, and of rich and abundant resource. Relying upon his own exertions, with untiring assiduity he had carefully cultivated his uncommon powers until he had become one of the first mechanicians in this section of our country. While his personal friends weep over his untimely death, the public will deplore his loss as a serious misfortune to the community in which he resided.[40]

MARKHAM'S LEGACY: PAGE'S COTTON FACTORY AND THE LATER MIDDLEBURY COTTON INDUSTRY

Markham's death was also a loss to the cotton factory, which was in the midst of an expansion, increasing the number of looms to eighty and building a gas plant to light the factory. The factory had already changed hands several times, and did so yet again within a few years. Most owners ran into debt, mortgaging the factory to merchants in Boston and New York, and eventually selling at a loss. J. W. Stephens, a New York City merchant, bought the factory in 1824 for $23,000, mostly on credit. He employed Markham as agent and superintendent. Stephens ran deeply into debt, leased the factory back to the previous owner, and in 1829 defaulted.[41] Only one of the several owners of Page's cotton mill in its first fifteen years seems not to have run into financial difficulties, possibly because he made no improvements.[42]

---

[38] Jonathan Hagar Ledger, 1812–1823, Hagar Family Papers, 38.

[39] Deed, Isaac Kellogg to Isaac Markham, August 12, 1823, *Middlebury Land Records*, Vol. 9, page 98.

[40] Obituary, *National Standard*, Middlebury, Vermont, 20 September 1825.

[41] See *Middlebury Land Records*, Vol. 9, pp. 180, 182, 447, 484, 561, 573, 574, 581; Vol. 10, pp. 97, 145.

[42] Joseph Hough, David Page's clerk who became a merchant to whom Page owed money, acquired Page's interest in the cotton factory in 1820 (*Middlebury Land Records*, VIII, 3–10); Hough sold it to Stephens in 1824 and leased it back from Stephens in 1826 (*Middlebury Land Records*, IX, 561).

The town history could as well have been talking about Stephens when it described the owner of the second cotton factory in town: "Mr. Warren, not content with the income which had made him independent, undertook to establish a cotton factory. ...But the business was then in its infancy in this country...and being left with no great abundance, Mr. Warren sold out his premises and returned to Massachusetts."[43] Stephens returned to New York City, whence came the next owner, who finally stabilized the factory, though not without some intrigue.

Benjamin Marshall (1782–1858) had the experience and business acumen to make a success of the factory. He was from a manufacturing family in Yorkshire, had entered the cotton trade in Manchester at the age of sixteen, emigrated to New York City in 1803, and established a business exporting raw cotton to England and importing finished textiles. Always seeking efficiencies in the trade, he and his partners established the Black Ball Line, the first packet line sailing across the Atlantic on a regular schedule. In 1825, perhaps motivated by the Tariff of 1824, he began shifting his interests from shipping to manufacturing. He built and bought cotton mills in several towns, and moved to Troy, New York, where he managed the Hudson River Print Works, one of the first textile printing mills in the country.[44]

Marshall had bought up several mortgages on the Middlebury factory, but the situation wasn't simple. In 1830 he wrote to a Middlebury lawyer that he was sending an agent who was studying law to work in the lawyer's office and gather information on the status of the mortgaged property:

> I wish to be particularly advised as to the value of the property and of every circumstance that may appear of interest. ...Mr. Burt [will be there] to gain information as to the intentions of our opponents and he will be enabled to fully explain my views on the subject of purchasing or settling our claim on the Middlebury factory and hope by special attention we shall be enabled either to get a cheap Factory entirely unencumbered or to get the value of our mortgage.[45]

By 1832, Marshall, along with his brother Joseph, had emerged as the principal owner of the factory. According to the *McLane Report* on manufacturing in the United States, the capital investment was $60,000, none of it borrowed, with an annual profit of 6%. There were 110 employees (25 men, 75 women, 10 boys) working twelve hours a day, year round, to produce 450,000 yards of cotton goods.[46] He added machinery, increased the number of looms to 100, and hired Joseph Gordon, the man who had brought the design for power looms to Middlebury, to run it.

The factory remained in Marshall's hands until his death in 1858. How long its prosperity continued after that is unknown. In 1882 the factory closed for good and its machinery was sold to parties in Alabama.[47]

---

[43] Swift, *Middlebury*, 288.
[44] "Benjamin Marshall" in *American National Biography*.
[45] Letter from Benjamin Marshall to Charles Linsley, Oct. 29, 1830 (#830579), Letter Collection, Sheldon Museum.
[46] *McLane Report* I, 878–881.
[47] *Middlebury Register*, February 24, 1882.

Chapter 2

Markham in the Context of the Nascent American Cotton-Manufacturing Industry

☆ ☆ ☆ ☆ ☆ ☆ ☆

ISAAC MARKHAM'S RELATIVELY SHORT LIFE, reaching from the 1790s to the 1820s, spanned the critical decades in the United States's efforts to industrialize. Pivotal to the establishment of an industrial sector in the infant republic were inventive mechanics such as Isaac Markham. Some indication of the wider contexts in which he developed his talents for machine building and machine designing will help in placing his work in a proper historical perspective.

In the 1790s it was by no means certain that the newly independent federal republic would be able to imitate its former parent and establish the revolutionary system of production that, by new combinations of labor, organization, and technology, was currently transforming some parts of the British industrial scene. One English manufacturer of woolens visiting America in 1794 later published his verdict that America, unable to prevent its rapidly expanding population from gravitating into its vast rural territories, would be dependent on John Bull for their cloth "for at least half a century to come."[1]

In Markham's lifetime this prediction was thoroughly confounded. The interplay of unforeseen economic and political circumstances conspired with developments more particularly conducive to textile manufacturing. Developments more specific and advantageous to textile manufacturing included a change in American attitudes toward manufacturing industry, changes in the supply of raw cotton, protection from foreign competition, knowledge of European advances in cotton manufacturing, the transatlantic migration of skilled workers, and American investments in the cotton manufactures.

Despite the lure of land settlement, rather than limit industrial expansion, rapid population growth (which between 1790 and 1820 more than doubled to nine million) became one driver of economic growth. Urban population, trebling to 693,000, deepened markets. Slowly developed improvements in highways—beyond intercity stagecoach routes along the Boston–Washington axis—and a few canals supplemented the arterial eastward-flowing riverine and coastal shipping networks. Transport and communication improvements, in extending markets, removed limits on new manufacturing technologies exploiting the division and specialization of labor. The potential to seize new market opportunities lay with America's well-established merchant class: Their city-based banks and insurance companies financed industrial investments and their academic institutions educated future entrepreneurs.

---

[1]Jeremy (ed.), *Wansey*, 141.

Growing through youth and manhood between the 1790s and the 1820s, Markham would have been keenly aware of the rumbling, if fluctuating, national debate over the merits of manufactures. The fathers of the infant American Republic stood at a crossroads in the 1790s and 1800s. Jeffersonians dreamed of a westward-expanding agricultural economy with its associated hopes for traditional virtues. For Jefferson in 1785:

> Those who labour in the earth are the chosen people of God, if ever he had a chosen people, whose breasts he has made his peculiar deposit for substantial and genuine virtue...let our workshops remain in Europe. ...The mobs of great cities add just so much to the support of pure government, as sores do to the strength of the human body.[2]

In opposition to the Jeffersonians, supporters of Alexander Hamilton wanted to complement political independence with economic independence. Hamilton's *Report on the Subject of Manufactures* (1791), which, as Secretary of the Treasury, he presented to the House of Representatives, both surveyed the state of manufacturing in America and served as a manifesto for proindustrialists. For them it was logical: there were greater possibilities for mechanization in manufacturing than in agriculture and therefore, deplorably, "The substitution of foreign for domestic manufactures is a transfer to foreign nations of the advantages accruing from the employment of machinery, in the modes in which it is capable of being employed, with most utility and to the greatest extent."[3]

A simple black-and-white choice between a moral agrarian ideal on the one hand and an immoral industrial vision on the other,[4] apparent to Jeffersonians in the late 1780s and early 1790s, has latterly been reinterpreted as much less clear-cut and more nuanced. John Nelson has emphasized the delay in bringing out the *Report on Manufactures*; its silence with respect to the position of the politically powerful merchants importing British manufactures, and the revenues they yielded; weaknesses in finding capital for investment in American manufactures; Hamilton's preference for using bounties to encourage larger enterprises (like his favored Society for Establishing Useful Manufactures); and his reluctance to protect American manufactures from English competition. On the other side, Jefferson's Secretary of the Treasury, Albert Gallatin, advocated government intervention in the economy that would have generally benefited all manufacturers, from small to large scale. Gallatin supported protective tariffs for American manufactures, a national transportation network, and a central bank. With these he reflected the argument for public works in a free-market economy made by Adam Smith, prophet of economic liberalism.[5] Gallatin's remarkable *Report on Public Roads and Canals* (1808) outlined a national transportation network of roads and canals spanning the country north–south from Maine to Georgia and east–west from the Atlantic to the Western rivers. Proposals to link Lake Champlain with the North River (the Hudson) reached into Markham's Vermont, bringing the industrial benefits (as Smith proposed) of new divisions of labor released by a widened extent of the market.[6] In all sections of the country, ambiguities such as these among the Jeffersonians helped to undermine, but not overcome, Americans' outright hostility to a European type of industrial system.

Economic and technological realities also played their part. In the first decade of the nineteenth century an international division of labor was beginning to appear in the global cotton industry. Raw

---

[2]Jefferson, *Notes on the State of Virginia*, query 19, in Folsom and Lubar (eds.), 16–17.
[3]Hamilton, *Report*, in Folsom and Lubar (eds.), 84.
[4]The debate over the disadvantages and merits of industrialization has been the subject of a considerable historical literature, for which see the Introduction in Folsom and Lubar (eds.), *Philosophy of Manufactures*.
[5]Nelson, *Liberty and Property*, 37–51, *passim*.
[6]Gallatin, *Report on Public Roads and Canals*, 8, 67–68, especially A. Smith, *Wealth of Nations*.

cotton reached manufacturers in Europe from agricultural economies in the tropics. Britain, with its fast-developing cotton industry, imported raw cotton, chiefly into Liverpool and, secondarily, London, supplied *inter alia* by numerous West Indian colonies; Turkey; Brazil; Surinam; the East Indies; and, largest of all, sending more than the rest put together, the United States. Not much cotton then came from India, partly because long-fiber, fine Indian cottons were retained for indigenous hand spinners; partly because short-fiber Indian cottons were unsuitable for machine spinning.[7] What placed the United States at the head of the world's cotton suppliers was the suitable climate of its southern states (albeit north of the tropics), the invention of the cotton gin, and the expansion of the institution of slavery.

For the global and American cotton industries the situation drastically changed after 1794. That year Eli Whitney patented a toothed cylinder gin, which other gin makers developed into the saw gin. Productivity with the saw gin was allegedly fifty times that of the older roller gin.[8] Thereafter roller gins were increasingly favored for long-fiber cottons; saw gins for short-fiber cottons. Thus, on the supply side, innovations in cotton gins and the institution of slavery in America's Southern plantations loosened restraints on flows of raw cotton into the global cotton economy. On the demand side, Lancashire's manufacturers relied on their organizational and technological competitive advantages to expand their export markets for cotton goods and on the Royal Navy for the opportunity during two decades of global war to exploit those advantages.[9] In this situation the United States's Southern planters increased their exports of cotton to Britain from nearly 0.2 million pounds weight in 1793 to 3.7 million in 1796, 15 million in 1800, 44.9 million in 1807, and 54.4 million in 1815.[10] For American manufacturers and engineers, access to supplies of the highest qualities of cotton without the burden of transatlantic transport costs would represent a significant competitive advantage. Not least it contributed to their scope for making changes to the imported British technology.

From the cotton fields, mechanization spread further in the South albeit on a domestic scale as Jefferson himself, still unreconciled to a thoroughly industrial manufacturing sector, privately but proudly admitted in 1812, "We have reduced the large and expensive machinery for most things to the compass of a private family, and every family of any size is now getting machines on a small scale for their household purposes."[11]

A major factor in tipping the balance toward manufacturing came as a result of strained relations with Britain, the fast-emerging first industrial nation. For American manufacturers protection from foreign, especially English, competition was critical. Infant industries needed captive markets, it was believed. These were imposed by the breakdown in Anglo-American relations, followed by war, between 1807 and 1814. Wartime separation from European markets persuaded a growing body of political opinion to overcome Jeffersonian anti-industrialism and to take steps to protect American manufactures. The tariff of 1816, arguably essential for the infant American cotton industry, was the government's first important protectionist measure.[12]

News of what exactly was going on in the first industrial nation, Britain (and was spreading to Europe), reached the United States through a number of movements and events that in sum may be denominated a transatlantic knowledge revolution. In the cotton industry in 1790 changes in British cotton manufacturing sprang from a series of foundational innovations. John Kay's fly shuttle of 1733,

---

[7]J. Thomson, "Cotton"; Edwards, *Growth of the British Cotton Trade*, chapter 5.
[8]Lakwete, *Inventing the Cotton Gin*, 48–49, 72–73, *passim.*
[9]Singleton, "Lancashire Cotton Industry," 61–63.
[10]Edwards, *Growth of the British Cotton Trade*, 251.
[11]Jefferson to Thaddeus Kosciusko, 28 June 1812, quoted in Folsom and Lubar (eds.), 28.
[12]Rosenbloom, "Path Dependence."

which had relaxed the constraints on weaving cloth but increased the demand for cotton yarn, was no more than a modification to the traditional handloom. The great spinning inventions—James Hargreaves' spinning jenny (1770), Richard Arkwright's water frame (1769), Samuel Crompton's spinning mule (1779)—were being modified by an increasing number of "improvements" (not all of them technically and commercially viable) in the 1790s and 1800s. The search for a viable power loom received a new fillip with the Rev. Dr. Edmund Cartwright's inventions of 1785–1788, but it would take more than another twenty years to develop a satisfactory production-line machine. Nevertheless the modern factory system, centered on machine spinning, was firmly established in Britain by the 1800s.

Americans learned about Europe's (and Britain's in particular) new technical knowledge in various ways. British newspapers and periodicals relayed knowledge about commercial aspects of the new technology. For example, when Richard Arkwright, pioneer of factory cotton spinning, died in 1792, the *Gentleman's Magazine* reported that he left wealth estimated at £500,000; thereby advertising the rewards awaiting new cotton entrepreneurs. Americans reprinted European titles when "vast" imports of books could not satisfy transatlantic demand.[13] By 1831 over 6,000 items of technical interest had been published in the Colonies and their successor states.[14] Indeed, between the 1790s and the Civil War a knowledge revolution transformed America. Old equilibria shifted, from the religious to the secular and commercial, from the elitist to the democratic, as reflected in publications, informal public interest, and the formation of institutions.[15] This interpretation has been reinforced by Daniel Walker Howe's recent synthesis of American history between 1815 and 1848. Howe emphasizes the period as "a time of a 'communications revolution,'" enabled by the arrival of the steam-powered rotary printing press, the reorganization of the U.S. postal system, and, in 1844, the introduction of Samuel Morse's electric telegraph.[16] Public information from New York City on average took 10 days to reach a Vermont town like Middlebury; by 1841 this time was cut to five days. With progressively growing ease, ideas and information spread across the Republic.[17] Technical knowledge also widened and deepened. Scientific knowledge, itself advancing, increasingly informed mechanic practices.[18]

The Markham family (see chapter 1) was schooled and raised in a world of books as well as practical trades. Although many American publications were brief and ephemeral works, they included two encyclopedias that had previously been published in Britain: Rees's *Cyclopaedia* and the *Edinburgh Encyclopedia*. The former (see Appendix 2) was the earliest reliable repository of descriptions of Britain's first wave of industrial technologies. Then there were patents. British specifications and/or drawings were spasmodically published in the London *Repertory of Arts* from 1794 onward. By 1820, of a range of patented textile inventions, a third (43 out of a possible 121) were published and presumably read by interested Americans.[19] Rarely, tools or machines were imported from Britain because, until after 1825, the British government imposed a blanket prohibition on the export of textile, metal-working, clock-making, leather-working, paper-making, and glass-manufacturing equipment.[20]

Until Parliament lifted restrictions in 1824, the emigration of skilled workers was also banned by the British government. However, determined individuals could and did evade Customs' scrutiny. During

[13]Jeremy (ed.), *Wansey*, 155.

[14]Rink, *Technical Americana*.

[15]R. D. Brown, *Knowledge Is Power*; Sinclair, *Philadelphia's Philosopher Mechanics*; R. Thomson, *Structures of Change*, 190–228, *passim*.

[16]Howe, *What Hath God Wrought*, 5. See review by Merritt Roe Smith in *Technology and Culture* 50, no. 1 (January 2009).

[17]Alan Pred, *Urban Growth and the Circulation of Information* (Cambridge, Massachusetts: Harvard University Press, 1973) quoted Howe, *What Hath God Wrought*, 224.

[18]Acutely aware of a growing and damaging gulf between theoretical sciences and practical mechanics, Olinthus Gregory published his *Treatise of Mechanics* in 1806 in an explicit attempt to bridge the two domains.

[19]Jeremy, *Transatlantic Industrial Revolution*, 55, 264–266 and author's data.

[20]Jeremy, "Damming the Flood."

the War of 1812 over 1,300 Britons were registered as alien immigrants working in American textile trades. Some 60 percent were operatives, 164 were manufacturers (on their own or others' account), 125 were general machine makers, and 60 were specifically textile-machine makers. From the mid-1820s many more immigrants arrived from Britain with textile skills, 5,000 arriving at Boston, New York, and Philadelphia in the eight years 1824–1831. Although half were handloom weavers, the 365 textile manufacturers and 152 textile-machine makers could have supplied up to half the actual expansion in the country's textile-manufacturing industry during the 1820s.[21] There was, in fact, an "international fraternity of mechanicians" in England, France, and the United States. Between 1820 and 1840 perhaps as many as 300 or 400 master mechanics, with over 5,000 journeymen and apprentices, were connected in networks across the English-speaking world, which they linked by word of mouth, letters, co-working, or moving among machine shops. Skilled in metalworking and familiar with the latest machine tools, these were the international carriers of new mechanical knowledge.[22]

Movements of skilled workers, both across the Atlantic and within their destination countries, would typically be channeled through networks. These might be family, trade, religious, or specifically organized emigration networks (to mention the most obvious kinds) for those crossing from England to the United States. Such transatlantic mechanisms have been richly documented by the late Charlotte Erickson and are further illustrated in the Hollingworth letters.[23] Within the United States concentrations of mills or workshops attracted the development of further trade networks. Machinists, with their metalworking, and machine-building skills, formed the most important regional and interregional networks. As David Meyer has shown, they collected in centers or "nests" for training, more experience, and exchange of ideas, forming networks of artisans along which knowledge could readily pass. One such nest was Almy and Brown's cotton mill at Pawtucket near Providence, Rhode Island, where in 1790 the English immigrant Samuel Slater had introduced the region to Arkwright technology from England. Slater's relatives and pupils fanned out into the localities of Providence, the Quinebaug Valley in Connecticut, upstate New York, southern New Hampshire, and Philadelphia. Through these networks machinists could move into mill management on the manufacturing side of the industry.[24] Ross Thomson has amplified this picture, drawing attention to the role of machinists in developing innovations and stressing their complementarity with institutions promoting the diffusion and generation of knowledge.[25] Although networks within America multiplied, intensified, and became interwoven within and between regions and industries, international exchanges did not cease. In textiles alone, after the early 1820s Americans added to mechanized spinning and power-loom weaving the British-sourced technologies of calico printing, the mechanized gig mill, and the self-acting spinning mule. In the opposite direction, the differential-geared cotton-roving frame, a hemp-spinning machine, and cotton ring spinning were taken to England in the late 1820s and 1830s.[26]

News of developments summed in the sensational size of the Arkwright fortune attracted American interest. Slater's capitalist backers at Pawtucket, the Quaker merchants Almy and Brown, were among a number of American entrepreneurs alerted by Arkwright's success. Slater introduced Arkwright's continuous-spinning technology to Rhode Island, transplanting organizational arrangements with which he had grown up. He employed a family-based workforce in factories equipped with both spinning mules for

---

[21]Jeremy, *Transatlantic Industrial Revolution*, 160–164, 174, 268–275.
[22] Wallace, *Rockdale*, 211, 219; Stapleton, *Transfer*.
[23]Erickson, *Invisible Immigrants*; Leavitt (ed.), *Hollingworth Letters*.
[24]Meyer, *Networked Machinists*.
[25]R. Thomson, *Structures of Change*, 1–33.
[26]Jeremy, *Transatlantic Industrial Revolution*, *passim*; Jeremy, *Technology and Power*, 14–15; Lozier, *Taunton and Mason*, 169–170, 245–271.

producing fine yarn and Arkwright's throstle frames for spinning coarser (lower count) yarns. By 1815 Slater had interests in six firms in southern New England and his spinning-mill model was widely emulated there and in states southward.[27]

Meanwhile the Hamiltonian proindustrialists were busy, engaging in international industrial espionage; recruiting skilled workers from England; and investing in experimental, often short-lived, manufacturing enterprises based on new technology.[28] The most influential of these was the Society for Establishing Useful Manufactures, brainchild of Hamilton and his associate Tench Coxe. Incorporated by the New Jersey legislature in 1791, its four-story mill opened in 1794 at the Great Falls of Passaic, near Newark, NJ, on a 700-acre site named Paterson in honor of the governor of New Jersey. As an experiment it was not a great success. Cotton manufacturing ceased in 1796, but the corporation was not dissolved and it was revived after 1814.[29] This mill has a precise significance in relation to Markham because a drawing of the Falls of Passaic (see Fig. 1.5, p. 14), preserved in his papers, suggests he visited New Jersey and it links him directly to the first generation of American industrial enterprise.

Much more successful than the Society for Establishing Useful Manufactures was another major American investment in cotton manufacturing. With foresight, powers of observation and memory, and no little cunning, the Boston merchant Francis Cabot Lowell, visiting Scotland and Lancashire in 1810–1812, gained sufficient knowledge of the power loom to carry its design back to Boston.[30] At nearby Waltham, he and other merchants heavily invested in a wartime cotton-manufacturing enterprise—much like other patriotic capitalists who were projecting cotton- and woolen-spinning factories on never-failing streams scattered between Delaware and beyond in the South and Maine in the North.

In northern New England, the organizational and technological patterns established by F. C. Lowell and the Boston Manufacturing Co. at Waltham, Massachusetts, in 1813 became the archetype for cotton manufacturing (Fig. 2.1). In essence, Lowell and his collaborators integrated power-loom weaving with machine spinning (in contrast to Slater's spinning-mill model). And they met the lack of skilled male workers, required in fine spinning to operate mules, by recruiting the unskilled and respectable daughters of the region's farming families to spin low-count yarns and weave coarse cloth. To do this the imported English technology was specially adapted by a series of labor-saving innovations (such as stop-motions).[31] Markham's technical outlook was shaped by men who were familiar with the Slater technology—likewise, he was influenced by Waltham technology because he spent some time working in the Boston Manufacturing Co. mills.

By 1820 the American textile industries were well established on lines recognizable to British eyes. The industry then had nearly 650 firms manufacturing cotton and wool, of which 386 worked in cotton spinning and weaving. Over 60 percent of the cotton firms were located in New England and another 20 percent in New York and Pennsylvania. Of the woolen firms, 44 percent were in New England and 21 percent in New York. From the government's returns of the 1820 Census of Manufactures, it appears that the cotton industry had as many as 357,000 spindles, and the woolen industry 56,000 spindles. Power looms were spreading in the cotton industry, with some 86 firms recording 1,665 power looms.

---

[27]Tucker, *Samuel Slater*, 89–111; Prude, *Coming of Industrial Order*, 34–64; see also Conrad, "Drive that Branch."

[28]For the multiplicity of means and ventures by which European technologies were taken across the Atlantic in this early industrial period see Stapleton, *Transfer*; Jeremy, *Transatlantic Industrial Revolution*; Jeremy, *Technology and Power*; Jeremy, *Artisans*.

[29]Davis, *Essays* I, 349–518; Cooke, *Tench Coxe*; Ben-Atar, *Trade Secrets*.

[30]Dalzell, *Enterprising Elite*.

[31]Ware, *Early New England Cotton Manufacture*; Dublin, *Women at Work*; Jeremy, *Transatlantic Industrial Revolution*, 180–203.

FIGURE 2.1 View of Boston Manufacturing Co. mills at Waltham, Massachusetts, by Isaac Ebenezer Markham, ca. 1819–1821, watercolor sketch (26 x 42.5 cm). This is the earliest known view of the Boston Manufacturing Co. mills at Waltham.

The view is taken from the north side of the mills, probably from near a line of single-storey structures that are visible in an 1830 painting viewing the mills from the south side, across the Charles River.

The building with the cupola (*to the right of the picture*) is the first mill; built 1813–1814, it measured 90 feet long by 45 feet wide and was four storeys above the basement. The second mill (*to the left of the picture*), built 1816–1818, measured 150 feet long by 40 feet wide. Its external but adjoining stair tower was a novel feature that was repeated on the south side of the mill. The 3,584 spindles in this mill became the unit of measurement of one mill power in northern New England. See Candee, "Architecture and Corporate Power in the Early Waltham System"; Jeremy, *Transatlantic Industrial Revolution*, 97.

Courtesy of the Sheldon Museum.

In contrast only seven woolen-manufacturing firms had power looms, 43 in all.[32] Because of its fiber structure, wool was more difficult to spin and weave by mechanical means than cotton. The spinning jenny and versions of the mule could produce woolen yarn.[33]

The average firm, having fewer than 1,000 spindles, was therefore small and at Middlebury, Markham worked in one of them. In 1820 it belonged to J. Hough & Co. and ran 840 spindles and 24 power looms with 8 men, 34 women, and 3 children. Its only local competitor, John Warren's factory, had 500

[32]Jeremy, *Transatlantic Industrial Revolution*, 276–283.

[33]In the jenny and the mule, drafting and twisting were simultaneous within one zone of the machine, so fibers were not pulled apart before they were twisted into yarn. See chapter 6, section 5.

spindles and 8 power looms run by 3 men, 20 women, and 4 children. These two mills produced sheeting and yarn—Hough's factory produced No 15 yarn, and Warren's produced No 14 yarn.[34] Their yarns supplied their own power looms but some presumably went to handloom weavers in rural households. Yet other yarns, doubled into thread, were bought by household consumers.[35]

The biggest American textile firms looked modest compared to the giants in Britain, where as many as 1,000 persons worked for the largest Manchester cotton firms.[36] In 1820, seven years after its incorporation, the Boston Manufacturing Co., the largest American cotton manufacturer, processed 450,000 pounds of New Orleans and Upland cotton per annum; employed 26 men, 225 women and girls, and 13 boys; and ran 5,376 spindles and 175 looms; even so, the Boston Manufacturing Co. employed just over a quarter of the number of workers the British firms required.[37] The Boston Manufacturing Co. consumed nearly two to four times the amount of cotton as each of its three nearest rivals, employed twice as many workers, though its spindleage was exceeded by that of J. G. Pierson—who operated the Ramapo Works at Hempstead, Rockland County, NY, in the Hudson Valley. This firm and the other two rivals of the Boston Manufacturing Co (the Matteawan Factory at Fishkill, Dutchess County, New York, also in the Hudson Valley, and Almy, Brown & Slater at Smithfield, Providence County, Rhode Island) each consumed approximately 100,000 pounds of cotton annually.[38] The Dover Cotton Factory, in Dover, New Hampshire, was above average in size, with 2,000 spindles; 50 power looms; employing 15 men, 105 women and girls, and 5 boys. It consumed three to four bales of cotton a week, which meant 60,000 to 75,000 lb. annually.[39] Without exception, all these companies used waterpower to drive their machinery, so numerous, large, and powerful were the rivers needed, invariably flowing eastward, on the Eastern seaboard of the United States. Until the late 1820s no attempts were made to power textile mills with steam engines, chiefly because they were uneconomic.[40]

The 1820 Census of Manufactures shows that the Boston Manufacturing Co. was producing nothing but shirtings and sheetings, as did the Dover Cotton Factory. The big firms in southern New England added stripes, plaids, and ginghams. The Schenks at Matteawan also made bedticks, tablecloths, diapers, and chambrays. Stripes and plaids, like shirtings, could be made into clothing. This left sheeting to be turned into bedding, curtains, or other household textiles.[41] After the tariff of 1816 this production would have gone into domestic markets—Eastern cities, Western states, and the slave population of the South.[42]

The relative profitability of the Waltham business encouraged its promoters to project similar large-scale enterprises on other waterpower sites in Massachusetts, beginning in 1822 with Lowell (named after the prime mover who died in 1817) on the Merrimack River. Sites in southern New Hampshire followed. The Appletons, Jacksons, Lowells, later the Lawrences and others, merchants and shipowners, members of the "Boston Brahmins,"shifted their investments from trade into manufacturing, establishing

[34]US Census of Manufactures, 1820, National Archives film M-279, VT doc 278.

[35]The household market for thread is hinted at by Markham's record of a thread fineness measurement formula. See Figure 6.86.

[36]Lloyd-Jones and Lewis, *Manchester and the Age of the Factory*; Lloyd-Jones and Lewis, "Size of Firms."

[37]US Census of Manufactures, 1820, National Archives film M-279, MA doc 33.

[38] J. G. Pierson had 5,820 spindles and 73 power looms, consuming 94,061 pounds of (mostly) Georgia Upland cotton annually. Almy, Brown & Slater ran 5,170 spindles and 70 looms and consumed 190,000 pounds of cotton. The Matteawan Factory ran only 1,600 spindles and 44 power looms but consumed 100,000 pounds of cotton. US Census of Manufactures, 1820, National Archives film M-279, NY docs 1409 and 1425 RI doc 4 (revised return); Jeremy, *Transatlantic Industrial Revolution*, 276.

[39]US Census of Manufactures, 1820, National Archives film M-279, NH doc 16. A bale of cotton weighed 3$^1/4$ to 3$^1/2$ cwt. J. Thomson, "Manufacture of Cotton."

[40]Hunter, *Steam Power*, 105–109; Jeremy, *Technology and Power*.

[41]US Census of Manufactures, 1820, film M-279, NH doc 16 MA doc 33 NY docs 1409 and 1425 RI doc 4 (revised return).

[42]With the protection of the 1816 tariff and its extensions in the 1820s, Samuel Slater could report in 1832 that American cotton manufacturers had "the whole domestic market for the coarser fabrics in their possession" (*McLane Report* I, 931).

a new industrial system pivoted on Boston. If Boston became the capital and credit hub of this industrial system—and the vital port link (rivaled by New York City) with Europe, it was buttressed by social networks visible in fine houses on Beacon Hill and the city's Congregational churches, as well as Harvard College and Massachusetts General Hospital.[43] This was the sort of world in which Markham, well below the Boston Brahmins in wealth and status, lived; not insignificantly contributing to the technology of the industrial base on which the new wealth of the Beacon Hill world rested.

His lifetime witnessed the years of infancy in an industry that unevenly grew from a dozen or so workshops equipped with no more than carding machines and jennies in 1790 (such as had been exhibited in the Independence Day parade of 1788) to an international giant, which by 1850 had 3.6 million cotton spindles (the third largest spinning industry in the world, behind Britain and France) and 95,700 power looms (second largest weaving industry in the world, behind Britain).[44]

During the narrow period 1810–1825, the major organizational and technological foundations of the American cotton industry were laid down: a vertically integrated organization coupled with labor-saving technology using high-quality cotton from the South and serving markets demanding utilitarian cotton goods of low quality for clothing and household textiles. Protection from foreign cotton imports, which had ended when war against Britain concluded in 1814, was resumed in 1816. Congress imposed a protective tariff of 6¼ cents a yard on all imported cotton cloth (whatever the real price or cost). This choked the postwar influx of cheap cottons from British India and effectively secured for American cotton manufacturers low-quality markets in New England, the Western states, and the South. [45] Prospects for promoters of industry in Vermont improved further after 1817. That year the New York state legislature authorized two canals: the Erie Canal, linking Albany on the Hudson River to Buffalo on Lake Erie; and the Champlain Canal, linking Albany to Lake Champlain. The former was completed in 1825 but the latter in 1823.[46] Markets would now be appreciably extended and, with increased demand, possibilities for manufacturing, machine shop, and inventive opportunities would multiply.

Given these propitious circumstances for the growth of the American textile industry in Markham's lifetime, what did he achieve as a machine maker? What explains his achievements? Three achievements stand out. He served as a catalyst to the introduction of power looms in northern New England. He was responsible for modifications that allowed the imported technology to meet local economic and social circumstances. And as a skilled machinist and machine maker he was an exponent of the new technique of projective, rather than perspective, machine drawing.

The references to Markham by the local historian Samuel Swift depend on the testimony of Joseph Gordon with whom, apparently, Swift had some conversation. As Swift tells it, Gordon brought the vital power loom plans from Scotland, relied on his own abilities as a machine maker, and used his first-hand knowledge of the operation of power looms. Markham simply "manufactured the iron part of the machinery."[47] Does this suggest a diminution of Markham's role? Understandably Gordon might inflate his own competence and contribution. What is unknown is the quality of the plans Gordon brought from Scotland. The fact that he had to rely on Markham for the iron parts itself indicates that Gordon's skills were limited. It is very possible that when he arrived in America, Gordon may not have been as

[43]Dalzell, *Enterprising Elite*.

[44]Information provided by the late Professor Douglas A. Farnie.

[45]Taussig, *Tariff History*, 30. See also Samuel Slater's letters to the Secretary of the Treasury, Louis McLane, March and 25 May 1832, *McLane Report*, I, 927-32, II, 876–877.

[46]Taylor, *Transportation Revolution*, 33–35.

[47]Swift, *Middlebury*, 331.

versatile as he later became. In the 1820s, Scottish cotton manufacturers, if not machine builders, tended to be specialized either in spinning or in weaving.[48]

Markham's skills, on the other hand, were far better proven. When in 1812, as a seventeen-year-old, Markham started to work in David Page, Jr.'s Middlebury cotton factory, he learned about machine building from John Houghton, whose prime skill derived from cabinetmaking (see chapter 1). A year or two later Markham learned metalworking skills from Jesse Holton. Holton was evidently more experienced and knowledgeable than Houghton and, originally a blacksmith, he had become a partner in a New Ipswich cotton factory, the machinery for which was constructed by a former employee of Samuel Slater (see chapter 1). In 1817 Markham obviously widened his repertoire after Joseph Gordon arrived. Two years later Markham was ready to move on. In readiness he obtained a handsome testimonial letter from his employer, David Page. It attested to his competence in building cotton and woolen machinery, including power looms, and identified his skills as those of draftsman, machinist, and superintendent of the machine shop. Markham was capable of doing all the necessary iron and brass work (after forging) and his power looms so satisfied Page that he would not exchange them for any other of which he was aware. Page concluded "the drawings he has is his of his own execution they speake for themselves" (see chapter 1, footnote 31).

Before he was twenty-five, Markham had become an all-round textile-machine maker. More than that, he was clearly experimenting with modifications to the technology to which Gordon could lay no claim because the modifications related to machinery used during other stages in a production line that occurred prior to weaving. Four are prominent. The carding-machine arrangement for feeding cotton into the machine, which extended the feeding cloth to the ground, was presumably designed to minimize the strains of lifting heavy loads of cotton off the ground (Fig. 2.2). The other three mechanisms seem to have a common feature. In several machines Markham used twin-cone pulleys to effect smooth changes in rotary motion, which was possibly the distinctive feature in his dresser design (Figs. 2.3–2.9). His flywheel spinning motion aimed to improve the regularity and increase the force in the factory power transmission system (Fig. 2.10). Similarly, his batten motion (Fig. 2.11) was an attempt to find a mechanical simplicity consistent with production for basic styles of cloth in rural markets. No doubt many more contemporary mechanics were likewise searching for their own trade-offs between mechanical simplicity and complexity in power loom designs, in response to localized clothing styles, machine construction, maintenance and operation costs, and the quantity and quality of labor supplies. These three types of motions evidence a quest for modifications that would meet John Farey's criterion for efficient mechanical movement: the achievement of "great uniformity of motion."[49]

However, the goal in these mechanisms was not simply technical perfection. They had important implications for operative satisfaction and management relations with labor. The dressing machine that used Markham's twin-cone pulley and a Waltham stop-motion, for example, promised to make a big difference in the process of dressing, then recognized as the most unpleasant procedure in the cotton factory.[50] If it reduced the number of yarn breakages, the girls who tended the dressers would find their work less arduous, labor turnover would fall, and throughput would increase. Markham, who was critical of the highhandedness shown toward employees by the management of the Dover and Waltham companies (see chapter 1, p. 14), would have received special satisfaction from these labor-alleviating, as well as labor-saving, improvements.

[48]Jeremy, *Technology and Power*, 4–5.
[49]Farey, "Machinery."
[50]Jeremy, *Transatlantic Industrial Revolution*, 195–196.

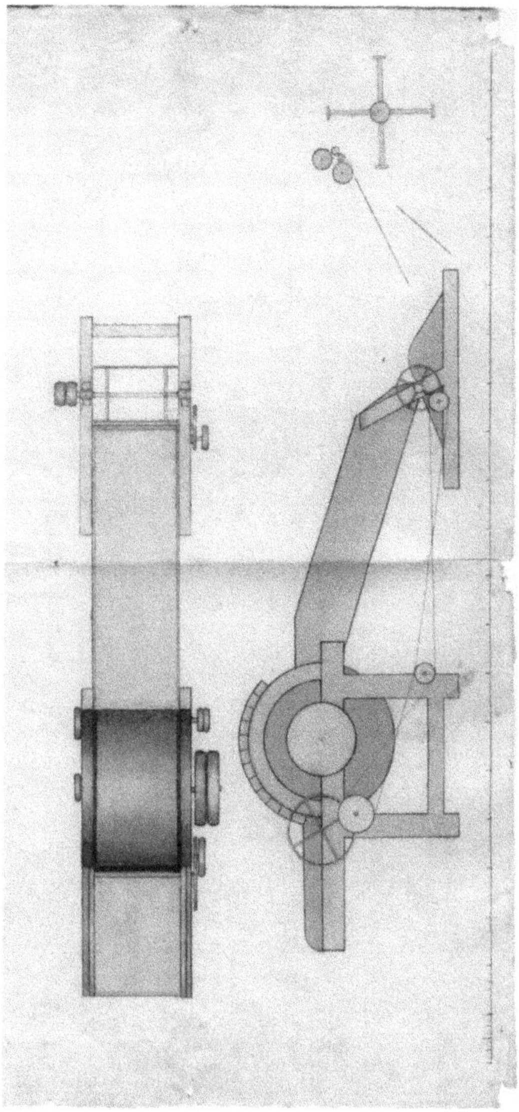

FIGURE. 2.2 Markham's cotton-carding machine-feed or delivery system. See also Figure 6.4 (p. 130) for further analysis.

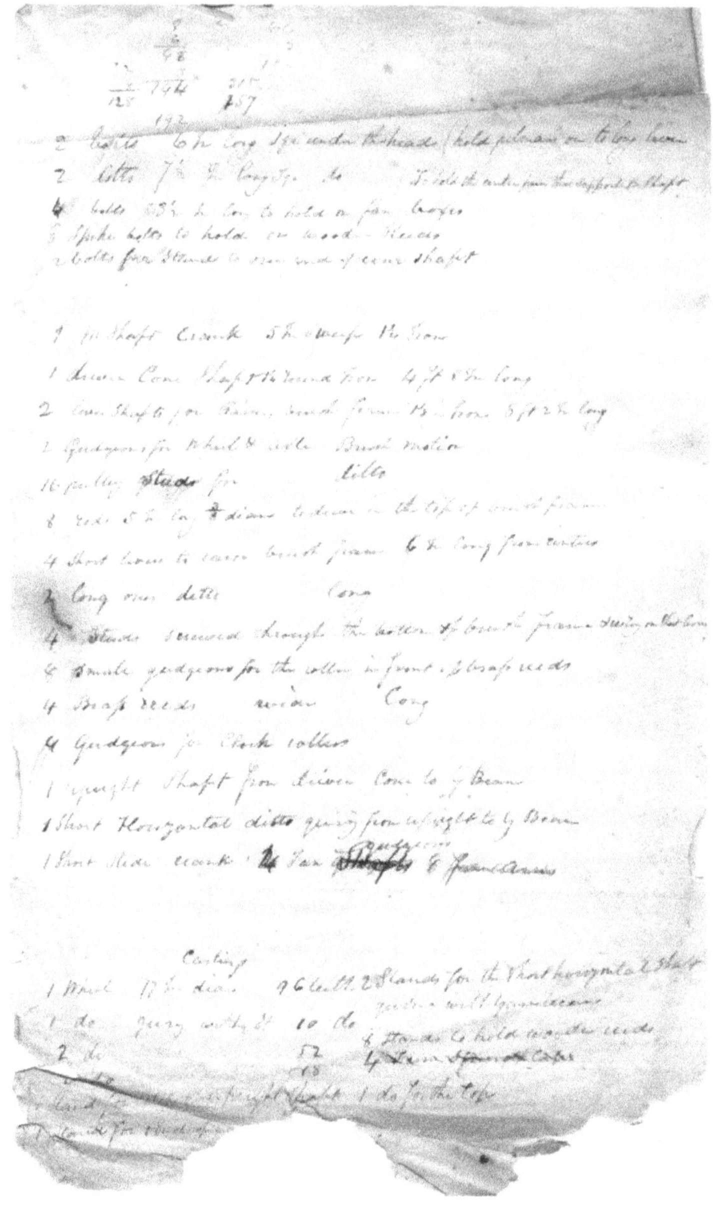

FIGURE 2.3 Loose sheet—inventory of patterns. Dressing machine. See also Figure 6.17A, 6.17B (pp. 160, 162) for further analysis.

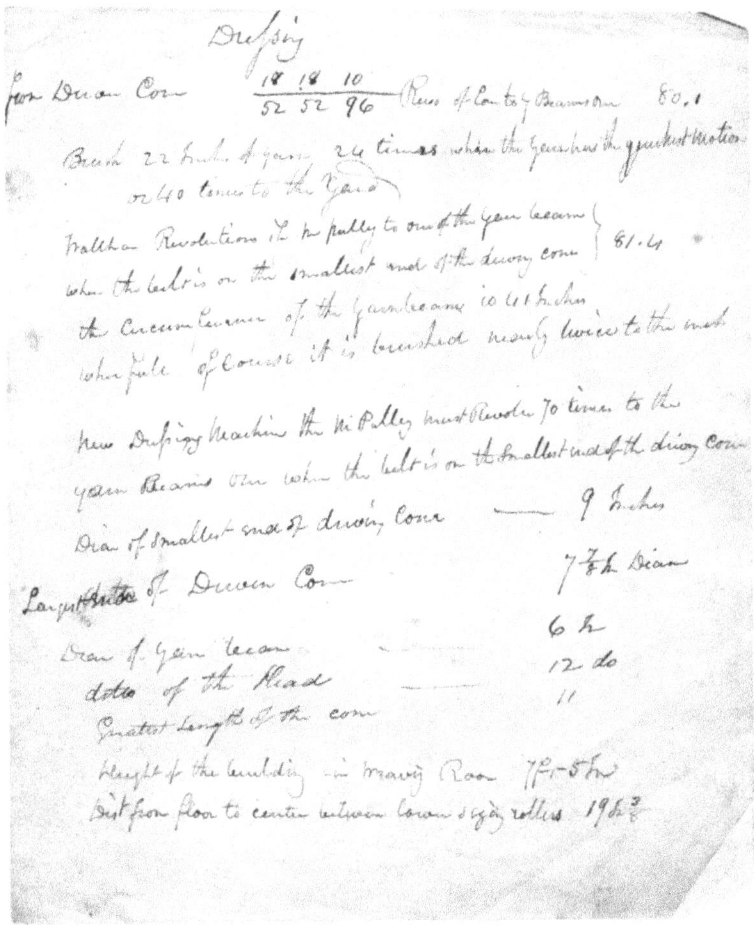

URE 2.4 Loose sheet. Dressing machine. See also Figure 6.22 (p. 174) for further analysis.

If Markham's mechanisms were wholly of his own invention there remains the puzzle of why he failed to gain patent protection for them. This is discussed more fully in chapter 3 in the section titled "Visual Aspects of the Generation of New Knowledge in Early–Anglo-American Industrial Engineering." Assuming that any one of his distinctive mechanisms was solely of Markham's devising, the most reasonable explanation seems to be that he did not regard it as important or profitable enough to warrant the undoubted struggle and costs of defending the invention in the courts.

Markham's skill as a machinist and machine maker would have been vastly increased by his skill as a draftsman using the new conventions of projective, rather than perspective, machine drawing. Competence in machine drawing allowed him to release the creative ideas in his head, try them out on paper, add alternatives, experiment before he or his employer was committed to the expense of casting, forging, and building prototypes. In this way solutions to technical problems could be given a preliminary testing. It moved mechanics and engineers toward the goals of creativity without limit, experimentation without cost, and construction without ambiguity. The ability to see differently, on which projective machine-drawing skill was predicated, distinguished the master machine builder and engineer from the machinist, confined as the latter would have been (as at Waltham and to a lesser extent in a small machine shop

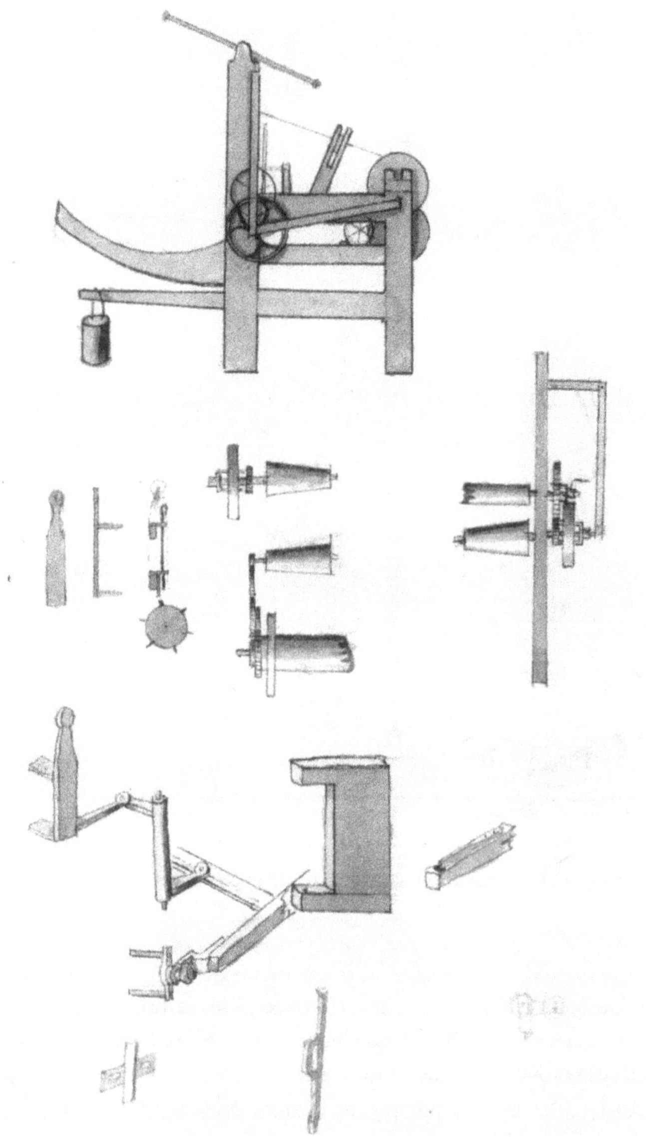

FIGURE 2.5 Power-loom mechanisms. See also Figure 6.32 (p. 194) for further analysis.

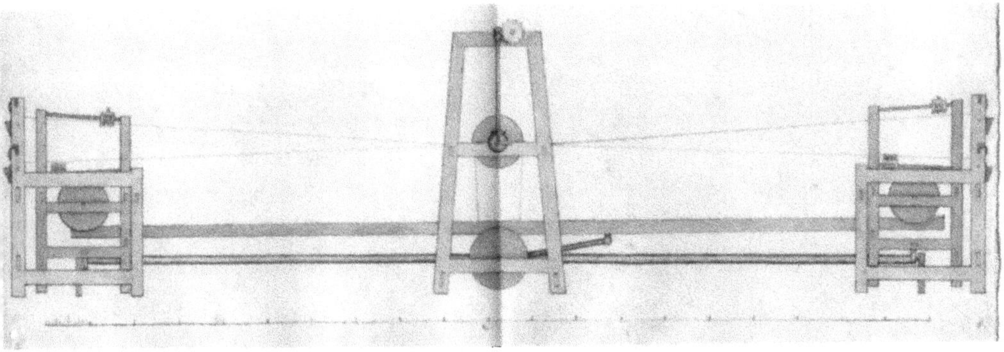

FIGURE 2.6 Dresser, side elevation. See also Figure 6.75 (p. 294) for further analysis.

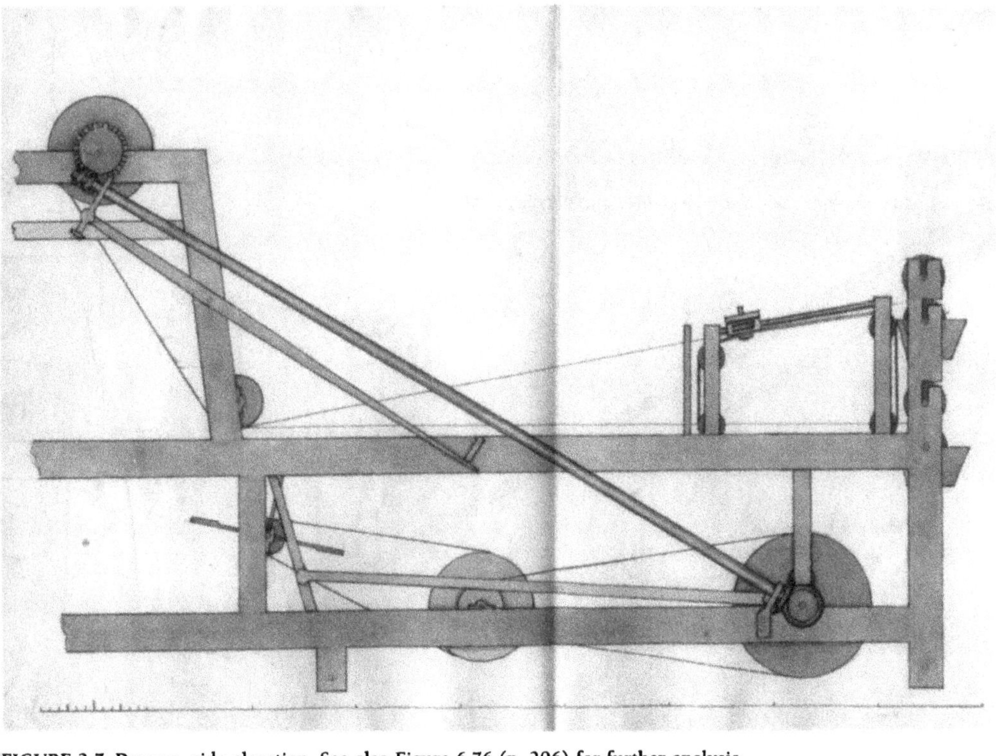

FIGURE 2.7 Dresser, side elevation. See also Figure 6.76 (p. 296) for further analysis.

like that at Middlebury) to the workshop production of a few parts. A decade or two after Markham's demise, projective machine drawing would be one of the gateways into the profession of engineer in the United States, as it already was in the French military world. The development and significance of machine drawing is examined more extensively and closely in chapter 3 and Appendix 5.

How can Markham's achievements be explained? His association with the introduction of an early power loom to New England can only be attributed to luck. He happened to be in the right place at the right time: in a workshop where a skilled immigrant (Gordon) turned up with the essential plans and

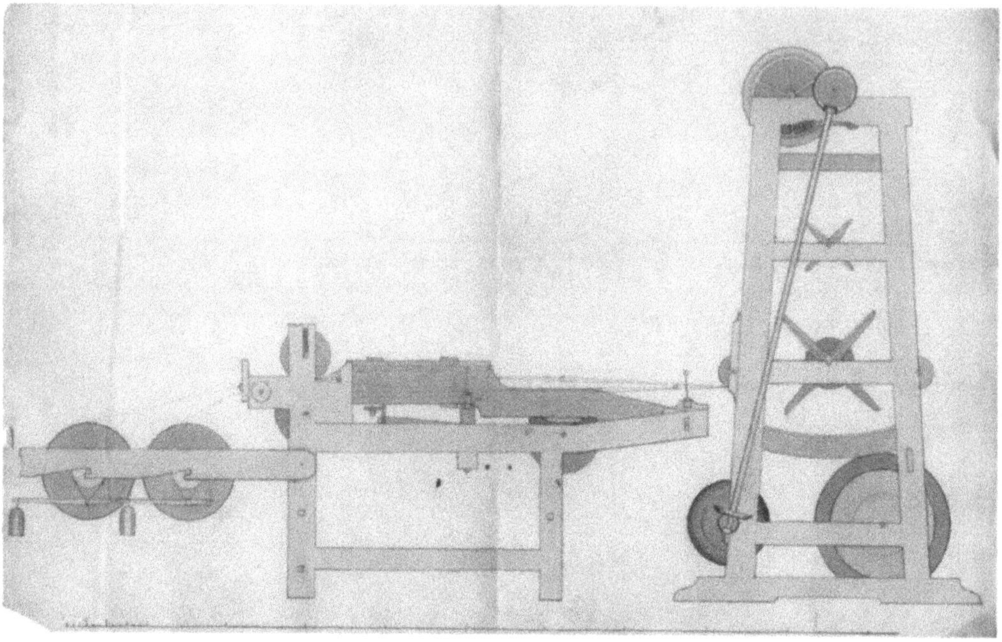

FIGURE 2.8 Dresser, side elevation. See also Figure 6.77 (p. 298) for further analysis.

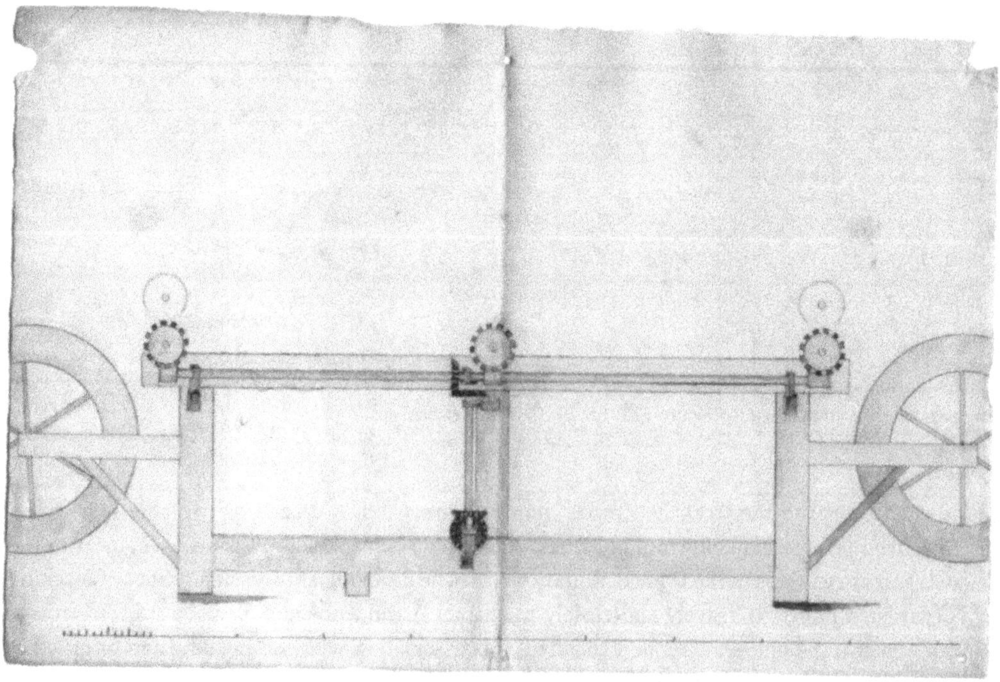

FIGURE 2.9 Dresser (?), side elevation. See also Figure 6.78 (p. 300) for further analysis.

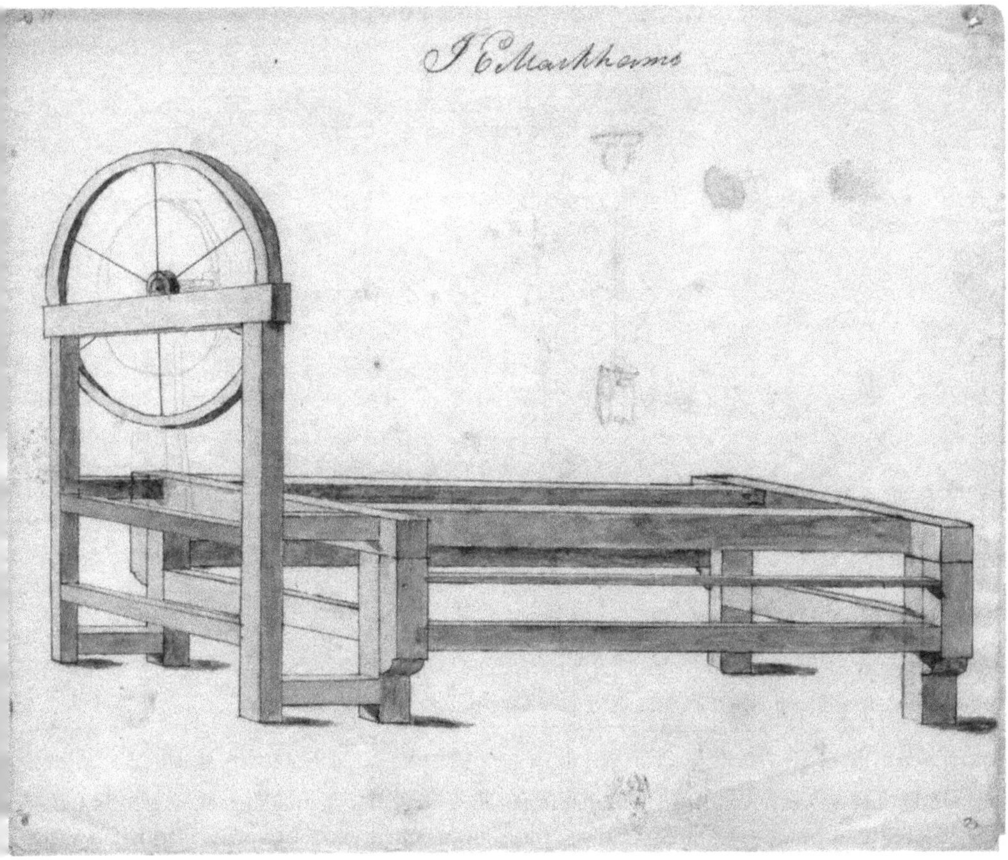

FIGURE 2.10 "I. E. Markham's" spinning machine, perspective sketch. See also Figure 6.65 (p. 273) for further analysis.

knowledge. Furthermore, that immigrant's arrival occurred only months after the design of the first reliable power loom was taken across the Atlantic and built in Rhode Island (by William Gilmour). Consequently, Markham participated in the pioneering transfer of power-loom technology from Britain to America.

As Louis Pasteur, the French chemist and bacteriologist, observed in 1854, "*le hazard ne favorise que les esprits prepares*" (chance favors only the prepared mind).[51] So it was with Markham. His achievements issued from a willingness to make the most of opportunities. However, those opportunities were molded partly by the networks within which he grew and matured, partly by the values that he imbibed and lived and worked by.

One important network, for which there is no detailed surviving evidence, was the network that existed inside the Middlebury Cotton Factory machine shop. It seems to have been a minimal network with, possibly, no more than two members at one time: Markham and one other machine maker. The latter—in turn Houghton, Holton, and then Gordon—probably were serially in charge until Markham proved his worth and succeeded them.

[51]*Oxford Dictionary of Quotations.*

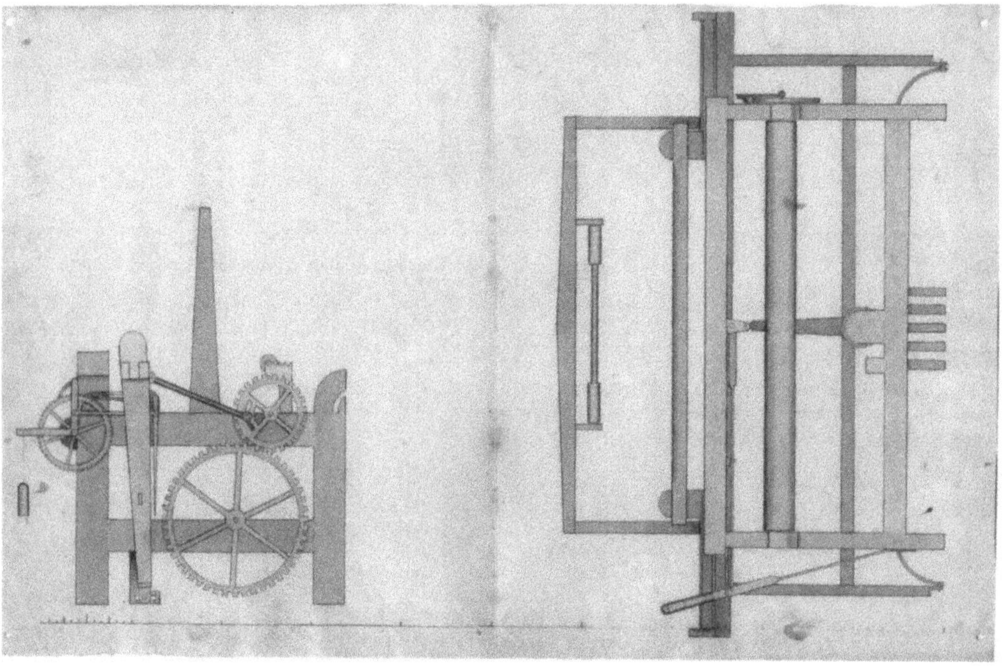

FIGURE 2.11 Power loom, side and front elevations. See also Figure 6.80 (p. 308) for further analysis.

The networks and values at work in Markham's family offered him confusing choices. His father plainly valued education because he hired a private tutor for his children. Yet Captain Markham's spells in prison for debt were hardly an encouragement to the pursuit of an independent business career. He was also an adventurer, risk-taker, entrepreneur, and experimenter, as exemplified by his forge and nail making. Isaac Markham's brother, David, chose a safer career and trained with a local lawyer. Isaac was obviously attracted and excited by what he saw and heard in his father's nail-making business and his tavern, enough to be drawn into the new and practical trade of machine making.

Other social networks opened up for Markham. When he was growing up, the local freemasons' lodge met in Markham's father's tavern. Freemasonry "had been an important force in the young republic. Its members had constituted a kind of Republican elite, with Benjamin Franklin and George Washington prominent among them."[52] Among the members of the Union Lodge at Middlebury was Daniel Chipman, an attorney who served in the state legislature and later in Congress. After Markham's father died in 1813, Chipman kept a close eye on the young Markhams and took the elder of them, David, to train in his office. Chipman had an economic interest in Page's cotton factory and it was here that Isaac Markham started work at the age of seventeen in 1812.

Chipman's influence in Isaac Markham's career became especially important because, until 1818, Daniel Chipman was the only resident of Middlebury who was listed as a subscriber to Abraham Rees's *Cyclopaedia*. This massive work ran into 45 volumes for the London edition and 47 volumes for the Philadelphia edition. Published alphabetically in parts, running at two a year for almost two decades,

[52]Howe, *What Hath God Wrought*, 267.

and completed in 1820, the article on the manufacture of cotton, with its detailed textual descriptions and excellent plates, threw open the window onto Britain's cotton-manufacturing technologies. It is inconceivable that Chipman, with patriotic and economic motives and Masonic commitment to mutual assistance, would refrain from sharing the textile-manufacturing installments of Rees with David Page and his machine makers in the Middlebury cotton factory (see Appendix 2). As only 2,000 copies of the *Cyclopaedia* were printed in America, or one set per 4,000 of population, Markham was specially privileged in his access to new knowledge from Britain.

His decision to move between the cotton mill machine shops of New England and possibly elsewhere, rather than staying in Middlebury, placed Markham in the networks and fraternity of tramping machinists and machine makers, a fraternity with international extensions, as indicated earlier. The larger textile mill companies had machine shops, concentrations of machinists, and access to more capital for machine making. Internalizing the machine-building function made economic sense, particularly if, like the Boston Manufacturing Co., the companies could create a market demand for machinery and sell to smaller manufacturers who needed to limit their overheads. In the summer of 1820 the Boston Manufacturing Co. employed 36 machinists and 4 blacksmiths. That year in the Hudson Valley, the Matteawan Co. had 16 machine makers. These were relatively high proportions of the workforce: approximately 13 percent at Waltham, and 10 percent at Matteawan, compared to 2 percent at Middlebury, assuming that Gordon was still there.[53]

Tramping among these large machine shops, machinists like Markham could expect to encounter different forms of organization and different technologies. The Boston Manufacturing Co.'s machine shop at Waltham, for example, was likely to have been far better equipped than the machine shop at Middlebury. In 1817 at Waltham there was $4,867 worth of equipment, including two cutting engines, one fluting engine, four roller engines, one polishing machine, and eight lathes for turning.[54] Markham's exposure to these various machine shops would also enlarge his acquaintance with machinist networks. But for Markham, at a critical point in his life, the family network took precedence over his machinist networks. In May 1821, dissatisfied with both the Dover and Waltham managements, he turned to his brother for advice on where to settle next (see chapter 1, p. 14).

These networks were undergirded by value systems. Again, only glimpses of those that surrounded Markham can be caught from the evidence. When he was a small child the Episcopal Church met for a time on his father's premises. After his death, Markham's funeral service was conducted by a Congregational minister, the Rev. Josiah Hopkins. Ten years younger than Markham, Hopkins came from the New Haven Congregational Church, eight miles north of Middlebury, where he had ministered since 1808. A native of Vermont, he combined a frontier earthiness with the compassionate nature of a pastor. Like Markham, he lacked college training and made up for it by private study. He was also a Freemason, though not a very active one. In an age of religious divisions Hopkins harmonized his congregation, bridging the old tensions between Halfway Covenant rationalism and New Light emotionalism. On the one hand he ran a theology class that attracted a following of young men—on the other hand his ministry saw two religious revivals, in 1816–1817 and 1822—Vermonters shared in the enthusiasms that were concentrated in upstate New York's "Burned-Over District."[55]

Hints of Markham's own churchmanship lurk in his obituary: His "extensive acquaintance with polite literature" and his "uniform temperance and sobriety of habits, and warmth and tenderness of affection"

[53]US Census of Manufactures, 1820, film M-279, NH doc 16 MA doc 33 NY doc 1425.
[54]Gibb, *Saco-Lowell Shops*, 49.
[55]Farnsworth and Rodgers, *New Haven in Vermont*, 120–124; Cross, *Burned-Over District*, 3–13; Howe, *What Hath God Wrought*, 164–176.

(see chapter 1, p. 15). Such traits would have found Hopkins's ministry congenial. At any rate the values of the Protestant Christian faith and ethic surrounded Markham, not least the ideals of good neighborliness, honesty, frugality, self-discipline, and hard work.

Freemasonry, still respectable and numbering among its members the most respected in the land, such as the Governor of New York, DeWitt Clinton (sponsor of the Erie Canal), promoted the values of the Enlightenment: reliance on reason, the pursuit of new knowledge, respect for science. Yet it also practiced "secrecy and mutual assistance," which after the infamous Morgan case of 1826, brought Freemasonry into disrepute in the United States.[56]

Markham's choice of occupation identified him as an industrializer and put him on the side of Hamilton rather than Jefferson, indicating he was committed to manufacturing, rather than mercantile or agricultural pursuits. His adoption of the latest improved techniques in power-loom weaving and in machine drawing show he had a determination to find and use the newest available knowledge in this field. His choice of career also implied a patriotism that would have led Markham to endorse the sentiments of the Schenck family, owners and operators of the Matteawan factory, when in 1820 they informed the Census of Manufactures returning officer, "No calculations was predicated on a continuance of the late war, but a firm belief that our country *must* manufacture for themselves."[57]

From the evidence assembled in Appendices 3 and 4, it is clear that Markham was of an independent cast of mind. He declined to settle down either at Dover, New Hampshire, or at Waltham. Nor did he slavishly copy either Waltham textile technology or the English textile technology transmitted through Rhode Island or Scottish immigrant, Gordon. His time at Waltham places him as a member of the Waltham mechanic network but one who brought his own ideas to the center and then moved back to the periphery to exercise independent creative thinking. That he returned to Middlebury to manage both factory and machine shop, attainments reached by the time he was thirty, may be linked to the expanding opportunities associated with the Canal boom then widening markets for East Coast manufacturers, implying that he held optimistic views for the future of the cotton business in his native community.

The qualities remembered by his obituarist echoed some of Markham's values. Some qualities bore on his moral relationships with others: "noble and generous feelings...independent and manly sentiments ...warm and unwavering friendship...undeviating and exact rectitude." Some qualities related to his capacity for work: "relying upon his own exertions, with an untiring assiduity." Some qualities induced his practical skills and ingenuity:

> [W]ithout the advantages of birth or fortune, by the force of talents and virtue alone [which rather overlooks Markham's early network advantages]...a mind of singular felicity of invention, originality and strength of conception, and of rich and abundant resource...carefully cultivated his uncommon powers until he had become one of the first mechanicians in this section of our country...talents of the highest order—mechanical and inventive powers of the richest cast.[58]

These qualities were external, evident to his friends and proven in the products of his workshop. Shaping those qualities were values that can only be guessed at: set by his religion, his Masonic code, his reading about Enlightenment values, or a mixture of these? The slenderness of the evidence denies absolute certainty, but these institutions would definitely have affected his life and his values. Clearly Markham had a very strong reputation in his community. In theory the values underlying a strong

[56]Howe, *What Hath God Wrought*, 267–268; Tyler, *Freedom's Ferment*, 351–358.
[57]1820s Census of Manufactures, New York returns, document 1425.
[58]*National Standard*, Middlebury, 20 September 1825.

reputation would have cemented his place in networks of machinists, induced customer loyalty, and enhanced the status of machinists/engineers. Rather more needs to be known about the values underpinning American machinist networks and structures of innovation.[59]

Markham's biography and his papers tell us much about the means by which American manufactures were advanced. They cast light on the international diffusion of textile machinery and skill; the inventive activities of New England textile machine builders; the workshop procedures of early-nineteenth-century machine builders; and, at a local level, the speed with which the industrial frontier spread into the remote rural parts of the infant American Republic. The drawings provide a rare opportunity, for the early years of the knowledge and communication revolution, to explore the nature of technical visual knowledge: The kind of knowledge that would have been at hand in the entrepreneurial machine-maker's workshop as he devised and assembled custom-built machines and incrementally pushed forward technological frontiers.

[59]Obituaries, despite the caveat *de mortuis nil nisi bonum est*, offer a way of exploring reputations and values. For related studies of nineteenth-century industrial values see Cochran, *Railroad Leaders*; Diamond, *Reputation of Businessmen*; MacLeod, *Heroes of Invention*; Sunderland, *Social Capital*.

Chapter 3

Visual Knowledge in the Workshop: The Wider Significance of Markham's Papers

☆ ☆ ☆ ☆ ☆ ☆ ☆

A VISUAL KNOWLEDGE APPROACH

In the last three or more decades the phenomenon of knowledge, under the impulse of the computer and biotechnology revolutions, has captured interest in many of the social sciences.[1] A knowledge revolution of a different kind, only recently explored as such, infused industrialization in the Western world during the late eighteenth and early nineteenth centuries.[2] So far, the visual knowledge dimension of this earlier transition has been relatively neglected. Markham's drawings may be understood as exemplifying an element of the visual side of the Western knowledge revolution that transformed the infant American Republic between the 1790s and the Civil War.[3] Small in quantity though Markham's drawings may be; large they appear when examined as pointers to wider aspects of the early industrial knowledge revolution.

How can visual knowledge be explored? The Markham drawings may be presented simply as examples of early textile machines, with technical commentary on how they presumably worked, and sometimes about their economic significance. That is the purpose of the editorial apparatus accompanying each drawing. However, more can be extracted from them.

Several kinds of knowledge attach to any tool, machine, or technological system. Each has information about design (purpose of device, operating principle, shape, components, materials, and so on), information about how to build a design (dimensions, materials, component interrelationships, skills, tools, etc.), and information about the operation of the built design (rules and routines to start, continue, stop, and maintain). Associated with each of these kinds of knowledge will be the underlying science, and also conditional information, about economic costs (like cost to build or cost to operate); social costs (like

---

[1]The breadth, complexity, and contested terrains of cognitive science have been historically and magisterially surveyed in Boden, *Mind as Machine*. Those disciplines influenced by the knowledge revolution can be glimpsed in a variety of texts: in economics, Cyert and March, *Behavioural Theory of the Firm*; in management, Easterby-Smith and Lyles (eds.), *Blackwell Handbook of Organisational Learning*; in economic history, Mokyr, *Gifts of Athena*. Eisenhardt and Santos, "Knowledge-Based View: A New Theory of Strategy?" in Pettigrew, Thomas, and Whittington (eds.), *Handbook of Strategy and Management*, 139, declare "The knowledge movement is sweeping through the field of strategy." In science and in the computer revolution, visualization and visual knowledge have played an increasingly important role; Barrow, *Cosmic Imagery*.

[2]Mokyr, *Gifts of Athena*.

[3]R. D. Brown, *Knowledge Is Power*. This interpretation has been reinforced by Daniel Walker Howe 's assessment of the role of expanding communications in America between 1815 and 1848, *What Hath God Wrought*.

kinds of worker needed or the impact on the community or the environment); and related technology costs (as implied in the location and skill levels of suppliers of metals and so on). Some of this technological knowledge may be distanced from the tool, machine, or technical system. It can be embodied in skilled persons or in "inscriptions" (Latour's word[4]), such as verbal directions, models, pictures, drawings, formulae, symbols, and so forth. Mechanical drawings convey information about the nature of the mechanisms depicted. They also disclose information about the manner of depiction. Living in an age of emergent science, Markham made drawings that therefore tell us much about how a mechanism worked and how a mechanism could be represented on paper (two dimensional, or 2D knowledge).

The territory where art and engineering overlapped was first surveyed by Francis D. Klingender in 1947. However, his observations on engineering drawing owed more to his aesthetic sensibilities than to technical and economic considerations.[5] In his history of engineering drawing, Peter Booker, Secretary of the Institution of Engineering Designers, emphasized that engineering drawings served both as an instrument of design and as a means of instructing machine builders.[6] Likewise, for Baynes and Pugh,

> Engineering drawing was a product of the Industrial Revolution. ...A dramatic and powerful tool...which made possible a new relationship between management and manufacture and separated the process of design from the process of construction.[7]

Hindle was one of the first American historians of technology to explore engineering drawings: "original drawings come very close to mental conceptions, and machinery components embody the concepts."[8] Latour, a sociologist, ingeniously endeavored to bring all graphic expressions in all scientific disciplines within one universe. He saw all two-dimensional papers and signs, with the knowledge they conveyed, as links between the anonymous multitude of workers and the relatively few individuals who make the big scientific and technological discoveries, or who head the great institutions of state.[9] Anatomizing the twentieth-century engineering-design process, Vincenti, an aeronautical engineer, identified plans and drawings as "design instrumentalities," interacting with concepts, criteria, theoretical tools, quantitative data, and practical considerations.[10] As an economic historian, Mokyr, analyzing the emergence of the knowledge economy, thought plans allowed the "streamlining of access to knowledge," thereby reducing the cost of knowledge.[11] Ferguson, coming from an engineering background, got closer to the essential nature of engineering plans and drawings. Although plans and drawings may store knowledge and reduce costs to accessing it, and serve as an intrinsic part of the complex modern engineering design process, they are, in his words, evidence of "the mind's eye" at work.[12] John K. Brown concluded that after 1850 the techniques of machine drawing gave engineer-managers enlarged capacity for trying new designs, transmitting technical knowledge throughout the engineering community, extending control over the production process, enlarging markets by cross-industry standardization, and regulating admission to the engineering profession.[13]

---

[4]Latour, "Drawing Things Together," 21–22.
[5]Klingender, *Art and the Industrial Revolution*; compare his 72–77 with sources cited here and in Appendix 5.
[6]Booker, *History of Engineering Drawing*, 188.
[7]Baynes and Pugh, *Art of the Engineer*, 11.
[8]Hindle, *Emulation and Invention*, 59, Appendix 5.
[9]Latour, "Drawing Things Together."
[10]Vincenti, *What Engineers Know*, 208.
[11]Mokyr, *Gifts of Athena*, 61.
[12]Ferguson, *Engineering and the Mind's Eye*.
[13]Brown, "Design Plans."

Art for aristocratic pleasure and art for technical use faced similar dilemmas in perception and representation, as well as reception. Over the course of millennia, human drawing and painting consisted largely of 2D representations of the world. Then in the early fifteenth century the perspective techniques, particularly the single-viewpoint perspective, discovered by Brunelleschi and codified by Alberti in Florence, became ascendant in the studios of Western painters, a situation lasting until the late nineteenth century.[14] With the twentieth-century shift back to 2D representations of the world, which was more challenging for the perception of the image, art historians could not agree whether nature or nurture shaped graphic representation. Famously, Erwin Panofsky argued that each historical period shaped its own perspective. Conversely, Ernst Gombrich believed that the persistence of the rule-driven-perspective forms of representation satisfied the "gravitation towards the schematic" because "It is even harder to see the visible world as a two-dimensional field than it is to see one's own image on the mirror's surface."[15] Doubt whether art was based on social construction or on physiology cast ambiguity over understandings of visual engineering knowledge. Since Gombrich's day, art historians have explored overlaps of art and science but none that I have seen has incorporated the long-term development of mechanical drawing.[16]

In the past thirty years or so, researchers in evolutionary and cognitive psychology have intensified investigations into the working of the human mind. One large and complex subfield of psychology is visual perception. "How we see" is not fully known and many findings are disputed.[17] However, a popular text offers several interim assessments that seem to have a bearing on Markham's world. One conclusion is that although the human eye registers on its retina the two dimensions of the material world, the third dimension, depth, is gauged by the human brain and is stored and interpreted through the millions of signals received by the retinas. This provides the human mind with stereoscopic vision, a phenomenon first discovered by Charles Wheatstone in 1838.[18] A second finding is that stereoscopic vision has to be learned.[19] A third finding is that, though most people see stereoscopically, our 2D perceptions are much sharper than our depth of vision is.[20] However, as Pinker emphasizes, mental images are frequently subject to mental reorganization; they are ambiguous if not inaccurate; and they cannot serve as concepts for the meanings of words.[21]

How does all this inform our understanding of Markham's workshop drawings? Essentially, it provides three starting points. First, in his lifetime the portrayal of a 3D (three-dimensional) object on a 2D surface was an exercise conventionally and habitually performed by following the Renaissance rules of perspective. Perspective was a powerful and enduring technique because, in presenting the human eye with the illusion of depth, it satisfied the human mind's need to assume physicality in the real world. Second, faced with the new world of industrial machinery, artisans were torn between the traditional perspective view of the world on the one hand, and, on the other, the deficiencies of that view in foreshortening or hiding surfaces and dimensions, vital data for the machine builder. Third, seeking a satisfactory solution to their problem, machine makers had to find ways of thinking about and looking at machines that ran

[14]Honour and Fleming, *World History of Art*, 9–11; Edgerton, *Renaissance Discovery of Linear Perspective*, 3–15; Kubovy, *Psychology of Perspective*.

[15]Gombrich, *Art and Illusion*, 205n, 247, 278; Snyder, "Picturing Vision" in Mitchell (ed.), *Language of Images*; Edgerton, *Renaissance Discovery*, 155–165.

[16]Kemp, *Science of Art*; Kemp, *Seen/Unseen*; Jones and Galison (eds.), *Picturing Science*.

[17]Professor Mark Mon-Williams, a cognitive psychologist, cautions me against easy acceptance of contested conclusions reported in popular texts, but thinks that the historical evidence on mechanical drawing will be of interest to vision scientists.

[18]Pinker, *How the Mind Works*, 218–237.

[19]Pinker, *How the Mind Works*, 237–241.

[20]Pinker, *How the Mind Works*, 259: because 2D is recorded by the human retina but the depth of vision by the workings of the mind, "we have an almost palpable sense of surfaces and the boundaries between them."

[21]Pinker, *How the Mind Works*, 287–298.

counter to the widely subscribed perspective schemata of representation.[22] This pointed them back to 2D representation (followed for centuries by architects and others). This meant focusing on two-dimensional surfaces rather than allowing the mind to think about the default choice of stereoscopic vision. The mental 2D image then had to be projected on paper as a 2D representation. Nor should the difficulties in doing this be underestimated, as art with an average class of teenagers demonstrates. Weaknesses inherent in mental images introduced increasing mathematical or geometrical discipline to representation skills, sometimes accompanied by verbal explanation. The next section describes the representational alternatives tried by those attempting to capture a visual record of the new world of machines before the mid-1820s.

SEEING IN A DIFFERENT WAY: THE CHANGING NATURE OF VISUAL REPRESENTATION IN EUROPEAN ENGINEERING BEFORE THE MID-1820S

Markham lived in a period of drastic and rapid change in the engineering world, not least in the ways in which visual knowledge might be perceived and represented. This section inquires into visual representation prior to 1825. How did Europeans portray machinery in general before and during Markham's lifetime? How were textile machines depicted for machine makers? What handbooks were available to guide machine makers?

Under the influence of the dominant-perspective paradigm, a stream of "theatres of machines" was published between the sixteenth and eighteenth centuries. The line began with Jacques Besson (1578), included Agostino Ramelli (1588), and culminated in Jacob Leupold's monumental work, *Theatrum Machinarum* (10 volumes, 1724–1739).[23] The typical drawing in these volumes deployed the perspective view.[24]

Some divergence came in France where, under the influence of the *philosophes*, at least six compendia of knowledge were produced.[25] The *Encyclopédie* (33 volumes, Paris, 1751–1780), edited by Denis Diderot and Jean d'Alembert, was the largest, with 3,100 plates. Its purpose was to rationalize and publicize the artisan's world of work, by categorizing and illustrating tools and machines, thereby undermining the old restrictive and secretive artisanal corporations.[26] The *Encyclopédie's* plates usually consisted of a mixture of perspective views of workshops, machine plans, and elevations of large machines (like handlooms) and close-ups of components and instruments, often shaded to convey the impression of depth.[27]

---

[22]This is the central premise on which this essay on visual knowledge is based. It was gratifying to discover that Ken Alder, in his work on eighteenth-century French military technology, identified the same visual shift over a decade ago in an article I did not see until a week or two before this manuscript went to press. However, Alder did not try to test this shift in the way attempted here. My thanks to Roe Smith for pointing me to Alder, "Making Things the Same."

[23]Ferguson, *Engineering and the Mind's Eye*, 115–130.

[24]I checked Gnudi and Ferguson (eds), *Various and Ingenious Machines of Agostino Ramelli*, which consists almost entirely of perspective views; and Leupold, *Theatrum Machinarum*, vol. 5, *Schauplatz des Grundes mechanischer Wissenschaften*: a mixture of perspective views, some derived, like Plate LII, Savery's steam engine; and visually curious orthographic views (parallel lines indicating depth of view) like the friction machines in Plate XXX, Fig. 1, and many places elsewhere. Yet Leupold also included 2D sections, suggesting a weakening of the perspective tradition.

[25]Ferguson, *Bibliography*, 57–60.

[26]Alder, *Engineering the Revolution*, 135–139; Alder, "Making Things the Same."

[27]Diderot et al. (eds), *Encyclopédie*. I checked the copy in the John Rylands University Library of Manchester, for drawings in Plates vol. XI (1772), the last plates volume to be published, and looked closely at those relating to "*Tisserand*" (weaving), eight in number, to reach this assessment; which cannot therefore be an exhaustive one.

The 1,700 plates in another giant French encyclopedia, the *Descriptions des Arts et Métiers* (45 volumes, Paris, 1761–1788), sponsored by the Academie des Sciences, comprised half-page views of workshops accompanied by half-page (or more) illustrations of related implements or machine parts.[28] The former were perspective views, usually on the upper part of the page; below, the detailed drawings were often 2D drawings made to appear 3D by means of shading. In both these encyclopedias the important aid (for a machine builder) of scale (in *pieds*) accompanied whole-machine views, but not usually the close-ups of implements and parts.

Although French artisans were being educated by the *philosophes*' great encyclopedias, an elite of French military engineers from the 1760s was being trained, like architects, to use mathematical and geometrical techniques to represent fortifications, buildings, siege engines, naval ships, and the like. As Alder has shown, staff in the engineering schools rejected perspective drawings as inadequate for engineers instructing subordinates. One of them, the Revolutionary Gaspard Monge, pioneered "multiple, orthogonal, projective views": "a '[universal] language necessary to all those who work in the mechanical arts' because it allowed one 'to represent with exactitude on drawings which have two dimensions, those objects which have three, and which can be rigorously defined.'" [29] In modern engineering terms, "Only orthographic projection, in which two or more views are projected on mutually perpendicular (orthogonal) planes, is able to show the true size and shape of an object without foreshortening or distorting some of the lines. All of the pictorial representations, such as isometric or perspective projections, show the object as it appears to the eye, but accurate dimensions cannot be scaled from them."[30] Orthographic projection minimized ambiguity in the transfer of visual information passing from designer to constructor.[31]

Monge was little known (if at all) among British mechanical engineers in Markham's lifetime. Instead, industrial engineering provided the impetus for a shift away from the perspective paradigm and toward a more customary use of 2D drawings derived from much older architectural practice. This is clear from the conventions used in English machine drawing published in books between 1780 and 1825. Some 27 of these are analyzed in Table 3.1.

Several points emerge from examination of the table. First, there were different kinds of books for mechanics. There were handbooks, outlining scientific principles (as then understood), like those of Banks, Emerson, Imison,[32] Brunton (who provided useful tables applicable to " the steam engine, water wheel, force pump, and mechanics in general" (to quote the subtitle), and the ambitious volumes of Gregory—who was anxious to bridge the gap between scientific and practical approaches. Then there were those that took individual machines and described and depicted them carefully. The works of Bailey, Blunt, Buchanan, Smeaton, Smith, White,[33] and the contributors to Rees were of this sort. Markham,

[28]Academie Royale des Sciences, *Descriptions*. I checked the copy in the John Rylands University Library of Manchester, for drawings in volume XXV (1763) for the article "*Art de la Teinture en Soie*" (the art of dyeing silk) by M. Macqueur (six folio plates): again, this cannot be an exhaustive appraisal.

[29]Alder, *Engineering the Revolution*, 139 and *passim*; Alder, "Making Things the Same"; Booker, *History of Engineering Drawing*, 86–113; J. K. Brown, "Design Plans."

[30]Belofsky, "Engineering Drawing," 25.

[31]How much, or rather how little, machine-drawing information is necessary for a mechanic to build a replica machine depends on several factors: the mechanic's prior knowledge, the complexity of the machine, and the machine shop resources available. It also depended on the kind of mechanical drawings available. As noted in Appendix 5, mechanical engineers came to rely on three kinds of drawing: the preliminary picture, which established the form of the machine; workshop drawings, which gave the exact shapes and dimensions of the machine's components; and the finished or general layout drawing, showing where all the components fitted together. All three are needed for making an accurate copy of a machine. (I am grateful to Michael Middleton, CEng for this professional opinion.)

[32]Discussed in Musson and Robinson, *Science and Technology*, 101–118 and *passim* for their milieu. Emerson and Imison are in the *Oxford Dictionary of National Biography (ODNB)*.

[33]Of these three, only Smeaton has an entry in the *ODNB*. For White, see Dickinson, 'James White'; Jeremy, "Technological Diffusion."

Table 3.1 Drawing Conventions in Twenty-seven English Books for Mechanics 1780–1825 (588 Plates)

	Short title	Author	Publication date	Number of plates	Perspective	Elevation	Plan	Diagram	Scale
Chambers	Cyclopaedia, sv Mechanics		1781	7	7	1		6	
Bailey	106 Copper Plates	p	1782	106	64	26	20		13
Imison	School of Arts	p	1785	16	13	6		3	
Imison	Treatise on the Mechanical Powers	p	1794	2	2	1		2	
Banks	Treatise on Mills		1795	3	1	1		1	
Smeaton	Experimental Enquiry	p	1796	5	4	2		1	3
Imison	School of Arts	p	1796	24	20	6		13	
Donne	Essay on Mechanical Geometry		1796		No machine drawings whatsoever				
Emerson	Principles of Mechanics		1800	43	41	6	1	13	1
Banks	On the Power of Machines		1803	3	1	1		2	
Gregory	Treatise of Mechanics		1806	55	42	34	11	34	2
Duncan	Practical Essays on Weaving	p	1808	14		9	5	7	
Blunt	Essay on Mechanical Drawing	p	1811	63	33			30	
Rees	Cyclopaedia, sv Manufacture of Cotton (Thomson)	p	1812	14		14	7		
Martin	Circle of Mechanical Arts, sv Cotton		1813	2	1	2			
Buchanan	Practical Essays on Mill-work	p	1814	15	2	15	6		
Smith	The Mechanic		1816	104	78	87	55	9	5
Buchanan	Practical Treatise on Propelling Vessels by Steam	p	1816	16	3	15	6	2	1
Rees	Cyclopaedia, sv Weaving (Duncan)	p	1818	2	2				
Rees	Cyclopaedia, sv Woollen Manufacture (Bakewell)	p	1818	5	5				
White	Memoir on...Toothed Wheels	p	1819	2		1		2	
Gray	Treatise on Spinning Machinery	p	1819	10	1	10	10		10
Farish	On Isometrical Perspective		1820	3	3			1	
White	Century of Inventions	p	1822	50	13	47	18	50	10
Fenwick	Essays on Practical Mechanics	p	1822	3	2	3		1	
Buchanan	Practical Essays on Mill-work	p	1823	19	4	17	5	1	
Brunton	Compendium of Mechanics	p	1824	2				2	
	totals, 1780-1825			588	342	304	144	180	45
	p = practical men	18		100%	58%	24%	52%	31%	8%

Notes: Because many plates, and often the figures within the plates, incorporate a variety of drawing conventions, these percentages do not add up.
Perspective includes emphatic shadows and shading indicative of depth.
Sections are usually subsumed under elevations.

incidentally, was familiar with Smeaton's work (Fig. 3.1) and the contribution to Rees (see Appendix 2), and may well have been aware of other authors represented in this table.

For the most part, these authors were interested in revealing and communicating machine designs. Bailey, Blunt, the authors of the Rees articles, Smith, and White all presented galleries of machines and mechanisms. Interestingly, very few images were of textile machines. These four authors (excluding the great work edited by Rees) presented over 300 plates, each plate often had numerous figures illustrating a wide range of machines. Smith, for example, offered all manner of mechanisms for the mine, the workshop, the farm, and the home.

A third feature of these books is that, in some, their drawings were intended to aid machine construction. Those of Bailey, Buchanan, Smeaton, Smith, and White were in this category. Thus, in compiling "delineations" of models in the collections of the Society for the Encouragement of Arts, Manufactures,

le 3.1 *(continued)*

	Short title	Author	Publication date	Number of plates	Perspective	Elevation	Plan	Diagram	Scale
mbers	Cyclopaedia, sv Mechanics		1781	7	7	1		6	
ey	106 Copper Plates	p	1782	106	64	26	20		13
son	School of Arts	p	1785	16	13	6		3	
son	Treatise on the Mechanical Powers	p	1794	2	2	1		2	
ks	Treatise on Mills		1795	3	1	1		1	
aton	Experimental Enquiry	p	1796	5	4	2		1	3
son	School of Arts	p	1796	24	20	6		13	
ne	Essay on Mechanical Geometry		1796		No machine drawings whatsoever				
rson	Principles of Mechanics		1800	43	41	6	1	13	1
ks	On the Power of Machines		1803	3	1	1		2	
gory	Treatise of Mechanics		1806	55	42	34	11	34	2
can	Practical Essays on Weaving	p	1808	14		9	5	7	
it	Essay on Mechanical Drawing	p	1811	63	33			30	
	totals, 1780–1812			**341**	**228**	**93**	**37**	**112**	**19**
	p = practical men	7							
				100%	68%	28%	11%	33%	6%
s	Cyclopaedia, sv Manufacture of Cotton (Thomson)	p	1812	14		14	7		
tin	Circle of Mechanical Arts, sv Cotton		1813	2	1	2			
hanan	Practical Essays on Mill-work	p	1814	15	2	15	6		
th	The Mechanic		1816	104	78	87	55	9	5
hanan	Practical Treatise on Propelling Vessels by Steam	p	1816	16	3	15	6	2	1
s	Cyclopaedia, sv Weaving (Duncan)	p	1818	2	2				
s	Cyclopaedia, sv Woollen Manufacture (Bakewell)	p	1818	5	5				
te	Memoir on...Toothed Wheels	p	1819	2		1		2	
y	Treatise on Spinning Machinery	p	1819	10	1	10	10		10
sh	On Isometrical Perspective		1820	3	3			1	
te	Century of Inventions	p	1822	50	13	47	18	50	10
wick	Essays on Practical Mechanics	p	1822	3	2	3		1	
hanan	Practical Essays on Mill-work	p	1823	19	4	17	5	1	
nton	Compendium of Mechanics	p	1824	2				2	
	totals, 1813–1825			247	114	211	107	68	26
	p = practical men	11							
				100%	48%	89%	45%	29%	11%

and Commerce in London, William Bailey asserted their purpose to be "for the benefit of the public [and] and to enable the artizan [*sic*] to construct it [the machine] from the description."[34] In the first two decades of the nineteenth century the best of these mechanical drawings and descriptions, though none had scale, were the series of essays by Robertson Buchanan, a Scottish millwright-engineer.[35] His

[34]Bailey, *One Hundred and Six Copper Plates*, Preface.
[35]Moss, "Robertson Buchanan."

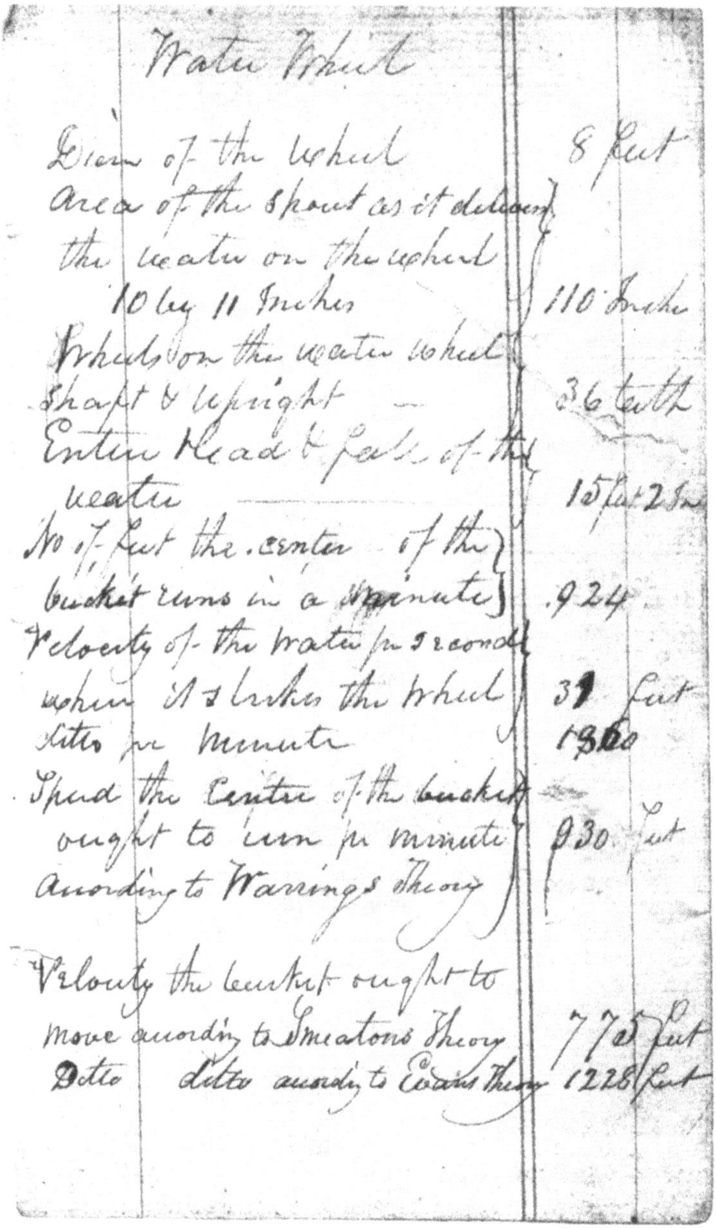

Water Wheel

Diam of the wheel 8 feet
Area of the spout as it delivers the water on the wheel 10 by 11 Inches 110 Inches
Wheels on the water wheel Shaft & upright 36 teeth
Entire Head & fall of the water 15 feet 2 In
No of feet the center of the bucket runs in a minute 924
Velocity of the water per second when it strikes the wheel 31 feet
ditto per minute 1860
Speed the Center of the bucket ought to run per minute according to Warings Theory 930 feet
Velocity the bucket ought to move according to Smeatons Theory 775 feet
Ditto ditto according to Evans Theory 1228 feet

FIGURE 3.1 Notebook page. See also Figure 6.1 (p. 124) for further analysis.

studies of millwork were reprinted posthumously and massively amplified in the early 1840s with a third edition.

However, the dominant finding from investigating forms of representation in these 27 books is that they reveal a decisive shift away from the traditional perspective view and toward conventions in mechanical drawing, namely, the elevation (which here includes sections) and the plan, more fit for the purposes of the machine maker. In Britain, perspective in machine drawing peaked in 1811 with Blunt's book. Although authored by a self-proclaimed "engineer draftsman" and purporting to demonstrate principles in mechanical drawing, in the long run it was thoroughly regressive. All the 30 or so actual machines depicted were represented in perspective.

The changing balance between perspective and nonperspective conventions in machine drawing becomes clear by separating the 27 texts into two groups: those pre- and those post-Blunt. Whereas 13 texts (including Blunt's) prior to 1812 showed perspective in 68 percent of their 341 plates, in the 14 texts spanning 1812 to 1824 that proportion fell to 48 percent of 247 plates. Conversely, whereas only 28 percent of the pre-Blunt texts offered elevation and section views, that proportion rose to 89 percent in the post-Blunt texts. As for plans: only 11 percent of pre-Blunt plates included this convention, 45 percent of the post-Blunt plates contained plans. Noticeably, across the two subperiods, a growing number of authors were practical men.

The persistence of perspective is slightly puzzling because machine makers in the 1790s would have been aware of all kinds of representational tools except orthogonal projection. At century's end, Edgeworth's guidance for the practical education of children declared, "To understand prints of machines, a previous knowledge of what is meant by an elevation, a profile, a section, a perspective view, and a (vue d'oiseau) bird's eye view, is necessary."[36] Duncan, in 1808, explained that perspective views only concealed many essential parts and distorted others so he deliberately used ground plans, elevations, and sections.[37] Bailey's delineations of machines, almost a quarter of them with sections and plans, were easily the best examples of machine draftsmanship published in England before the plates in Rees's *Cyclopaedia* appeared. Why, then, the preference for perspective in texts intended for both the general public and the practical machine maker?

The answer may be twofold. First, the perspective view gave the most readily understood and communicable 3D-on-2D image of an object, and especially a complex object like a machine. It was therefore valuable in providing the form of a machine, the overall shape, which was readily understood by layperson and machine maker alike. If perspective was often necessary, it was never sufficient in mechanical drawing. Second, a skilled machine maker might at times have had sufficient knowledge to attempt to build a machine from the information supplied in the perspective drawing. He would have needed to rely on accumulated experience and skills of improvisation. Depending on the complexity of the perspective design being copied, several attempts may have been necessary. Sometimes the efforts would have failed.[38] Over time, he would have needed access to increasingly specialized resources of a machine shop, as testified by the cast-iron frames, gears, and so forth, of White's machines. One essential metric in any useful machine drawing was scale. Unless dimensions were known it would be extremely difficult, if not impossible, to replicate a wholly unfamiliar machine. It was evidently sufficiently important for Smeaton to give scale with one of his three perspective views (where it would have been ambiguous).

[36]Edgeworth, *Practical Education* (1798), 460.

[37]Duncan, *Art of Weaving*, 3.

[38]Without any kind of drawings, efforts over three years to assemble a spinning mule from its disassembled parts failed in Philadelphia in the 1780s. Jeremy, "British Textile Technology Transmission," 39–40. So much at that time depended on prior knowledge, however.

Toward the end of the period under review, Professor William Farish, evangelical clergyman and Cambridge scientist, suggested a modification to perspective views: isometrical perspective. Drawings of this kind, he reported, allowed his laboratory assistants to make an accurate model of any kind of machine from the set of basic mechanical components that he had designed for this purpose (like modern Meccano). Apparently he regularly employed this system in his student lectures. However, it is not clear how far isometrical perspective percolated into everyday use among machine makers.[39]

Experience outside textbooks is indicated by James Watt, who set a widely followed example when he allowed architectural representational practices to inform his machine drawings. With John Roebuck and, more famously, with Matthew Boulton, he developed a market for steam engines secured by his five patents, 1769–1785. Winning contracts for the construction of his steam-engine designs in factory locations, he supplied the engine erectors with elevations, sometimes with shading, for use with the very detailed "Directions for erecting and working the newly-invented Steam Engines" (1779).[40] Watt, of course, trained as a mathematical instrument maker, and was therefore accustomed to engineering precision.[41]

Business in pre-Revolutionary France, which took him there, surely gave Watt a keen awareness of the importance of machine drawing for practical artificers and scientific engineers alike. Both Watt and Boulton sent their respective sons to the Continent to learn the French and German languages, to further their mathematical skills, and to improve their scientific knowledge. They also learned drawing. For young Boulton this included architecture, perspective, and "fancy."[42] This last may have been topics from nature and the imagination. Along with geometry, "drawing education served as the core curriculum in French technical education" in the mid-decades of the eighteenth century.[43] Despite pre-Revolutionary cross-Channel exchanges, Monge's descriptive geometry was overlooked by English engineers like Watt. One reason for this on the French side was that it was taught at the *École du corps royal du génie* at Mézières, the finest engineering school in Europe in the 1780s, where "officers trained in Monge's method were forbidden to communicate it even to those engaged in other branches of government service."[44] It was not translated into English until 1851.[45]

How were textile machines depicted in published and unpublished sources just before and during Markham's lifetime? Neither of the two great French encyclopedias included the new spinning machinery developed in Britain because the throstle (an improvement on Arkwright's water frame) and the spinning mule, the dominant cotton-spinning machines in Britain and the United States in the early nineteenth century, were either not developed or had not spread to France before 1780.[46] However, one of the *philosophes'* publications, Roland de la Platière's *L'Art du Fabricant de Velours de Coton* (Paris, 1780), portrayed a spinning jenny (Fig. 3.2). Following the format and representational conventions of Diderot's *Encyclopédie* and the Académie's *Descriptions*, the plate showed a plan, a side elevation, and 3D view with shading to emphasize depth, but without scale: not all, but certainly some, of the elements that

[39]McConnell, "William Farish"; Farish, *Isometrical Perspective*; Booker, *History of Engineering Drawing*, 114–127.

[40]Dickinson and Jenkins, *James Watt*, 383–390. Something of a transition can be seen in the drawings in Tann, *Development of the Factory*, for example, on pp. 56–58 where a 1784 drawing shows perspective views set in a plan but a 1795 drawing of another seed-crushing mill uses 2D elevation and plan drawings.

[41]Hills, *James Watt* I, 46–58.

[42]Musson and Robinson, *Science and Technology*, 205–214.

[43]Alder, *Engineering the Revolution*, 138.

[44]Artz, *Development*, 98–101.

[45]Judging by the British Library Integrated Catalogue.

[46]Harris, *Industrial Espionage*, 361–389.

would unambiguously enable a constructor to build the machine.[47] As an example of engineering-drawing technique this revealed little improvement over the (unpublished) jenny design patented by James Hargreaves in England in 1770 (Fig. 3.3), but its drafting techniques were better than those of the jenny design of Christopher Tully, published in Pennsylvania in 1775 (Fig. 3.4). As a spinning mechanism, the French jenny of 1780 was an advance on the original English jenny of 1770 but was decidedly inferior to the American jenny of 1775.[48] Ironically, the first published view of the most important of cotton-spinning machines, Arkwright's water frame, appeared as a detail in an aristocratic status painting (Figs. 3.5, 3.6, 3.7, and 3.9). It is not impossible that Markham saw it because mezzotint copies (like all prints, "exactly repeatable pictorial statements"[49]) were produced and, passing into domestic and foreign markets, could have reached New England and its educated middle classes, one member of which was Markham's patron. A short digression on this portrait offers another dimension to the visual engineering world in which Markham moved.

The painting in question is Joseph Wright's lifesize portrait in oil of Sir Richard Arkwright (1732–1792), which now hangs in the City Museum and Art Gallery at Derby, England (Fig. 3.5). Painted in 1789 when Arkwright was fifty-seven or thereabouts,[50] it shows his large frame straddled across the corner of a plain padded chair beside a plain table with a rich red curtain as backdrop. In white silk cravat and stockings, old-fashioned short bob wig, untrimmed frock coat and breeches, he sits with his right arm akimbo, his left arm and fist resting on the table.[51] His striped waistcoat (possibly of cotton and silk) envelopes an ample stomach that in his times and at his age testified to prosperity, disease, or dissolution (in his case certainly the first, possibly the second). Slightly concealed by his left hand is a piece of machinery, placed there because it made him an immense fortune, totaling nearly £500,000 of property at his death (Fig. 3.6).

Arkwright commissioned the portrait, paid £52.10s for it, and was "much pleased" with it.[52] Clearly it flattered him. He sits in a domestic setting of comfort without grandeur, rather than standing amid the dust, noise, and discomfort of one of his cotton factories: He is the epitome of the meteoric self-made man. Aware of his poor family background, he wants to record two decades of extraordinary success without appearing to be excessively ostentatious or reaching beyond his nonetheless elevated social place.[53] His posture echoes his achievement. If his head is not proudly tilted upward, his face expresses the satisfaction of huge success. That derived from ruthless determination, entrepreneurial acquisitiveness,

---

[47]Harris, *Industrial Espionage*, 150–151 reproduces an identical jenny from the papers of a French Inspector of Industry but dated 1779 when there were about 1,200 jennies in France (*ibid.*, 156). In all likelihood this was the design taken by two French machine makers from Paris to Barcelona in 1784. See J.K.J. Thomson, "Transferring the Spinning Jenny to Barcelona."

[48]Among various differences, two are prominent. The operative's driving wheel in Hargreaves's 1770 patent was tilted, so that the operative worked it manually, like stirring porridge; the French arrangement of 1780 placed the operative's driving wheel in a horizontal plane and to the side of the machine, still requiring small, circular manual motions; ergonomically best of all was the American jenny of 1775 in which the driving wheel was positioned on a horizontal axle, so that the operative could use body weight as well as arm strength to propel the driving wheel. The other main difference was the manner in which power from the driving wheel was relayed to the spindles. The original English jenny used a tilted-axle drum (itself directly rotated by the operative) from which bands ran to each of the spindles; the 1780 French jenny separated the vertical-axle drum from the operative's driving wheel; and the 1775 American jenny replaced the drum with a long tin roller on a horizontal axle, from which bands individually drove each of the spindles. Again the American design was superior. The long tin roller would appear again in the English throstle, the improved water frame of Arkwright (see Fig. 3.10). For the English jenny see Aspin and Chapman, *James Hargreaves*. For the characteristics of the Tully jenny see Jeremy, *Transatlantic Industrial Revolution*, 16.

[49]Classic insight of Ivins, *Prints*, *passim*.

[50]Fitton, *Arkwrights*, 202.

[51]For costume and style, see Ribeiro, *Visual History of Costume*, 106, 124.

[52]Fitton, *Arkwrights*, 202; Egerton, *Wright of Derby*, 197.

[53]Crouzet, *First Industrialists*, 31–32, 88, 134, 147.

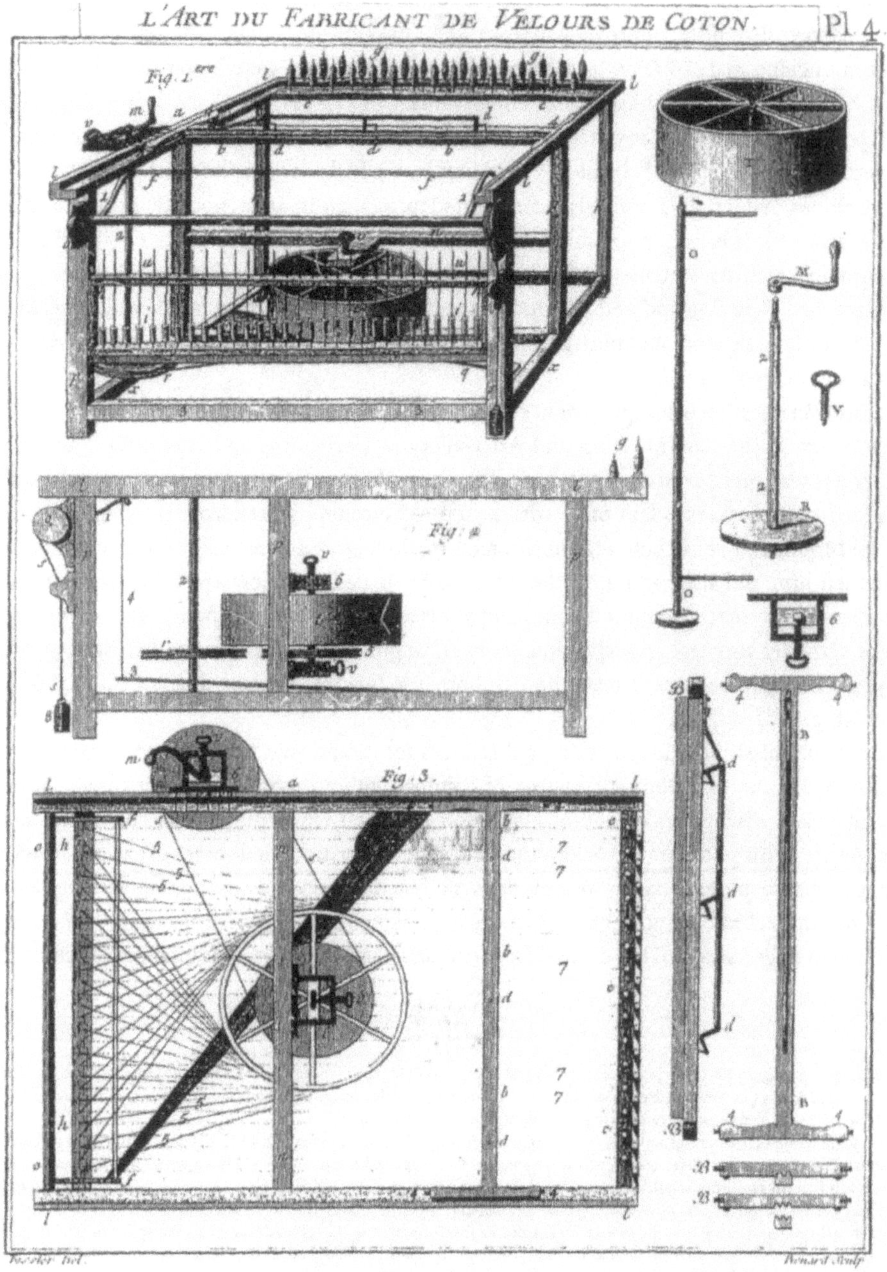

FIGURE 3.2 French spinning jenny, perspective view, with plan and side elevations; in Jean Marie Roland de la Platière (1734–179 *L'Art du Fabricant de Velours de Coton* (Paris: Moutard, 1780), Plate 4. See chapter 3, "Seeing in a Different Way" for technical comment.

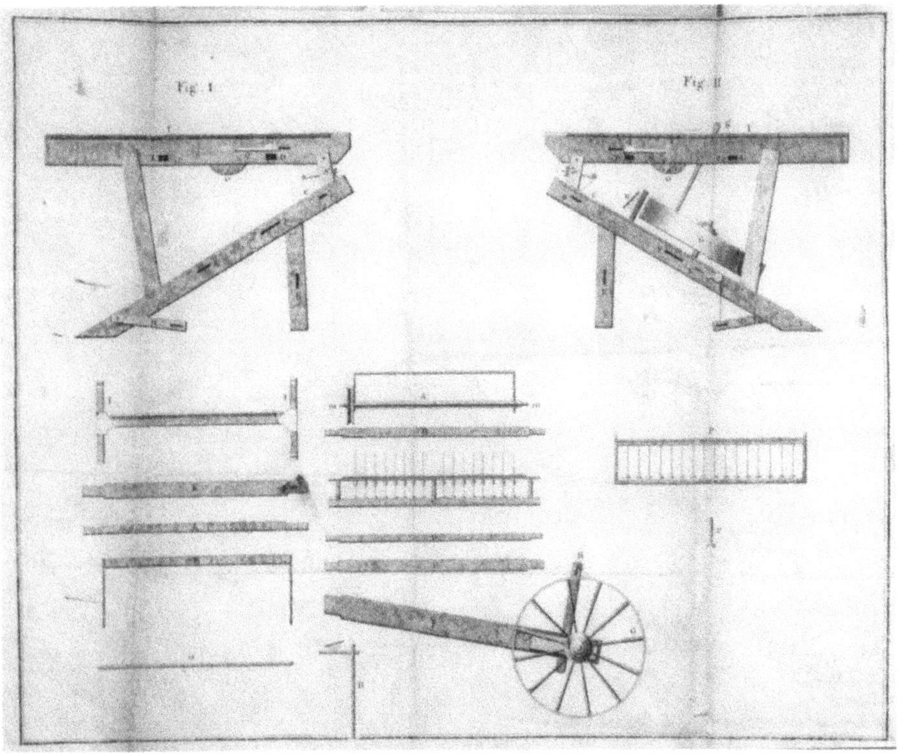

FIGURE 3.3 James Hargreaves (1720–1778), English patent for the spinning jenny, 12 June 1770. *Top*, side elevations; *bottom*, plan of driving wheel. Pen and wash on skin (33.1 x 41.4 cm). National Archives C 210/11.GB Patent no. 962. See chapter 6, section 5 for technical details.

Courtesy of the National Archives, London.

and managerial control, all (perhaps unwittingly) symbolized in the clenched fist. His eyes search a distant horizon, thinking perhaps of new commercial partners with whom to widen his Midlands and Scottish industrial empire, to challenge his competitors' 140 or more cotton mills.[54]

Wonderfully realistic though the portrayal of the machinery on the table as well as the sitter certainly is, it would not have been very helpful to somebody unfamiliar with the technology of Arkwright's cotton factories who was hoping to learn from this portrait the key to making a fabulous fortune. Unquestionably there is technical information in the painting about the machine on the table (Fig. 3.7). But it is really very minor. The object is only part of the water frame, the revolutionary machine Arkwright commercially exploited, being one of the roller drafting heads, the novel element that enabled continuous high-productivity cotton spinning. Executed in the superbly realistic style learned first from the genius Renaissance painters, to the initiated it shows a perfectly recognizable part of the water frame drawn according to the rules of perspective. Capable of spinning four yarns simultaneously, it is apparently complete in its elements: the front pairs of drafting rollers, the upper ones pressed down on the lower by the weighted

[54]Fitton, *Arkwrights*, 50–90.

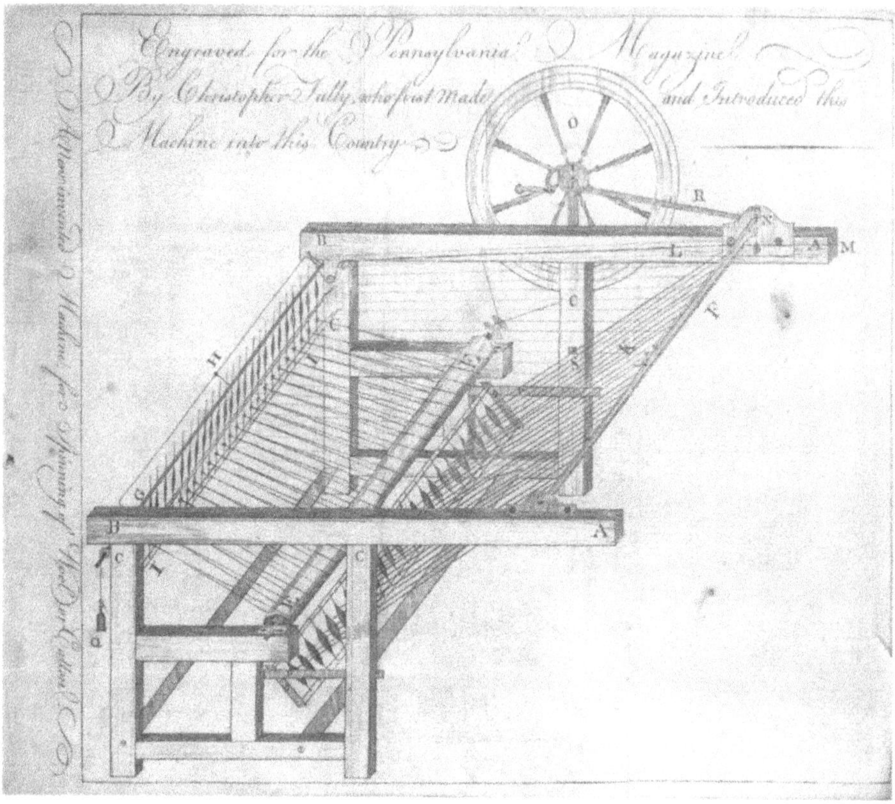

FIGURE 3.4 Spinning jenny, perspective view, built by Christopher Tully, who, as he claimed in his text, "Introduced the Machine into this Country"; *Pennsylvania Magazine* 1 (April 1775), 157–158. This 24-spindle machine was 6 feet deep, possibly as wide, and with a 3-foot high frame. If the draftsmanship conformed to the dominant perspective paradigm, the machine's design incorporated significant improvements over the original Hargreaves patent design. Prominent were the driving wheel on a horizontal axle (O); power transmission to spindles from a hollow tin layshaft (E); a wheeled clasp (F); a trigger in the middle of the clasp to lower the faller wire (H) on its axis (G), to wind the yarn onto the spindles (I); a weighted cord-and-pulley (Q) to raise the faller wire off the spindle points; friction-reducing glass bases under the spindles; tensioning devices for the driving wheel and the lay shaft; and an adjustable roving box.

Whether these improvements reflected English advances or whether they were made in America by Tully and/or others is unknown. See chapter 3: "Seeing in a Different Way" for technical comment.

saddle bar in the middle, and below pairs of guide rollers (colored a warm sienna). At the end of the unit, furthest from Arkwright, is a large brass gear, part of the "clockwork" that, driven by gears and belts, powered the rollers.

In reality much more was there than is seen in the painting. Hidden behind the water-frame rollers shown in the Arkwright painting would be two more pairs of rollers similarly constructed. Beyond that there would be a rack supporting large bobbins holding cotton roving that would be passed through the drafting rollers. And, of course, there would be a machine frame to support other elements, especially the bobbins and flyers that continuously twist and wind up the spun yarn. There would be other gears, as well as pulleys and belts conveying power within the machine (as seen in Figs. 3.8 and 3.9, pp. 57 & 58). The concealment of the flutings in the drafting rollers is intriguing. Was it to deceive imitators? (Unlikely, because the 1769 patent for the rollers was overturned in 1785.) Was it artistic license, a

URE 3.5 Sir Richard Arkwright (1732–1792) by Joseph Wright; 1789, oil (95 x 60 inches, or 241.3 x 152.4 cm). (Egerton, *Wright* *erby*, 256).

rtesy of Derby Museum and Art Gallery.

splash of white to relieve the compositional triangle of legs and waistcoat? Or was the drafting head recently taken from a working machine, with cotton lint adhering to the flutings?

After Sir Richard Arkwright's death the painting was kept in the family and his son, Richard Arkwright II (1755–1843), authorized a print of this portrait of his father to be made by John Raphael Smith (see Fig. 3.7). Published in mezzotint on 5 May 1801,[55] it appeared after Smith published a catalogue of his prints (ca.1798) but this shows that most of his prints cost under 10 shillings (then equivalent to two weeks' wages of a farm laborer in England). As many as 200 or 300 mezzotint copies could be made from a single plate at this time and London print retailers (often the printers themselves) kept large stocks, running into hundreds or thousands of prints in the late eighteenth century.[56] Smith's was an accurate copy, as might be expected of "Wright's most prolific interpreter" and a foremost exponent of mezzotint printing.[57] It was this mezzotint that Markham just might have seen. Much less flattering and well illustrating the social construction of art, was a later engraving by J. Jenkins, published in 1833, after Markham's death.

The Jenkins stipple print was used by Baines in his *Cotton Manufacture* (1835) (Fig. 3.11). It registered two modifications, hitherto unnoticed. Arkwright now appeared somewhat coarser, fleshier in the face, and with more pronounced waistcoat creases that swelled his stomach. This engraving, rather than the original painting, must have provoked Thomas Carlyle's judgment: "not a beautiful man; no romantic-hero with haughty eyes, Apollo-lip, and gesture like the herald Mercury; a plain almost gross, bag-cheeked, potbellied Lancashire man, with the air of painful reflection, yet also of copious free digestion."[58] The other significant change in the Jenkins engraving was made to the position of the roller drafting head. Still placed on the table beside Arkwright, it was all but hidden behind the sitter's left arm. There was a message here. Baines, among others (but in contrast to Arkwright's Scottish defenders), believed that Arkwright had stolen his inventions from other men. By making this engraving the frontispiece of his book, Baines was announcing to the world that the roller drafting head owed little to Arkwright's ingenuity, that Arkwright purloined it, refining only details.

As this consideration of the history of Joseph Wright's portrait of Sir Richard Arkwright has demonstrated, the artist's view of life was conditional on two factors that made their depictions of machinery largely irrelevant to the machine maker. Patrons influenced artists and culture influenced patrons. In the late eighteenth century, machinery was the instrument and symbol of "trade" and "trade" implied an inferior status in aristocratic society. Illustrations of machinery in high-culture paintings commissioned by wealthy patrons were invariably accessorized, even distorted. Consequently, no matter the artist's wishes, the oil painting or its print version was unlikely to satisfy the requirements of the mechanic. Committed to the dominant paradigm of perspective (reinforced at the end of the eighteenth century by the craze for the picturesque[59]), high-culture paintings or drawings could never be more than hints to a novice constructor.

In Britain in the 1790s and 1800s "workshop drawing" evidently meant different things to different classes of machine makers. Practical machine builders, the craftsmen and mechanics, relied on sketches of elevations that edged away from the perspective paradigm. This may be seen in some of the earliest

---

[55]Reproduced in Egerton, *Wright of Derby*, 256; Wallis, *Joseph Wright*, 9.

[56]D'Oench, *"Copper into Gold,"* ix–11, 237, 259–264; Clayton, *English Print*, 211–218; Burnett, *History of the Cost of Living*, 248–249.

[57]Clayton, "Engraving," 27.

[58]Egerton, *Wright of Derby*, 256–257; Thomas Carlyle, "Chartism" in *Miscellaneous Essays*, quoted by Fitton, *Arkwrights*, 202; Baines, *Cotton Manufacture*, 194–195. For the controversy over Arkwright's reputation, see Jeremy, "British and American Entrepreneurial Values."

[59]Klingender, *Art and the Industrial Revolution*, 83–103; Moir, *Discovery of Britain*.

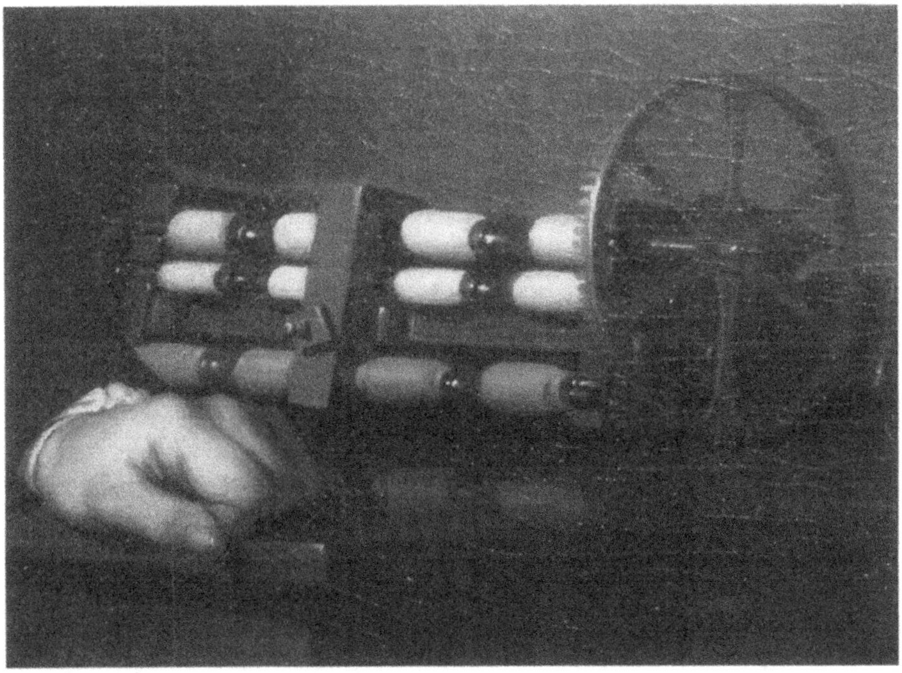

;URE 3.6 **Detail of water-frame rollers from Sir Richard Arkwright by Joseph Wright, 1789.**

urtesy of Derby Museum and Art Gallery.

surviving drawings of the new industrial textile machines. The first of these mechanic drawings (Fig. 3.12) is an elevation (called a "plan" by the draftsman) of a slubbing billy made by a member of the Scholfield family in New England ca. 1798. The Scholfields emigrated (as a family network) from Yorkshire to Boston beginning in 1793 and were responsible for diffusing woolen carding and spinning technology in New England in the 1800s.[60] The billy was used to piece together short lengths of carded wool in readiness for spinning on a jenny; piecing was done by hand on a feeding cloth (to the right of the drawing). Scholfield's practices were confirmed by the example of John Duncan in his essays on weaving: "to convey a just idea" of the construction of the weaver's various instruments he rejected perspective drawings in favor of plans using elevations and sections, "the modes of representation used by engineers, architects, and other artificers."[61]

The elevation of the Scholfields's billy is a working drawing because it is done with some geometric care, has scale relationship to the built machine, and is supported by functional and dimensional information keyed to the drawing. Standing alone, however, this elevation offered no idea of the billy's third dimension. That is provided by another sheet (Fig. 3.13), which shows three more machine surfaces with similar features: depicting the plan of the moving carriage, the side section opposite the driving wheel, and the feeding cloth. The treatment of the carriage and its wheels hints at the need for a perspective view, but the urge is otherwise rejected. No information is provided about materials, either wood or

[60]For the Scholfields see Jeremy, *Transatlantic Industrial Revolution*, 118–128.
[61]Duncan, *Essays on Weaving*, 3.

FIGURE 3.7 Sir Richard Arkwright, mezzotint of Joseph Wright's portrait, by John Raphael Smith; 5 May 1801 (608 x 455 mm, image size). (Egerton, *Wright of Derby*, 256).

Courtesy of Derby Museum and Art Gallery.

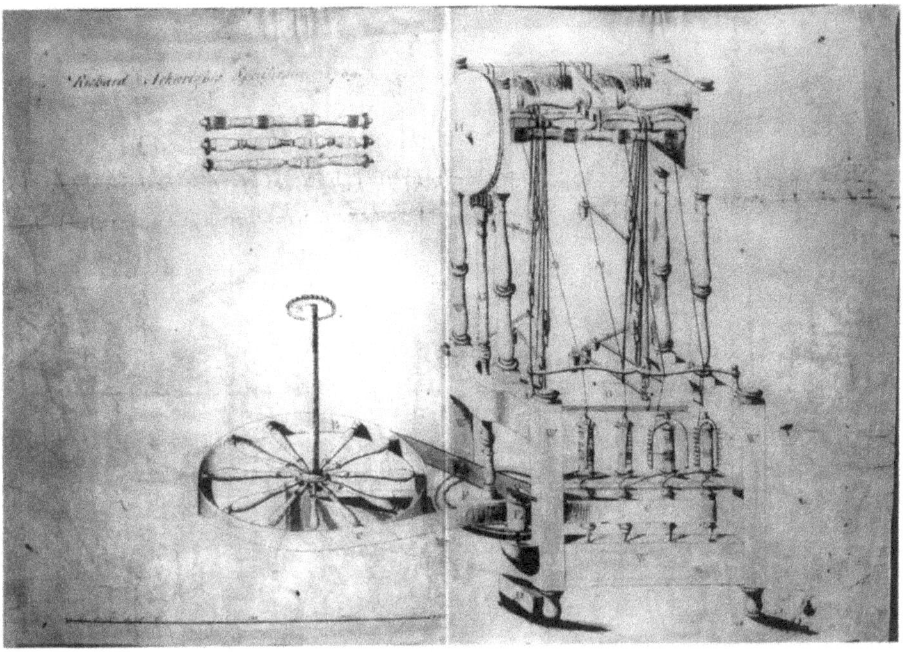

GURE 3.8 Richard Arkwright, English patent for water-frame spinning, 3 July 1769; perspective views; pen and wash on skin (31.5 47 cm). National Archives C 73/13. GB Patent no. 931. This was the pivotal machine in the cotton-spinning system that launched modern factory into the world. See chapter 3: "Seeing in a Different Way" and chapter 6, section 5 for technical comment.

urtesy of the National Archives, London.

metal or textile. Only a mechanic familiar with the machine would make sense of these drawings. He could see 3D solid forms beyond the 2D; supply information about materials; and, above all, provide a mental reconstruction of the whole machine essential to any attempt to build it.

A viewer never having seen a real slubbing billy might well have had great difficulty in understanding what these drawings were depicting. The overall design, however, becomes clearer from the 3D perspective drawing published in Rees's *Cyclopaedia* in 1818 (Fig. 3.14). Here the feeding cloth is on the left. (Also, the manually powered driving wheel, mounted on supports fixed to the main frame of the Scholfield billy, is now set on the moving carriage.) Like Joseph Wright's painting of the Arkwright roller drafting head, the depiction is vivid because it is 3D. Because it uses a 3D perspective, with the third dimension inaccurately presented, it is only useful to a machine builder as a machine form drawing. But, of course, it is better than no visual image at all.

For practical members of the emerging British engineering profession, like Watt, "workshop drawing" meant a careful geometrical rendering reliant on plans and elevations supplemented perhaps by perspective views. Even this understanding was in doubt in the 1790s and 1800s.

It might be expected that London draftsmen, responsible for making the drawings that accompanied most patent specifications, would have abandoned perspective views, especially after Lord Justice Mansfield's judgment of 1778, that the specification should be sufficiently detailed to enable anyone skilled in the trade relating to the invention to understand it.[62] By implication, the uninitiated, for whom the

[62]MacLeod, *Inventing*, 49. For an example of an obscure patent drawing see John Wood's patent of 1803 for spinning mule improvements, seen at Figure 3.19 (p. 67).

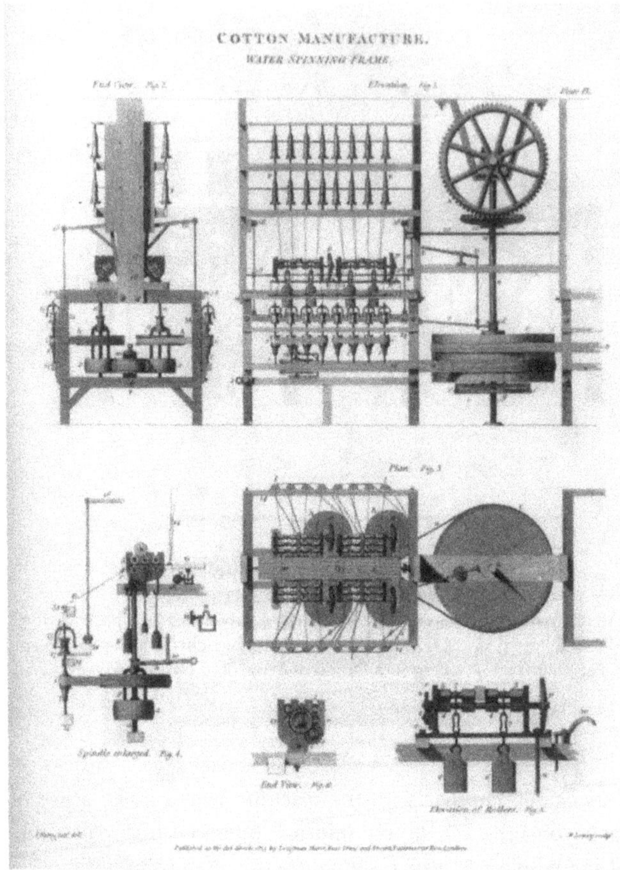

FIGURE 3.9 Cotton-water-spinning frame, from Rees (ed.), *Cyclopaedia (1812)*. See also Figure A3.8 (p. 355) for further analysis.

perspective view catered, need no longer be regarded as the audience. Patent draftsmen could even have adopted French engineering drawing techniques, particularly after the publication of Monge's book in 1795.

Yet a rapid scan of the more important textile patents reveals a persisting vacillation between perspective and 2D views, but never following Monge's method. Before 1812, when less than 20 percent of patent drawings were published, some used perspective views: Arkwright's 1769 water frame (see Fig. 3.8; Garlick's 1797 roving frame (Fig. 3.15). In an effort to be helpful no doubt, the draftsman of Arkwright's patent drawing included a scale, surely realizing that it only helped with an elevation, which the drawing did not show. On the other hand, more followed the new reliance on elevations: Hargreaves's 1770 spinning jenny (see Fig. 3.3, p. 51); Miller's 1796 power loom (Fig. 3.16); Johnson and Radcliffe's 1803 dresser (Fig. 3.17); their 1804 dresser, which also gave a scale (Fig. 3.22); and Marsland's 1806 variable batten speed motion. Nevertheless perspective views persisted after 1812, as in Raynor's patent of 1813 for a roving frame (Fig. 3.18) or Jones's 1824 patent for a cloth brusher.[63]

[63]Illustrated in Jeremy, *Transatlantic Industrial Revolution*, 58, 61, 64, 65–67, 132.

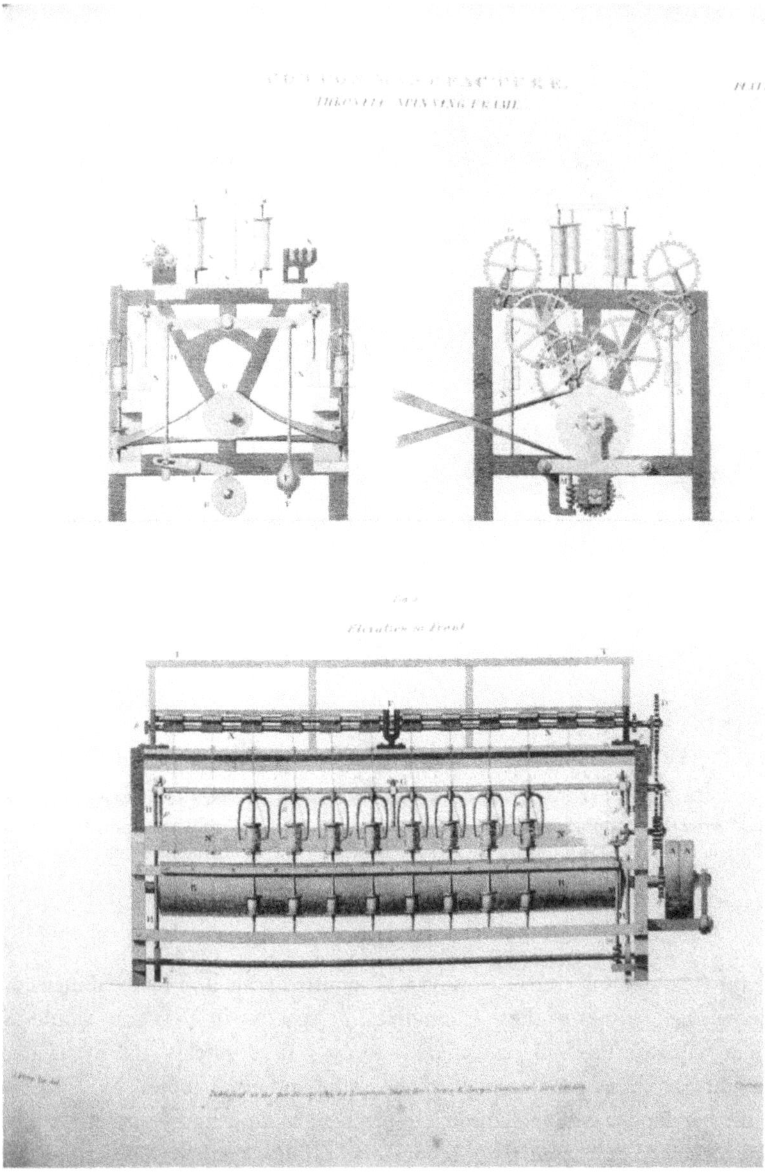

FIGURE 3.10 Cotton-throstle-spinning frame, from Rees (ed.), *Cyclopaedia* (1812). See also Figure A3.9 (p. 356) for further analysis.

If some craftsmen used 2D elevations, some "engineers" preferred 3D-perspective views. Numbers of machine drawings in Rees's influential *Cyclopaedia* preserved the perspective view, like the slubbing billy and the spinning jenny (see Figs. 3.14 and 3.20; and see Table 3.1, p. 44). As already emphasized, the perspective view, satisfying human stereoscopic vision, gives a really vivid impression of what a machine looks like. For a machine maker they supplied the first step in design, providing an overall idea of form. Essential supplements were elevations and plans of components and the total layout. A preponderance of these accompanied the textile articles in Rees's *Cyclopaedia*, for example, the water

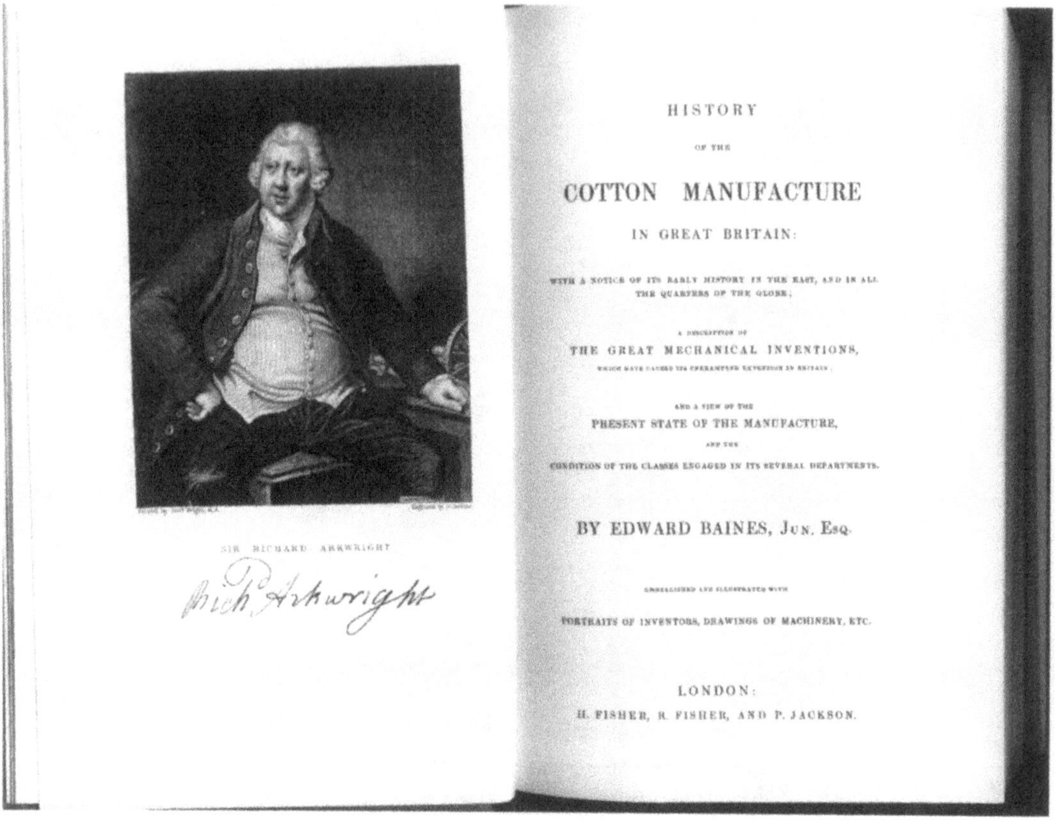
SIR RICHARD ARKWRIGHT

Richd Arkwright

HISTORY

OF THE

COTTON MANUFACTURE

IN GREAT BRITAIN:

WITH A NOTICE OF ITS EARLY HISTORY IN THE EAST, AND IN ALL THE QUARTERS OF THE GLOBE;

THE GREAT MECHANICAL INVENTIONS,

AND A VIEW OF THE

PRESENT STATE OF THE MANUFACTURE,

AND THE

CONDITION OF THE CLASSES ENGAGED IN ITS SEVERAL DEPARTMENTS.

BY EDWARD BAINES, JUN. ESQ.

PORTRAITS OF INVENTORS, DRAWINGS OF MACHINERY, ETC.

LONDON:

H. FISHER, R. FISHER, AND P. JACKSON.

FIGURE 3.11 Sir Richard Arkwright, stipple copy of Smith's engraving of Joseph Wright's portrait, by J. Jenkins; 1833 (115 x 92 m image size). (Egerton, *Wright of Derby*, 256). Chosen by Edward Baines for the Frontispiece of his *History of the Cotton Manufacture in Great Britain* (1835).

frame and the throstle (see Fig. 3.10). However, all omitted scale and hence dimension. As records of engineering knowledge, neither of these alternatives, perspective or elevation, would satisfy a machine maker unfamiliar with the machine and anxious to copy it as quickly and as accurately as possible. However, he might have been good enough to make trial-and-error copies.

Rare were the handbooks wholly dedicated to mechanical drawing in Markham's lifetime. Two of the earliest in English showed no awareness of Monge's radical new method of descriptive geometry. On the contrary, whereas one, Benjamin Donne's, *An Essay on Mechanical Geometry* (Bristol, 1796), offered a diet of geometric theorems and problems with the promise of three-dimensional models to exemplify geometrical objects and facilitate education in geometry, the other was very traditional.

Because Charles Blunt styled himself "engineer draftsman," it might be expected that his *An Essay on Mechanical Drawing* (London, 1811) would point the way to new 2D solutions to the three-dimensional representational problem. Instead his ninety-one-page volume, published by the inventive art publisher Rudolph Ackermann (1764–1834),[64] offered commentary on 63 plates, 30 of which illustrated aspects

[64] Ford, "Rudolph Ackermann."

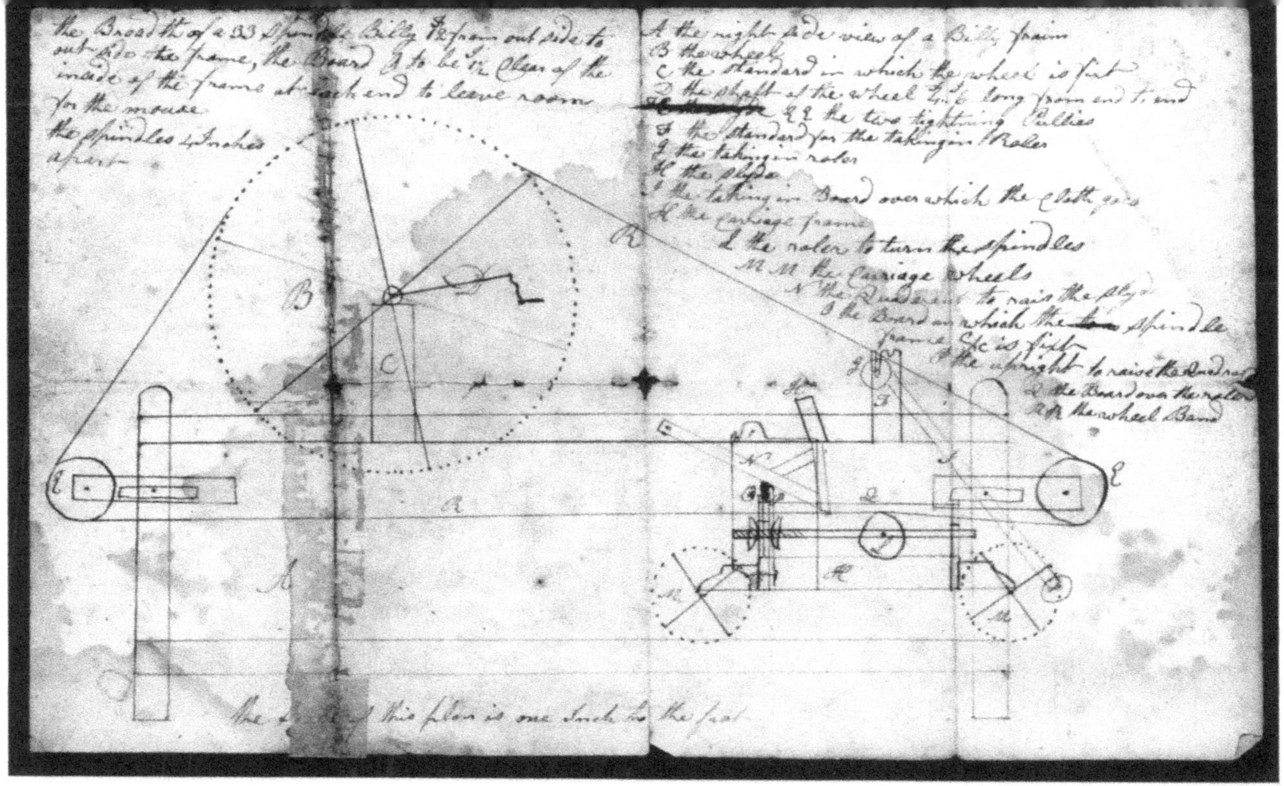

FIGURE. 3.12 Slubbing billy, drawn by an unknown member of the Scholfield family; 1790s, side elevation, ink drawing (18.7 x 29.8 cm). Scholfield Papers. This, the earliest known drawing of a slubbing billy (reportedly invented in 1786), shows the simplest version of the machine. Of 33 spindles, it measured just over 13 feet side to side and almost 10-feet deep (assuming this elevation relates to the drawings at Fig. 3.13). It allowed two or more children, piecers, to splice short lengths of cardings (taken from the woolen-carding machine) on a revolving apron (G, I). These slubbings were then drawn by the retreating carriage (apparently pushed and pulled to and fro by the machine operator, the slubber) into the spinning zone, between the slide or clasp (H) on the frame and the spindle tips (at O) on the carriage. Here they were drafted and twisted to form a continuous length of semi-spun woolen yarn and then wound onto the spindles. The hand-turned driving wheel (B) powered the long tin roller (L), on the moving carriage, which drove the spindles.

When the carriage was closest to the piecing apron, the quadrant (N) (presumably hand operated) opened the clasp (possibly closed by gravity), allowing slubbings to be pulled into the spinning zone.

The intriguing reference to a "mouse" suggests that an object of some kind ran along the side of the machine frame. Probably this alluded to a block-and-cord device, possibly weighted, which enabled the slubber, after drafting and twisting, to depress a faller wire allowing the slubbed yarns to meet the spindles at right angles and so to be wound onto them. Afterwards slubbed yarns might be spun into finer yarns on a mule (French, *Samuel Crompton*, 90; Wadsworth and Mann, *Cotton Trade*, 496n; Mann, *Cloth Industry*, 200, 288). See chapter 3 "Seeing in a Different Way" for further comment.

Courtesy of Connecticut State Library.

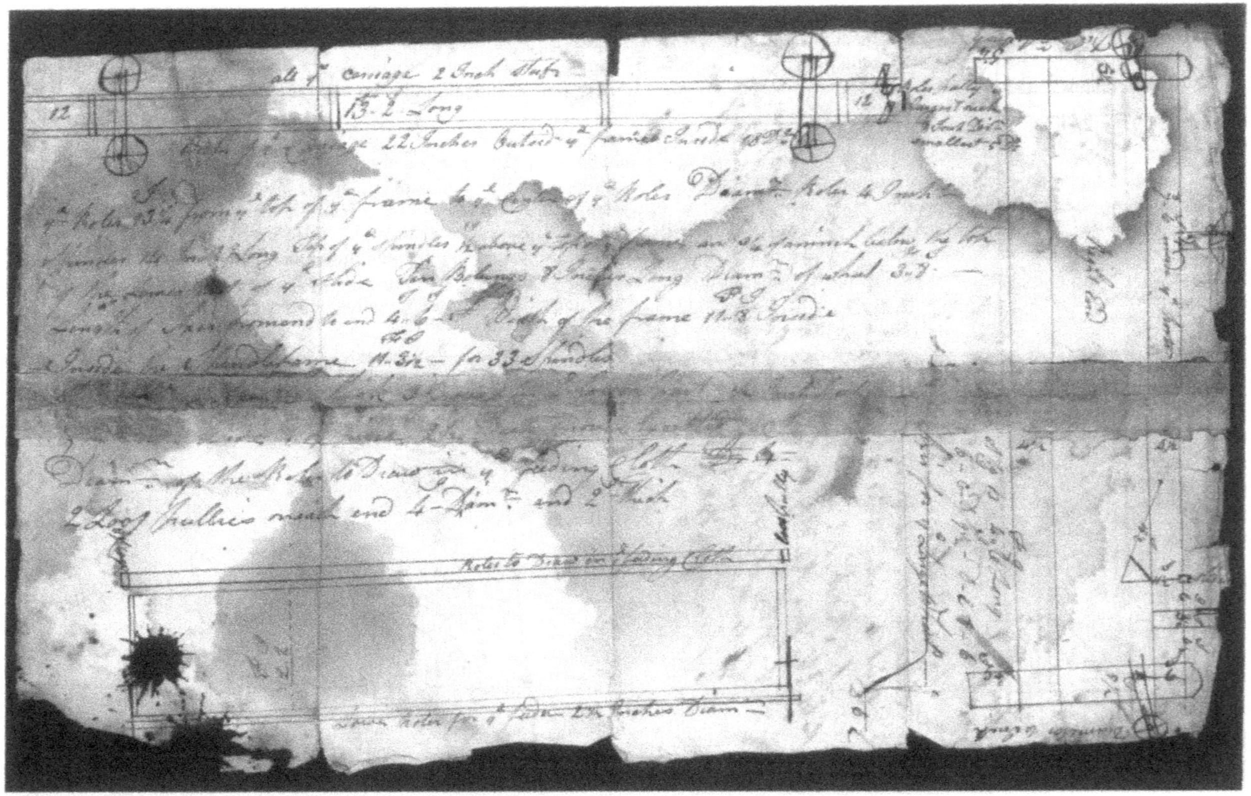

FIGURE. 3.13 Slubbing billy, drawn by an unknown member of the Scholfield family; 1790s, plan and side elevations, ink drawing (20.3 x 33 cm). Scholfield Papers. Since this drawing also shows a 33 spindle billy, it is assumed that it relates to the elevation shown at Figure 3.12. Forms and dimensions also seem to correspond. The top drawing shows the plan of the moving carriage. The bottom drawing depicts the feeding cloth arrangement. The drawing on the right details the right side of the machine. The draftsman, clearly aware of the merits of plan and elevation, fell back on perspective when depicting axles and wheels. See chapter 3 "Seeing in a Different Way" for further comment.

Courtesy of Connecticut State Library.

URE 3.14 Slubbing billy, perspective view of a design "now universally employed"; from Rees (ed.), *Cyclopaedia* (1818), Plates, 6. In two respects this differed from the Scholfield billy. The driving wheel was mounted on the moving carriage, rather than ng fixed to the main frame. Second, a system of cord-and-weights synchronized the speed of the carriage with that of the apron— s avoiding excessive strain on the delicate cardings as they came from the apron (Bakewell, "Woollen Manufacture").

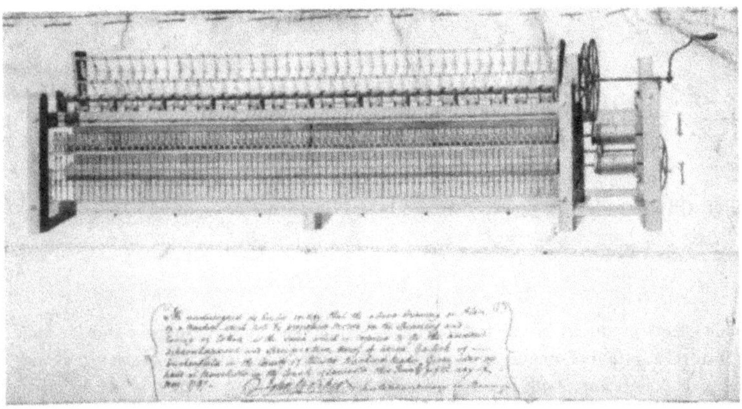

URE 3.15 Aaron Garlick, GB Patent for the double speeder, 7 February 1797. See also Figure A3.12 (p. 359) for further analysis.

ırtesy of the National Archives, London.

of practical but elementary geometry; exercises for making perspective views of blocks, spheres, and the like; and principles of shading and shadowing. The other 33 plates provided views of real machines, ranging from the Woburn apparatus for weighing livestock to a pigment-grinding machine for paint making, and Stanhope's printing press. In every single case the perspective view was presented.

Yet Blunt clearly and correctly realized the urgent need for sound courses in engineering drawing in the country's schools:

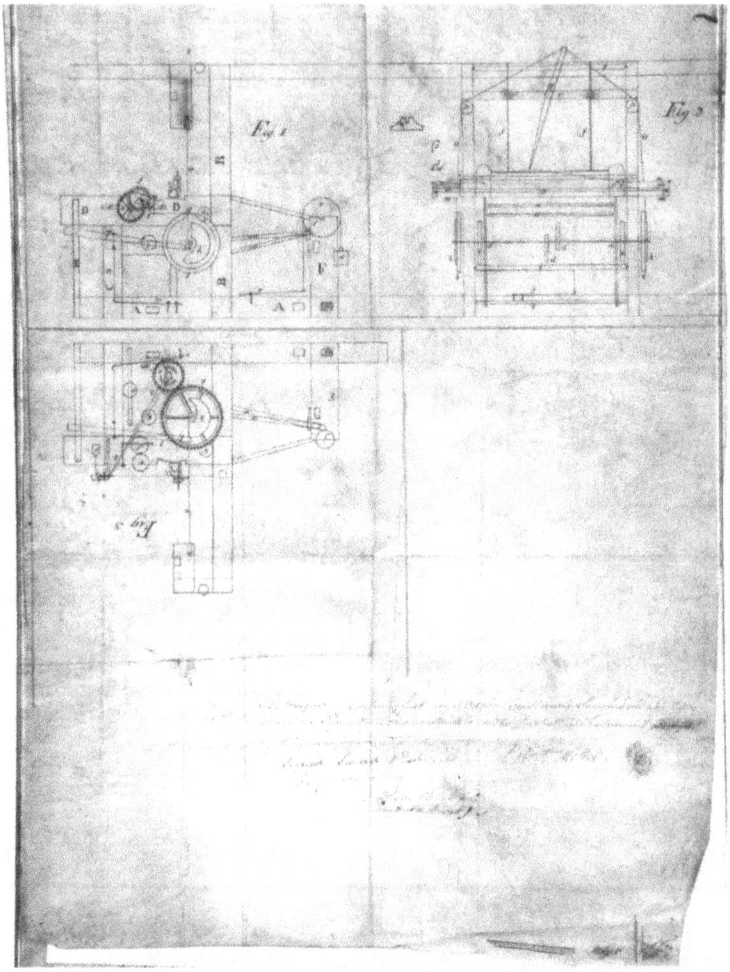

FIGURE 3.16 Robert Miller, GB Patent for the power loom, 28 June 1796. See also Figure A3.16 (p. 363) for further analysis. Courtesy of the National Archives, London.

> The stupendous effects produced in our manufacturing operations, by the extensive introduction of machinery, either in aid or in substitution of manual labour, has within a few years, rendered accurate drawings of mechanical subjects indispensably necessary, both to the practical mechanic and the projector; while the circulation of useful knowledge, which distinguishes the period in which we live, has produced an increased demand for the journals of science and a record of those improvements, which at once places this branch of drawings above the rank of a merely ornamental art, by rendering a knowledge of its practice not only essential to those concerned in the projection and execution of engineering operations, but a matter of considerable importance to the whole intelligent community.[65]

Thus in Britain's textile industry during the vital decades of the 1790s–1820s, there is evidence that practical machine makers worked from rough 2D elevation drawings, whereas the better trained engineers

[65]Blunt, *Essay on Mechanical Drawing*, "Introduction."

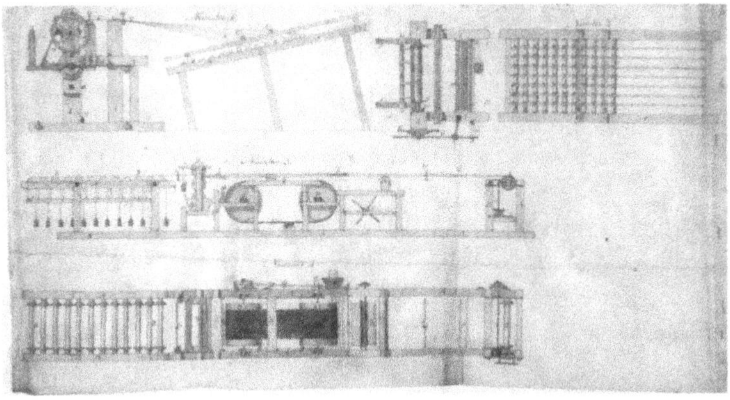

GURE 3.17 Thomas Johnson, GB Patent for warping machine and for dresser, 28 February 1803. See also Figure A3.14 (p. 361) for rther analysis.

urtesy of the National Archives, London.

were widely prompted to stick to the revered 3D-perspective paradigm, albeit beautifully drawn with the aid of Euclidean geometry and newly invented drawing instruments.[66] Exceptionally, it appears, Watt and the steam engine makers together with some of the patent draftsmen cultivated the habit of 2D elevations, also relying on careful geometry. In the late 1790s and the 1800s the growth of an engineering community in Britain's major industrial districts, with London at the fore, led to improvements in engineering design capabilities.[67] Steam-engine technology required heavy machine tools to manufacture the new all-metal production machines of workshops and factories. In contrast to bobbins or spindles and other small, light components in the textile industry, heavy iron machine tools and manufacturing machines involved multiple production stages and demanded careful work on the drawing board in order to avoid unnecessary waste of materials and labor time.

Yet changes in machine drawing came slowly and patchily. Recalling his successful attempt in 1829 as a twenty-year-old to gain a place with Henry Maudslay, premier London engineer of the day, James Nasmyth remembered how he had hugely impressed the great man with his specimens of a working model of a high-pressure steam engine and a selection of working drawings. Nasmyth commented, "Mechanical drawing is the alphabet of the engineer. Without this, the workman is merely 'a hand.' With it he indicates the possession of 'a head'. ...I was the more desirous of exhibiting the ability which I possessed in mechanical draughtsmanship, as I knew it to be a somewhat rare and much valued acquirement."[68]

Departing from perspective demanded learning new ways of looking. As Professor William Farish in 1820 observed of his isometrical perspective system (a halfway house between classical perspective and Monge's First Angle projection[69]), "though liable in a slight degree, to the last-mentioned inconvenience [the appearance of incorrectness], *till the eye becomes used to it,* I found much better adapted to the exhibition of machinery "[this author's italics].

---

[66]Woolrich, "John Farey."
[67]Brown, "Design Plans"; Woolrich, "London Engineering Industry" and other essays in Cantrell and Cookson (eds.); Rolt, *Tools for the Job.*
[68]Smiles (ed.), *Nasmyth*, 125.
[69]Farish, *Isometrical Perspective*, 3. See Appendix 5, this volume.

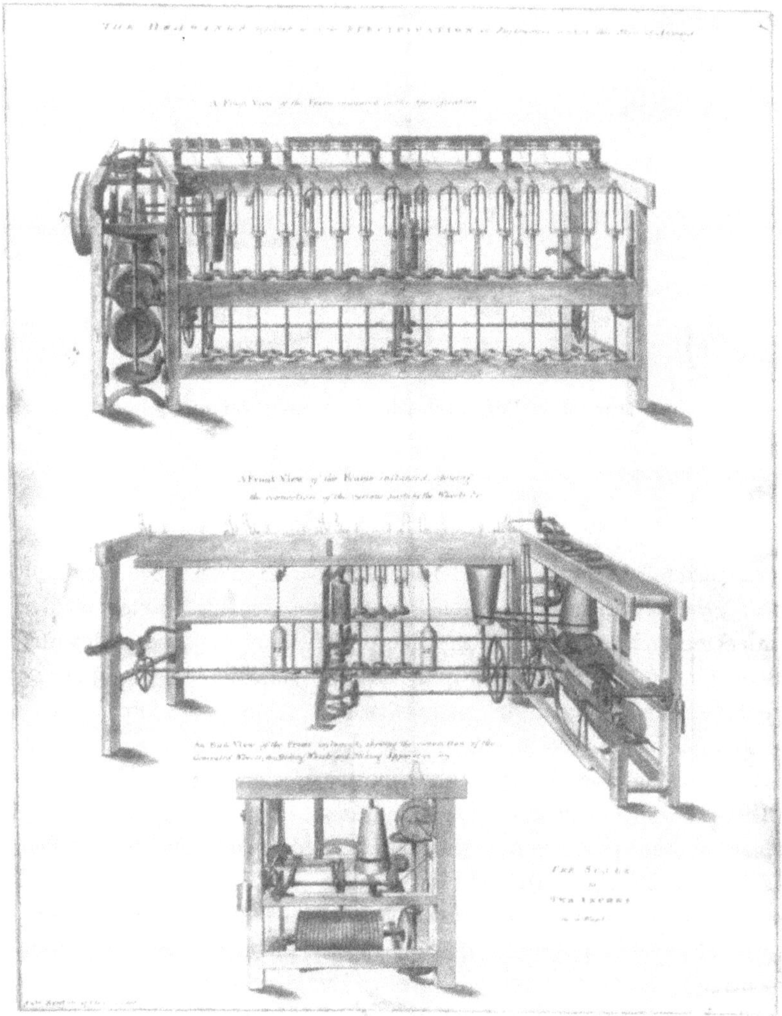

FIGURE 3.18 Joseph Raynor, GB Patent for the double speeder, 1 January 1813. See also Figure A3.13 (p. 360) for further analys
Courtesy of the National Archives, London.

THE DIFFUSION OF VISUAL ENGINEERING KNOWLEDGE IN THE EARLY AMERICAN REPUBLIC

Between 1790 and 1820 the United States cotton-spinning industry grew from about 10 manufactories or workshops to 386 firms; from less than 2,000 jenny spindles to nearly 357,000 cotton-factory spindles; from about 50 workers to around 12,600 workers (including about 7,000 children).[70] The underlyin

[70]Jeremy, *Transatlantic Industrial Revolution*, 19, 276–279.

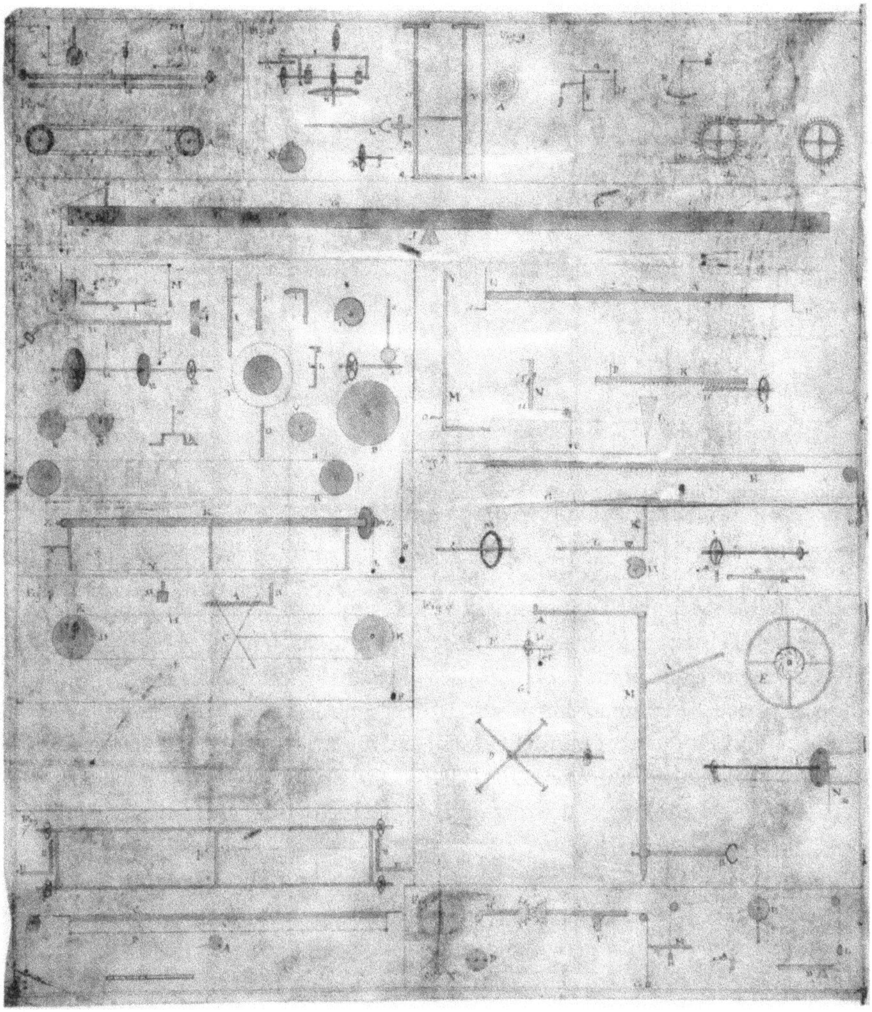

GURE 3.19 John Wood, English patent for spinning-mule improvements, 14 June 1803; various perspective, plan, and elevation ews; pen and wash on skin (62.5 x 56 cm). National Archives C 210/98. GB Patent no. 2711. Not published in *RAMA*.

This patent drawing illustrates how difficult some of these early technical drawings were to understand, and therefore how mited they might be as vehicles of technology transfer. As can be seen, the mule involved the use of much "clock work." A mule inner might make sense of this patent, but only if he or she were accustomed to using a hand mule, that is, one that partly used eam or waterpower, rather than being mostly automatic (the first satisfactory automatic mule was covered by Richard Roberts's atents of 1825 and 1830).

ourtesy of the National Archives, London.

technological forms on which this expansion was based (the carding machine, drawing frame, roving frame, water frame [and its modification, the throstle], as well as the spinning mule) were brought from Britain. So far as is known, in this period very few full-size machines were imported from Britain to America or there copied (not least because machinery export was prohibited in Britain and especially so

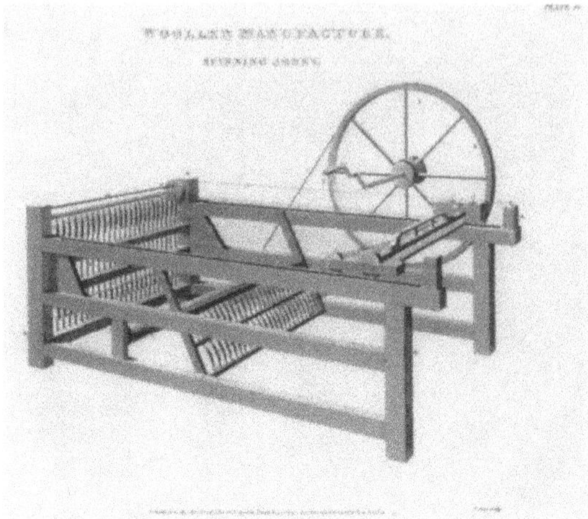

FIGURE 3.20 Woolen-spinning jenny, from Rees (ed.), *Cyclopaedia* (1818). See also Figure A3.7 (p. 354) for further analysis.

in wartime).[71] This implies that an industrial technological system was transferred across the Atlantic either by 2D forms of visual engineering knowledge or by artisans. It has long been known that aggregated immigrant workers were one of the prime vehicles of technological diffusion to the United States in the early industrial period. How important were 2D images? Answers to this question may be found in examining the utility and extent of transfer (inward and internal to the United States) of the engineering-drawing techniques imported into the United States; the nature of the textile engineering knowledge transmitted; and, last, the extent of interdependency between visual and artisan vehicles of knowledge transfer.

By 1814, the year Markham commenced his workshop portfolio, European sources offered three engineering-drawing techniques: perspective, the 2D elevation and plan, and Monge's orthogonal projection. The last, the most fitted for design and construction purposes, was not systematically introduced into the United States until 1817 when Major Sylvanus Thayer, the new Superintendent of the United States Military Academy at West Point, NY, modeled the curriculum on Napoleonic lines. He imported from France over a thousand technical books and supported a French engineer, École Polytechnique-educated Claudius Crozet, to teach military engineering, including descriptive geometry.[72] In Markham's lifetime, although Crozet published his own treatise on descriptive geometry, West Point engineering-drawing knowledge never percolated into the civilian practice of American mechanical engineering.[73]

---

[71]In the 1780s a handful of machine models were smuggled into America. Two important ones required the addition of much artisan skill to assemble or become workable. One of them was a spinning mule imported into the United States in 1784 in parts, foiling American efforts to reassemble it, which were then shipped back to England in 1787. A mahogany and brass model of a water frame, which William Pollard, first American patentee of the Arkwright water frame, secured from the importer in 1788–1789 proved unworkable and needed months of adjustment effort to become operable. Jeremy, "British Textile Technology Transmission," 40; Wallace and Jeremy "William Pollard," 407; Jeremy, "Damming the Flood."

For models as representations of engineering vision see Ferguson, *Engineering and the Mind's Eye*, 44–45, 102–107. After the first US Patent Act of 1790, models of patented machines were deposited and inspected in the patent section of the Department of State, after 1802, this occurred in a separate Patent Office. Ferguson and Baer, *Little Machines*.

[72]Forman, *West Point*, 42, 51, 79. Presumably for security reasons, I could not electronically access the US Military Academy (West Point) Library but did find the New York Public Library held early or first editions of both Monge, *Géométrie Descriptive* and Claudius Crozet, *Treatise*: transferred possibly from the US Military Academy?

[73]Crozet, *Treatise*; Stevens, *Grammar of the Machine*, 32; Belofsky, "Engineering Drawing."

Perspective drawings were by far the most common technique employed in 2D visual sources imported or internally spread through the American colonies and in the early American Republic.[74] The earliest perspective drawing of a textile machine published in an American source is Christopher Tully's spinning jenny in the *Pennsylvania Magazine* of 1775 (see Fig. 3.4, p. 52). Tully claimed to be the first maker and introducer of the machine to the American colonies. His device gives a vivid impression of a machine derived from Hargreaves's English jenny but is improved in important ways.[75] The use of perspective was reinforced somewhat after Abraham Rees's *Cyclopaedia* began appearing in London in 1802 and in the United States in 1806 (see Tables 3.1, 3.3, and Appendix 2, Table A2.1, p. 336). For the general reader, of course, the perspective view was more immediately vivid because it provided the third dimension, satisfying cyclopean vision. Perspective would always be useful in conveying an impression of form. Although sometimes necessary, it would never be sufficient for the machine maker.

A much better route to accuracy and minimized ambiguity in engineering drawing, though lacking the ultimate fitness-for-purpose of Monge's descriptive geometry, were elevation-and-plan drawings. In the early American Republic two individuals above all others were instrumental in pioneering the shift away from perspective and toward the 2D elevation-and-plan machine-drawing conventions. These two, one native-born, the other an immigrant, came from very different technical traditions. Oliver Evans (1755–1819) was a flour miller and millwright; Benjamin Henry Latrobe (1764–1820), an architect and civil engineer.

The essence of Evans's career is well known: born in Newport, Delaware; son of a farmer; apprenticed to a wheelwright; an autodidact with a questioning and inventive mind.[76] His prime work and influence were encapsulated and disseminated through his classic volume *The Young Mill-Wright and Miller's Guide* (1795). It went through 15 editions, the last in 1860 and was remarkable in combining practical experimental knowledge with scientific theory. For his science Evans was indebted to a Philadelphia schoolmaster and Quaker, William Waring. In the book, after outlining the laws of mechanics and hydraulics as then understood, Evans described in detail his automatic flour mill, a revolutionary mechanical system incorporating the principles and techniques of flow production.

The automatic milling system was not the only astonishing feature of *The Young Mill-Wright and Miller's Guide*. Also groundbreaking was the draftsmanship in the 26 plates of the volume. Concerned with both the structure of the mill and the machinery it housed, Evans and his collaborator, Thomas Ellicott, a millwright, took the logical step of applying architects' and builders' rules of 2D representation to the mechanic's problem of depicting machinery. This proved so effective that Evans and Ellicott (whether they originated the idea or not) used elevation/section and plan in the *Young Mill-Wright's Guide*. In so doing, they massively contributed to the education of American mechanics in learning to see differently. This shift in drawing conventions is demonstrated in the analysis of Evans's and Ellicott's 26 drawings presented in Table 3.2.

In addition to elevation/section and plan in well over a quarter of the 118 figures in the 26 plates, compared to only 36 percent of figures with evidences of perspective, Evans and Ellicott provided scale for nearly two thirds of the figures.[77] Although Evans's drawings were crude compared to those of Latrobe,

---

[74]See the illustrations in Hindle, *Emulation*, 60, 62, 65, 67, 69–71, 73–74, 77; York, *Mechanical Metamorphosis*, 12, 50, 191; Barnhill, "Dissemination of Technology."

[75]*Pennsylvania Magazine* 1 (April 1775), 157–158; Jeremy, *Transatlantic Industrial Revolution*, 16. The *Pennsylvania Magazine* drawing was accompanied by a detailed verbal description that gave four key dimensions and identified construction materials.

[76]Bathe, *Oliver Evans*; Ferguson, *Oliver Evans*; Ferguson, Foreword, Evans, *Young Mill-Wright*.

[77]For his part of the book, Evans provided a note on scales at the end of the list of contents (ranging between two and 18 feet to the inch); Ellicott, in the fifth part of the book, gave his scales on page 63 (mostly an eighth of an inch to a foot).

Table 3.2 Drawing Conventions in Oliver Evans, *Young Mill-Wright's Guide*, 1795 (1st) edn

Plate number	No of figures	Short title	Part	Text page	Perspective	Shading	Elevation	Section	Plan	Diagram	Scal
1	10	Science of mechanics	1	39	4	5				10	
2	9	Watermill for grain, parts and powers thereof	1	34	4	7				4	3
3	8	Hydraulics	1	93	2	7			1	1	6
4	8	Waterwheels: undershot, pictured back, overshot, breast, tub wheel	2	9		7			2	1	6
5	8	Gears and cogs	2	48	3					2	
6	12	Mill for grinding flour into parcels; elevator	3	73	3	12		1	1		7
7	6	Grist mill for small parcels; hopper boy; drill	3	75	2	6		1			5
8	1	Evans's automatic flour mill; hoppers	3	71	1	1		1			1
9	5	Evans's improved mill; wheat elevator	3	161	4	5					
10	3	Mill for hulling and cleaning rice; log dam construction	2	132	2	3		1			3
11	7	Circle of motion; gudgeons; millstones	2	37	4	7				1	5
12	1	Machine for experimenting on waterwheels	1	128	1	1				1	
1	3	Undershot waterwheel	5	31				3			3
2	3	Breast waterwheel	5	18				3			3
3	2	Pitch back waterwheels	5	19				2			2
4	2	Overshot waterwheels	5	20				2			2
5	7	Forebays; cogwheels; trundle for wallower	5	13		7		6	1		7
6	1	Groundfloor plan; keys to sills; spur blocks and head blocks; sieve	5	40		1			1		1
7	1	Second floor plan; millstone construction; bolting chest; reel	5	50		1			1		1
8	1	End view of old-fashioned grist mill	5	66		1		1			1
9	1	Side view of old fashioned grist mill showing wheel and millstones with crane	5	49		1		1			1
10	1	Improved grist milll, Occoquam, Virginia	5	viii	1	1		1			1
11	1	Sawmill	5	80	1	1		1			1
12	9	Fulling mill; pitman irons	5	84	8	9		1			9
13	2	Threshing machine	5	Ap 2	2	2					2
14	6	Improved waterwheel for sawmill	5	Ap 9	1	6		6			6
	118		Totals		43	91	0	31	7	20	76
Percent of 118 figures (some conventions overlap)					36%	77%	0	26%	6%	17%	64%

The distinction between elevation and section is often unclear in these drawings.

when set in their contemporary British context they appear remarkable in their subordination (though not elimination) of perspective in favor of elevations, plans, and scale. One of Ellicott's motives in deploying the new draftsmanship, perhaps his prime one, was to share his practical knowledge with other mill and machine builders:

> Perhaps some are of opinion that draughts are useless pictures of things, serving only to please the fancy. That is not what I intend by them; but to give the reader true ideas of the machines, &c described, or to be made. ...The great use of draughting mills, &c to build by, is by conveying our ideas more plain, than is possible to be done by writing or words, which may be misconstrued or forgotten; but a draught, well drawn, speaks for itself, when once understood by the artist; who, by applying his dividers to the draught and to the scale, finds the length, breadth and height of the building, or the dimensions of any piece of timber, and its place in the building &c.[78]

Their improvements in milling were reported by Evans and Ellicott in the *Young Mill-Wright* in order to encourage merchants and others to build mills on the improved principles and thereby, with Evans's patents in force, to accrue royalties from the inventions as well as the book. In Markham's lifetime the *Young Mill-Wright* went through four editions (1795, 1807, 1818, 1821) and he certainly read and used it (see Fig 3.1, p. 46 and Fig. 6.1, p. 124).

The new engineering drawing techniques were decisively demonstrated in America by English-born Benjamin Henry Latrobe. After emigrating to the United States in 1795, Latrobe (who had trained under John Smeaton and William Jessop) diffused British standards and significantly raised the quality of American mechanical drawing through the creation of outstanding engineering drawings and by teaching others his methods.[79] His solution (derived from British architectural practice) to the problems posed by engineering drawing was to use 2D drawings for each face of a machine or machine part, accurately drawn to scale. His drawing techniques included the use of preliminary sketches, measured workshop drawings, scale, plans, sections, elevations, details, and color coding.[80] This could be seen in his scheme for the Philadelphia waterworks on the Schuylkill River (1799–1803), which included a steam engine . Although the first drawings for the steam engine were made by the draftsman of the Soho Works—near Newark, NJ (where parts were cast)—they were authorized by Latrobe and used by his assistants. Latrobe, famously, became architect of the Capitol in Washington, DC, as well as other buildings, and designer of other engineering works.[81]

Elevation-and-plan drawings, better drafted than those in Evans's *Young Mill-Wright*, appeared in the plates of Rees's *Cyclopaedia*, as Table 3.3 indicates.

Slowly, though not overwhelmingly, elevations, shaded and unshaded, displaced perspective as the preferred drawing technique of American patent draftsmen, or so the 54 patents in Table 3.4 suggest.

One of the earliest examples of American machine drawing using elevations was William Pollard's patent of 1791, which is the earliest graphic evidence of the introduction of Arkwright's spinning frame to the United States (Fig. 3.21). Although the machine has no scale, the heart-wheel dimensions are recorded. Interestingly, it shows Arkwright's spinning and roving machines on the same base and apparently sharing the same machine frame[82]: this is an advance in engineering drawing (see Appendix 5).

---

[78]Ellicott continued, "I have heard a man say, he believed his mill was £500 better, by having employed an experienced artist, to draw him a draught to build it by. And I know by experience the great utility of them. Every master builder ought, at least, to understand them." Evans, *Young Mill-Wright*, fifth part, 63–64.

[79]Stapleton (ed.), *The Engineering Drawings of Benjamin Henry Latrobe*, 69.

[80]Stapleton (ed.), *Engineering Drawings of Benjamin Henry Latrobe*, 62–63.

[81]This paragraph is based on Stapleton (ed.), *Engineering Drawings of Benjamin Henry Latrobe*, especially 3–71, 144–199.

[82]US National Archives, Reconstructed Patents, RG 241, 30 December 1791 patent, headed "William Pollard, ass[ignee] of Richard Arkwright Spinning and Roving Cotton." Wallace and Jeremy, "William Pollard," 412; and Jeremy, *Transatlantic Industrial Revolution*, 77.

Table 3.3 Drawing Conventions in Textile Articles in Abraham Rees (ed.), *The Cyclopaedia* (London, 1802–1820; Philadelphia, 1806–1820)

"Manufacture of Cotton" article (London 4 Nov 1812; Philadelphia 8 July 1813)

Plate	Subject of plate	London date of plate	Perspective	Elevation shaded	Elevation
1	Calico printing	1808; also 1809		x	
2	Batting machine	1809		x	
3	Devilling	n/a		x	
4	Carding	1809; also 1810		x	
5	Drawing frame	1808		x	
6	Roving can frame	1810		x	
7	Roving frame termed "double speeder"	1811		x	
8	Roving frame	1811		x	
9	Water spinning frame	1813		x	
10	Throstle spinning frame	1813		x	
11	Mule spinning	1814		x	
12	Reeling	1812		x	
13	Doubling machine and twisting machine	1814		x	
14	Messrs Strutt's Cotton mills at Belper, Derbyshire	1814			x

"Weaving"article (London 29 May 1818; Philadelphia 25 Sept 1818)

Plate	Subject of plate	London date of plate	Perspective	Elevation shaded	Elevation
1	Weaving: Mr Austin's engine loom	1815	x		
2	Weaving: Duff's draw boy	1819	x		

"Woollen Manufacture"' article (London 30 July 1818; Philadelphia 24 Nov 1818)

Plate	Subject of plate	London date of plate	Perspective	Elevation shaded	Elevation
1	Slubbing machine or billy	1811	x		
2	Spinning jenny	1811	x		
3	Shearing machine	1811	x		
4	Carding engine	1809	x		
5	Gig mill	1815	x		

Sources for dates: *Philosophical Magazine and Journal* 56 (1820), 218–224 for London dates; Professor Robert D. Arner for Philadelphia dates.

Although Rees's *Cyclopaedia* and the *Repertory of Arts* were the paramount public sources for the transatlantic transfer of visually represented knowledge about Britain's industrial technology before 1830 (Appendix 2), there were private visual communications as well. About these little is known because at the time they were secret and sporadic, partly for fear of Britain's prohibitive laws (which did not specially

le 3.4 Drawing Conventions in 54 American Textile Patents Surviving for the Period 1790–1825

e	Perspective	Elevation shaded	Elevation	No drawing
0–1812	3	4	6	—
3–1824	16	1	18	—
5	4	—	—	2
al	23	5	24	2

ce: Author's collection of American textile patent drawings and specifications. All were reconstructions or reissues.

mention plans or drawings, but their possession might be incriminating), partly to gain competitive advantage. In the textile industry, "models and patterns [drawings?] of a number of machines and engines now used in the Kingdom of Great Britain...for manufacturing cotton" were the target of an attempt to bring Arkwright spinning technology to the United States in 1787–1788, aborted by English Customs.[83] Possibly the Scholfield family took their billy sketches with them from Yorkshire to New England in the 1790s (see Figs. 3.12, and 3.13, pp. 61, 62, respectively). When in 1818 Paul Moody patented his dressing frame he admitted that he had not seen the English prototype for his design, patented by Thomas Johnson in 1804 (see Fig. 3.22 and Appendix 3), but "I have seen the specification and drawings accompanying his patent, from which I constructed a machine resembling very nearly, as I suppose, that which was invented by the said Johnson."[84] How he got hold of that patent specification and drawing was never disclosed. The Scotsman Joseph Gordon arrived at Middlebury in 1817 reportedly with drawings of textile machinery used in Scotland. Gordon was certainly familiar with Scottish power looms, though whether drawings were in his pocket when he crossed the Atlantic is not known.[85] Zachariah Allen, the Rhode Island woolen manufacturer, visited England in 1825 and made clandestine copies of machinery used to manufacture woolens that he saw in Leeds; he took them back to Providence where, adding a little modification, he built and patented them.[86] These were rare examples of drawings as vehicles for the transatlantic transfer of textile technology. Significantly, those drawings that have survived (the Scholfield billy, Johnson's patent, and Allen's sketchbook), all used the new drawing conventions of machine elevation and, in the Scholfields's case, the plan technique was used. In these early industrial decades transfer could not exclusively rely on plans because they needed to be accompanied by skilled workers carrying tacit knowledge. Perhaps that is why plans were not itemized in Britain's prohibitory laws.

What of the extent of internal diffusion of the engineering-drawing techniques imported into the United States? Here knowledge networks were crucial. Some centered on prominent architects' offices; others facilitated the spread of visual and verbal information conveyed by patent specifications or technical publications. Factory-based machine shops, like that of Markham at Middlebury, Vermont, also played a part though we know virtually nothing about it. Heading some of these network mechanisms were Oliver Evans and other machine makers in Philadelphia and New York during the late 1790s and 1800s, the drawing offices of Benjamin Henry Latrobe, and the social systems by which Rees's *Cyclopaedia* was introduced to the United States.[87]

Evans moved his inventive attention to other endeavors, chief of which was his high-pressure steam engine. Between 1806 and 1812 he promoted the Mars works, "a full-scale array of engineering shops"

---

[83]Cooke, *Tench Coxe*, 107–108; Wallace and Jeremy, "William Pollard."

[84]Paul Moody's dresser patent specification (no drawing) in Boston Manufacturing Co. versus Jonathan Fisk et al., May 1821 file, in the United States, First Circuit Court records.

[85]Swift, *Middlebury*, 331.

[86]Jeremy, "Transatlantic Industrial Espionage," 104–112.

[87]Meyer, *Networked Machinists*, 39–40, 44–47; R. Thomson, *Structures of Change*, 33–47.

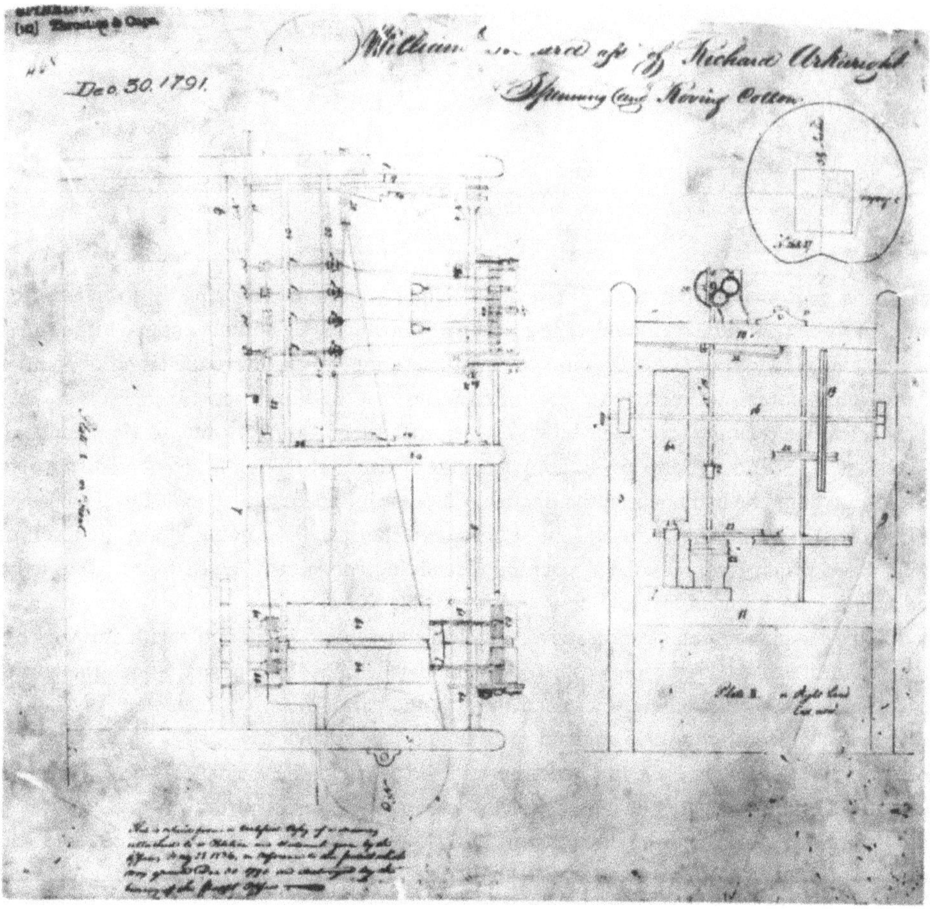

FIGURE 3.21 William Pollard (d. 1801), United States patent for cotton roving and spinning, 30 December 1791. Restored Patent Drawings (X-Drawings), un-numbered, Records of the Patent and Trademark Office, Record Group 241, National Archives, Washington, DC.

Plate 1, on the left, shows a front elevation of the machine (when the figure is turned sideways). To the right, rovings from c presumably taken from a can frame, were drawn up and between drafting rollers. To the left of Plate 1 yarn is shown coming dow from the weighted drafting rollers. Plate 2 shows a right-hand, side elevation of the machine. Rather than two machines, Plate 1 probably shows the front and back elevations drawn on a common base. The heart-shaped cam, to the upper right of the sheet, activated the bobbin-lifter rail patented by Coniah Wood in England in 1772. As seen by the endorsement on the drawing, Pollard claimed to be an assignee of Richard Arkwright. See chapter 3, "The Diffusion of Visual Engineering Knowledge" for technical comment.

Courtesy of the US National Archives, Washington, DC.

in Philadelphia, where steam-engine construction was supervised by an engineer from England.[88] The Mars works specialized in high-pressure steam engines and acted as the hub of a network of mechanics in the developing steam-engine industry. Evans influenced and was influenced by other machine makers in the city. His earlier advocacy and use of machine drawings intensified with his production of steam

[88]Ferguson, *Oliver Evans*, 45.

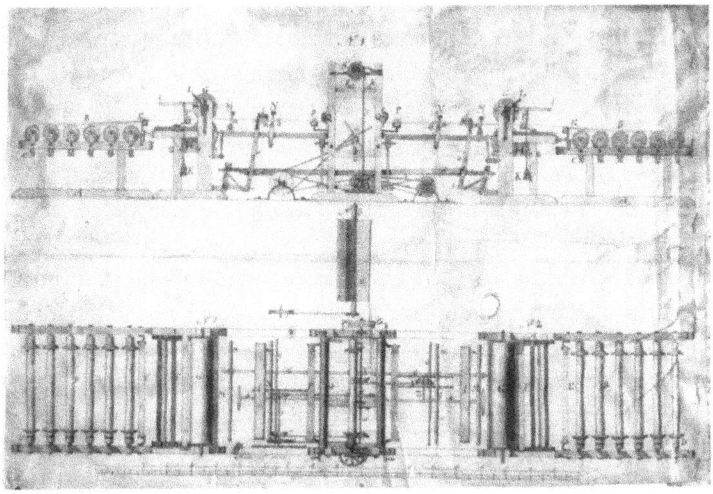

FIGURE 3.22 Thomas Johnson, GB Patent for the dresser, 2 June 1804. See also Figure A3.15 (p. 362) for further analysis.

Courtesy of the National Archives, London.

engines to avoid wastages of materials and labor time, though evidence of this was removed by Evans himself and the fiery destruction of the Mars works in 1819.[89]

Latrobe also promoted manufacture of steam engines, but of the Boulton and Watt low-pressure type. The Soho Works with which he was associated were on the Passaic River near New York City and functioned contemporaneously with the Mars works. By 1820 Philadelphia and New York were leading producers of heavy industrial machinery, increasingly installed in factories and in steamboats in the Midwest. A snatch of evidence illustrates how they influenced machine building design techniques in other industries. By including Latrobe-approved drawings (without the scale seen in the original workshop masters) of the improved iron boiler of the Schuylkill Engine (1803), in its *Transactions* in 1809, the American Philosophical Society publicly offered an example of 2D sectional drawings with the highest imprimatur in the land. In all likelihood they were heeded by educated proprietors of emergent small-scale factory machine shops that were beginning to proliferate on riparian sites between Virginia and Vermont. More privately, Latrobe's associates, pupils, and indirect imitators trickled the technique into the community of mechanics and machine builders.[90]

The knowledge networks by which Rees's *Cyclopaedia* was imported to the United States and then distributed throughout the country are more precisely known (see Appendix 2). And they almost certainly included Markham. By importing, sharing, or funding the publication of British books, pamphlets, or articles, America's middle-class professional elite was instrumental in providing mechanics and incipient engineers with access to technical literature in the young Republic. Eventually the twin passions of the United States's educated urban elite, the promotion of national economic and industrial independence, and the expansion of their own business investments, would produce the formation of the New York Mechanic and Scientific Institution (1822) and in Philadelphia the Franklin Institute (1824).[91]

89Pursell, *Early Stationary Steam Engines*, 44–50; Meyer, *Networked Machinists*, 44–45, 47.

90Stapleton (ed.), *Engineering Drawings of Benjamin Henry Latrobe*, 162–163; Pursell, *Early Stationary Steam Engines*, 34–43; Meyer, *Networked Machinists*, 39–40, 46.

91Sinclair, *Philadelphia's Philosopher Mechanics*, 1–107, *passim*.

Prior to these developments, the American mechanic world depended on the cascade publication of Rees's *Cyclopaedia*, modeled in part on the French encyclopedias, for its knowledge of British manufacturing technology. This huge publishing project lasted eighteen years in Britain (1802–1820) and fourteen years in the United States (1806–1820), where it was reprinted. In London Rees mobilized over a hundred contributors to write 39 million words accompanied by six volumes of plates (see Appendix 2). Aiming at printing 2,500 copies (in London 6,000 copies were issued), the American *Cyclopaedia* project was necessarily dependent on cash flows. It left Samuel Bradford, the lead publisher, bankrupt and caused misery (and further bankruptcy) for his successors.[92]

To spread his risks, Bradford and his successors created two distribution networks, which also doubled as knowledge networks. Although these have not been analyzed as thoroughly as David Meyer has dissected antebellum industrial networks, sufficient traces remain to establish elements of their stratified, cascade structure.[93] By this means the *Cyclopaedia* was brought to the United States and then distributed throughout the country. The strata in this multilayered network comprised publishers, subscribers, libraries, and all those with access to copies belonging to subscribers and libraries. Markham stood in this last group.

Samuel Bradford, as prime mover, obviously acted as the network node in the publishers' stratum. On one side he had suppliers—printers, artists, engravers, and so on—on the other were his customers, other printers who would take batches, and subscribers who ordered single copies. As Table 3.5 shows, Bradford was linked to 20 publishers, mostly in the seaboard states between Maine and Georgia. The result was a coverage, no doubt roughly calculable from Census figures, of the most heavily populated, urbanized parts of early-nineteenth-century America, as Table 3.6 suggests.

Comparing numbers of publishers to regional proportions of subscribers, it appears that publishers in the middle states, primarily Pennsylvania, were most successful in recruiting subscribers to Rees. Those in northern New England attained proportionate parity between subscribers and population. Publishers in the South were less successful and those in southern New England least effective of all. This profile was hardly surprising because the project was launched in Philadelphia. A somewhat different picture emerges when proportions of subscribers are compared to proportions of the population. By this measure it is evident that the greatest interest in Rees's "Scientific Dictionary" was stirred in northern New England, probably reflecting the region's strong educational traditions and the sort of cultural ethos that encouraged both the theoretical and practical approaches to new knowledge for which New England's universities and mechanics became equally famous. This of course overlooks the statistical effect of the slave population in the South where the proportion of Rees's subscribers was much less than the region's share of the total U.S. population.[94] Relative lack of interest was evident also in southern New England, which is less easily explicable.

Six of the subscribers lived in Vermont. From the evidence assembled in Appendix 2, it is most likely that of the two who lived in Middlebury, the lawyer Daniel Chipman would have shared his copy of Rees's *Cyclopaedia* with Isaac Markham. Chipman had economic and emotional stakes in the cotton factory at Middlebury; Markham's brother trained in Chipman's office as an attorney, and Markham and Chipman were in the same Masonic lodge. This circumstantial evidence makes it almost certain that

---

[92]We are grateful to Professor D. Arner of the University of Cincinnati, who is working on a study of Samuel Bradford and Philadelphia publishing, for generously sharing some of his groundbreaking information with us.

[93]Meyer, *Networked Machinists*. Clearly more can be learned about publishing networks either by applying Meyer's model or by using other models of social networks. See Van Wijk et al., "Knowledge and Networks."

[94]Jeremy, *Transatlantic Industrial Revolution*, 70.

able 3.5 **Publishers and Associates for Abraham Rees (ed.),** ***The Cyclopaedia*** **(Philadelphia, 1806–1820)**

ublisher or Associate	Place	State	Region
radford, Samuel F.	Philadelphia	Pennsylvania	Middle states
urray, Fairman & Co	Philadelphia	Pennsylvania	Middle states
esier, Peter	New York	New York	Middle states
ucas, Fielding	Baltimore	Maryland	Middle states
lake & Cunningham	Boston	Massachusetts	New England, N
ushing & Appleton	Salem	Massachusetts	New England, N
Vhipple, Charles	Newburyport	Massachusetts	New England, N
dams, John W.	Portsmouth	New Hampshire	New England, N
atten, Stephen	Portland	Maine	New England, N
owe, Hezekiah	New Haven	Connecticut	New England, S
apine, Daniel	Washington	District of Columbia	South
illigan, Joseph	Georgetown	District of Columbia	South
tewart, John A.	Alexandria	Virginia	South
enderson, D.	Fredericksburg	Virginia	South
itzwhylsonn, William H.	Richmond	Virginia	South
ottom [*sic*], Richard	Petersburg	Virginia	South
off, John	Charleston	South Carolina	South
Villiams, William T.	Savannah	Georgia	South
obby & Bunce	Augusta	Georgia	South
ramer, Spear & Eichbaum	Pittsburg	Pennsylvania	West
almer, James W.	Lexington	Kentucky	West

ource: Title page of Rees (ed.), *Cyclopaedia* (Philadelphia edition).

able 3.6 **Publishers and Subscribers of Abraham Rees (ed.),** ***The Cyclopaedia*** **(Philadelphia, 1806–1820), Compared to opulation, 1820**

egion	Publishers (number and %)		Subscribers (%)	Population (%)
orthern New England (ME, VT, NH, MA)	5	24	23	13.5
outhern New England (RI, CT)	1	5	1.5	3.7
iddle states (NY, NJ, PA)	4	19	33.5	28.0
outh (below Mason–Dixon Line) & West	11	52	42	54.8
bsolute figures and percentages	21	100%	1,851 (100%)	9,638,453 (100%)

ources: Jeremy, *Transatlantic Industrial Revolution*, 70; title page of Rees (ed.), *Cyclopaedia* (Philadelphia edition).

Markham would have been shown the articles and plates in Rees relating to the cotton and woolen manufacturing industries. This in turn fed his ideas both for adopting and for adapting British designs.

The importance of the 2D images of European (chiefly British) textile technology transferred to the United States depended not only on the engineering-drawing techniques employed but also on what kind of manufacturing information the images contained. As shown in Appendix 2, the single most important source of visual information about Britain's new manufacturing technologies, not least textile technologies, was Abraham Rees (ed.), *The Cyclopaedia*. None of the other two dozen texts for millwrights and mechanical engineers published in Britain during the forty-five years before Markham's death (see Table 3.1) include either such complete descriptions or such a range of manufacturing equipment as did the articles in Rees. For the first time in a single work of wide comprehension, the window onto Britain's new industrial world of machines and techniques was flung open. For textile machines, Rees was unrivalled.

The Rees contributors abstained from loading their factual machine descriptions and their unpeopled machine drawings with political or personal biases. Detached from the human aspects of their operatives, managers, or owners, these drawings were intended simply to explain how the new technology worked. The key textile industry articles were "Manufacture of Cotton" and "Woollen Manufacture." The former appeared in 1812 in London and in 1813 in Philadelphia; the latter appeared in both countries in 1818. They were written by practitioners, covered typical industrial forms in Britain, and included most of the most recent machinery available at the time of writing. The visual knowledge conveyed by the plates shared these features (reliability, representativeness, and recency). Judging by the drawing techniques (perspective, elevation, and plan views), the illustrations and their accompanying texts evidently aimed to enlighten both the scientific minded and the general reader. Lacking scale, they would not have satisfied a machine maker. Nevertheless, they were important in revealing the design forms of the new machines and (in English) naming the parts of components. To these merits must be added one more, their visual objectivity. Whatever the new knowledge they brought to America, individual artisans could be problematic. English immigrants like handloom weavers did not always have experience relevant for factory work. Others claimed competencies they did not possess, like one of the early steam-engine builders in Philadelphia whose millwrighting knowledge was faulty.[95] To test self-asserted abilities, authoritative sources of knowledge were required. For this reason Rees's published verbal descriptions and visual depictions of new machines held great importance. They offered benchmarks by which all self-proclaimed possessors of new technical knowledge could be judged—until a new wave of innovations left the printed authorities behind.

This brings the evaluation of visual evidence in technology diffusion to the last test: the extent of interdependency between visual and artisan vehicles of knowledge transfer. Even with accompanying textual descriptions, the visual information conveyed in patents or in Rees's *Cyclopaedia* was insufficient to build any of the machines accurately. Dimensions, angles, tolerances, numbers of gear teeth, and surface qualities were usually missing. Construction materials were usually unidentified. Operating routines, such as adjustments necessary to accommodate different qualities of cotton, were not specified. Maintenance and repair procedures were not detailed. Above all, in Rees's *Cyclopaedia* drawings, no scale appeared and only fitfully came in patent drawings. Without scale, machine builders could never be sure that they had accurately constructed a fully operable machine nor could they run it without the assistance of a mechanic familiar with the dynamic relations among the moving parts. But, obviously, from the machine makers' viewpoint, elevation and perspective drawings were a great improvement over the artist's view of the machine as an accessory object subject to classical perspective rules of drawing. These early machine drawings like those found in Rees's *Cyclopaedia* were reasonable design drawings, good for establishing the form of a machine (see Appendix 5).

What no inscriptions or images could capture was "tacit knowledge." This belonged to experienced artisans. Tacit powers of knowing, a concept first used and explored by the scientist-philosopher Michael

---

[95]Jeremy, "Immigrant Textile Machine Makers," 230–234.

Polanyi, have generated a long-running debate over its limits and whether tacit knowledge could ever become explicit and learned. For Polanyi tacit knowing consisted of the interaction between two elements, the proximal at the body's center and the distal at its extremities, with one, usually the distal—being known, and their gestalt totality being known but not communicable. "The things that we know in this way include problems and hunches, physiognomies and skills, the use of tools, probes, and denotative language."[96] So behind drawings there is much tacit information. Images cannot convey abstract thoughts.[97]

Thus, the feel of a yarn, the sound of a correctly tuned throstle spinning frame, the appearance of a cloth surface, working within proper tolerances, and the like, triggered tacit knowledge in the mind of the experienced mechanic and produced the decision that it felt, sounded, or looked right, or otherwise. Like evaluating the degree of tightness in a power loom warp, these exercises of tacit knowledge could only be learned from trial and error and never accurately expressed in pictures or words.[98] The logical consequence of this, argued Tsoukas, is that tacit knowledge could never be made explicit: "[I]t can only be displayed, manifested, in what we do...it comes about not when the tacit becomes explicit, but when our skilled performance—our praxis—is punctuated in new ways through social interaction."[99] That being so, the admixture of skilled and novice, immigrant and native, older and younger artisans in nodes and networks of machine shops provided the social interaction that developed tacit knowledge and hence was critical for the transfer of new technologies. Visual knowledge could only be subsidiary.

Markham's background, life, and drawings illustrated the primacy of artisan knowledge over visual knowledge per se. Opportunities for artisans to engage in social interaction and thereby to disseminate and develop new knowledge came when the first industrial immigrants of the 1790s formed "nests" and then, through their trainees, networks of skilled machinists. As Meyer has plotted, the first nests were located in Providence RI, and vicinity, near Boston, and in Philadelphia.[100] A network proliferating from the first of these nests was created by Samuel Slater. A former Arkwright mill manager, his arrival in Pawtucket, RI, from England (via New York) in 1790 effectively marked the beginning of the American cotton industry. Slater's family and trainees and their associates developed across a region stretching northward a hundred miles to include Vermont and Middlebury. Markham grew up as this network was ramifying and eventually entered it himself.

In these early years, ca. 1790s–1812, the first generation of capitalist promoters of early-American manufactures and the native craftsmen they recruited learned most from networked immigrant knowledge. Onto this they grafted their own tacit knowledge. For them the information about machine spinning of cotton or wool in Rees's *Cyclopaedia* was published after immigrant-originating knowledge arrived. David Page, Jr., owner of the first Middlebury, VT, cotton mill, may have learned much from the Rees articles but between 1811 and 1817 he depended more on his first two machine makers. They had earlier worked at New Ipswich, NH, where the machinery of the earliest cotton mill had been built by an employee of Samuel Slater. By means of this network, from immigrants to native craftsmen, mechanized cotton-spinning technology was introduced at Middlebury (see chapter 1).

Much more direct dependence on immigrant knowledge brought powered cotton-weaving technology to Middlebury in 1817 when Joseph Gordon arrived from Scotland. Markham, who was still an assistant

---

[96]Polanyi, *Tacit Dimension*, 29.

[97]Pinker, *How the Mind Works*, 295–258.

[98]Michael Polanyi, the originator of this concept in the 1960s, described tacit knowledge as "arising from the fact that we can know more than we can tell" or we have a "power to know more than we can tell"; Polanyi, *The Tacit Dimension*, 4.

[99]Tsoukas, "Do We Really Understand Tacit Knowledge?" 426. For the opposite view that tacit knowledge can become explicit and communicable see Ambrosini and Bowman, "Tacit Knowledge."

[100]Meyer, *Networked Machinists*, 63–64.

mechanic, helped Gordon construct 20 power looms. A year or two later Markham himself went on the tramp, gaining experience in building textile machinery at Waltham, where the Boston Manufacturing Co. had established an alternative textile-manufacturing system to that brought by Slater to Rhode Island. The Boston Manufacturing Co. experience would have been important. This was the premier machine shop in New England. Promoted by a group of Boston merchants, the Boston Associates, its success,

> rested on a community of practice in Boston's environs which encompassed iron foundries, general machine shops, and shops that also built textile machinery. Hub members of this community consisted of some of the leading machinists in the East, including [Jacob] Perkins, [Paul] Moody, [General Shephard] Leach, and [Luther] Metcalfe. These individuals and their firms or shops maintained network bridges among themselves and to other machinists who had previously worked for them. Their community of practice structured the tacit exchanges of knowledge which Paul Moody (and probably several of his other senior mechanics) had in face-to-face meetings and sub-contracts with other machine shops.[101]

Although leading machinists and their associates located with relative permanence at the hub machine works, others seeking training, employment, or work, traveled among the shops. In his late teens, Markham was one of these migrant mechanics. He went from Middlebury, VT, to Dover, NH, and then to Waltham, MA, and possibly back to Dover, before settling back at Middlebury.

In assuming that visual knowledge was part of the knowledge that flowed through early-nineteenth-century textile-manufacturing networks, the level of difficulty presented by the jump from the perspective view to the 2D view is easily overlooked. It required artisans to see differently, as explained, and not all mechanics could cope with 2D perception and representation. It defeated at least one of Markham's mechanic contemporaries. John Brandt, later master mechanic of the New York & Erie Railroad, found it almost impossible to make or to read workshop drawings. He thought full-scale and 3D, making his machine prototypes full size in wood even when he was over forty years of age in the late 1820s. It took more than a year for him to understand scale machine drawings and for years he worked from full-scale models.[102] Because he was extremely ingenious and inventive, he presumably had a serious conceptual problem.

Markham's surviving drawings include kinds that a modern mechanical engineer would recognize, albeit they are much less exact than their modern counterparts. There are the preliminary sketches in his notebook, recording experiments with various machine forms. The major corpus of 48 drawings listed in this edition (see Table A5.1, p. 379) show two kinds of machine coverage: whole machines (27) and components (21). Of the 48, all but five present elevations or plans. No longer rough scale sketches of a machine profile (like the Scholfield billy drawing) but carefully executed geometrical drawings to scale, they required simple drawing instruments: at least a long ruler and a pair of compasses.

Eight drawings adumbrated the final shift, after Markham's lifetime, to Monge's First Angle and the American Third Angle projections (see Appendix 5). These eight showed the relationship between machine faces or profiles by aligning the drawings of the faces on common bases, either shared or adjacent on a shared base, as identified in Table A5.1 (p. 379).

Nevertheless Markham deployed perspective views, one for a carding machine (Fig. 3.23, see also Fig. 6.42), one for his spinning machine (see Fig 2.10, p. 33), one for a possible warping frame (Fig. 6.74, p. 292), and two for his loom drawings (Figs. 3.24 and 3.25). The loom perspectives were made

[101]Meyer, *Networked Machinists*, 66.

[102]Ferguson (ed.), *Early Engineering Reminiscences*, 168–171. Brandt sounds as if he was a classic whittling boy, of the type noted in Cooper, *Shaping Invention*, 10, 13–17.

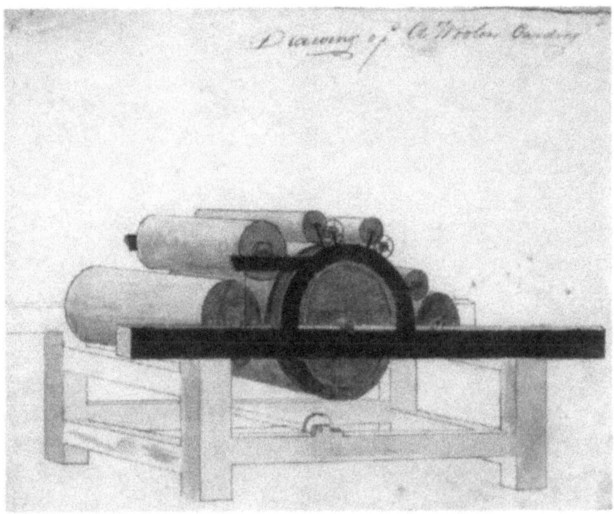

FIGURE 3.23 Woolen-carding engine, perspective sketch. See also Figure 6.42 (p. 221) for further analysis.

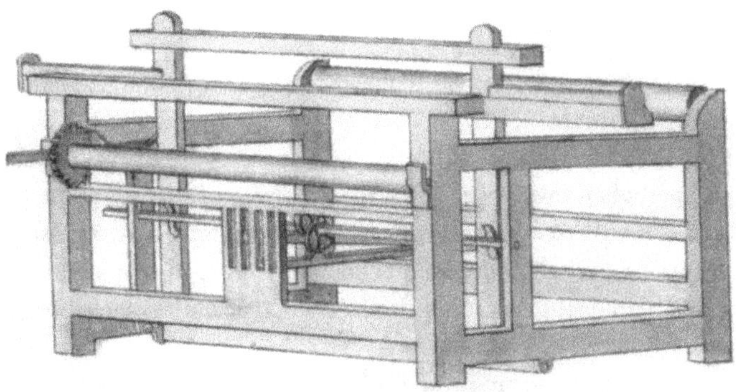

FIGURE 3.24 Power loom, perspective drawing, front view. See also Figure 6.81 (p. 311) for further analysis.

several years after his portfolio was begun in 1814, when the power loom was brought to Middlebury in 1817. Was this return to perspective regressive?

During 1813–1814 most probably, he would have read the text and seen the plates for the "Manufacture of Cotton" article in Rees's *Cyclopaedia*. They would have served as a benchmark for the machine forms of the new spinning technology. In addition the cotton machine plates provided or reinforced the idea of using 2D elevations, though without any indication of scale. Similar machines were increasingly found in states south of Vermont. Markham watched them being built and installed in Middlebury by experienced machine makers. Consequently, he had the clearest possible idea of what the spinning machines looked like. His mind could, therefore, fill in the depth of field from his 2D drawings and recall the relationships among the various machine surfaces.

FIGURE 3.25 Power loom, perspective drawing, front view. See also Figure 6.82 (p. 312) for further analysis.

The situation with the power loom was somewhat different. No up-to-date drawings of the best British power loom design (prior to Roberts's of 1822), that of Horrocks (1813; Fig. 3.26), were included in Rees's *Cyclopaedia*. Remarkably, the specification and drawing of Horrocks's loom reached the public domain the following year when they appeared in the *Repertory of Arts*.[103] It is not impossible, of course, that Gordon and Markham were acquainted with the patent design of the Horrocks power loom. Given the great interest of Britain's French and Continental competitors in this source, it is just as likely that American merchants and manufacturers would also have known about it. So far no evidence in Middlebury indicates this, however.

So the Horrocks design, in the form of the Scotch loom, was introduced to the United States by the immigrant William Gilmour in 1816. It had more moving parts and hence complexity than the Arkwright spinning frame. And this seems to have been the design that Joseph Gordon followed. Markham watched Joseph Gordon build 20 power looms in the Middlebury factory in 1817, so he should have been very familiar with the loom's design. However, he may well have needed a reminder of the machine form. As the evolutionary psychologists have demonstrated, "people cannot reconstruct an image of an entire visual scene. Images are fragmentary"[104] and visual memories are not accurate.[105] In addition, and as likely, Markham may have wanted to use the perspective drawings in marketing his skills as a mechanic. He would need them when dealing with a capitalist who had never before seen a power loom. Use of the perspective view was not regressive. It was helpful in recording machine forms, and for this reason has continued in use to the present: sometimes necessary, never sufficient.

[103] *Repertory of Arts, Manufactures, and Agriculture*, 2nd ser, 25 (June 1814).
[104] Pinker, *How the Mind Works*, 294.
[105] Pinker, *How the Mind Works*, 295.

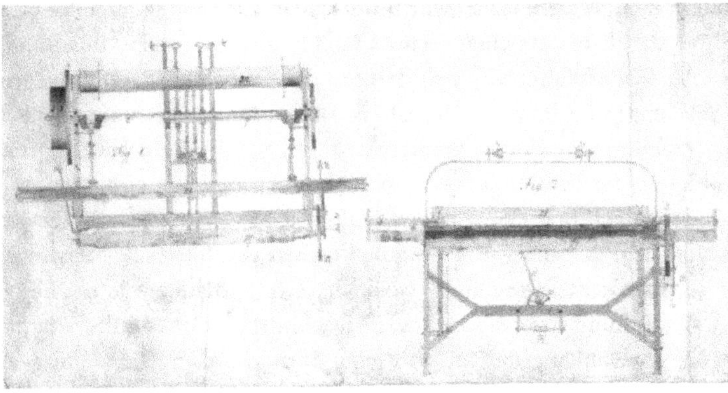

FIGURE 3.26 William Horrocks, GB Patent for the power loom, 31 July 1813. See also Figure A3.17 (p. 364) for further analysis. Courtesy of the National Archives, London.

Although Markham's drawings tell us a lot, there is much explicit knowledge about which they are silent, which in effect they hide. Two types of explicit information are unrecorded on the drawings. There is information that more properly belongs in an accompanying notebook, knowledge that can be expressed verbally. For example, technical details, like kinds of materials, economic information such as costs and outputs, instructions to operatives, and so on, could not always be found alongside the shop drawing. These can be seen in the surviving fragments of Markham's notebooks included here (see chapter 6, "Notebooks and Loose Sheets of Writing" and "Sketchbook" sections, also see Appendix 5). Second, over time, using geometrical plans would require some mathematical competencies as well as what would increasingly come to be known as "scientific training."

Across the whole range of manufacturing industries in the United States during the first four decades of the new nation, the practice of mechanical drawing for workshop purposes seems to have been patchy. As seen, well ahead of other millwright-engineers was Oliver Evans and in an economy based on waterpower, his handbook, *Young Mill-Wright and Miller's Guide*, was one of the most popular technical texts. Markham used it (see Fig. 3.1, p. 46). Evans advocated a school to teach mechanical drawing. Yet, according to George Escol Sellers (1808–1899), Evans would not use scale drawings, believing they would be too difficult for ordinary workmen to understand. Instead, Sellers claimed, Evans developed his ideas from rough pencil sketches and then, with the aid of a limited range of drawing instruments ("two-foot rule, straight edge, square and compass"), he made his drawings "full size on chalk boards." For some draftsman's skills he relied on the services of two copperplate engravers.[106] This recollection seems to contradict the use of elevation, plan, section, and scale apparent in Evans's *Young Mill-Wright's Guide* (1795) already analyzed.

Mechanical-drawing techniques in the textile industry apparently lagged behind, especially in remote rural machine shops.[107] In places there were certainly instances of the rough-and-ready approach to engineering drawing. In New England, William Gilmour, who introduced the power loom from Scotland in 1817, drew the design of the warper, dresser, and loom, full size on the floor of an empty mill room

[106]Ferguson (ed.), *Early Engineering Reminiscences*, 38.
[107]Brown, "Design Plans," 205.

and from this pattern moulds were made, which were then taken to the forge for casting or forging iron machine parts.[108] Yet Markham's drawings seem to modify this assessment. In addition, Zachariah Allen, a Rhode Island woolen manufacturer who visited England in 1825, made careful and probably surreptitious sketches of the machinery he found in Yorkshire. Dimensions were omitted but the drawings were supplemented by "exact iron patterns of the frames, keys, &c." He must have approximated dimensions when he returned home, for he built a close replica from at least one of his stolen designs.[109]The large size of the Boston Manufacturing Co. works at Waltham, MA, with over 250 employees in 1820, and where Markham worked at that time, surely dictated the use of dimensional drawings, though none have apparently survived. Its patented machinery would have had to adhere to the designs specified in the original patents[110] and its numbers of mechanics (37 machinists, 5 blacksmiths, 1 carpenter, and 9 laborers on its payroll in 1820 [111]) would need to build the equipment they supplied to other textile manufacturers in conformity to Boston Manufacturing Co. cost schedules.

Few modern studies of early-nineteenth-century American industries (apart from architecture and civil engineering) mention the use of machine or workshop drawings.[112] This could be because drawings have not survived. As observed at the outset of this volume, Markham's surviving scale drawings are extremely rare for the textile industry of this period. Another possibility is that the use of drawings in the machine shop was so common as not to excite comment: they were always present. This seems to be discounted by Sellers's recollections of Evans's practices. The third possibility is that workshop scale drawings were little used.

What purpose did paper scale drawings serve? A number of situations in early American industrial experience would require use of mechanical drawings. One was to preserve an archival record of the machines made by the workshop, either to make comparisons over time or between workshops, or as some kind of insurance, should the wooden patterns for components ever be destroyed. With drawings, the wooden pattern moulds could readily be reconstructed. It is important to note that the total layout drawing was the key to final assembly: without it, there were jigsaw pieces but no picture. One or another of these reasons may have shaped the preservation of Markham's drawings. Another purpose was to make a portable record that might easily be taken between factories or industrial plants by the machine maker owning the designs, for technical, commercial, or marketing reasons. Presumably Markham did something like this when he worked at the Dover, NH, cotton factory, either as employee or contractor.

Drawings of course were sometimes used to transmit design ideas along mechanic networks. Drawings were the form in which a new graining machine was secured from Paris for his gunpowder works on the Brandywine in Delaware by Elèuthere Irénée du Pont in 1804.[113] It is likely that drawings were used to transmit design ideas along the mechanic networks identified by Meyer, though apparently he found no evidence for this.

Again, drawings could be important when a machine maker placed work with suppliers. One example of this occurred at Harpers Ferry Armory in 1827. When John H. Hall, Superintendent of the Rifle Works,

[108]Bagnall, *Textile Industries* I, 548–549.
[109]Jeremy, *Transatlantic Industrial Revolution*, 133–136.
[110]Jeremy, *Transatlantic Industrial Revolution*, 185–198.
[111]Baker MSS, Harvard Business School Archives, Boston Manufacturing Co. Papers, vol. 82 (1820). These are the numbers of males on the payroll between 4 March and 26 August 1820 but who were not always employed concurrently. About three dozen other males were also on the payroll, chiefly as overseers, at the Boston Manufacturing Company's two mills at Waltham in 1820.
[112]I searched the following studies, with results that are incorporated in the text: Cooper, *Shaping Invention*; Ferguson (ed.), *Early Engineering Reminiscences*; Hindle and Lubar, *Engines of Change*; Hounshell, *From the American System*; Hoke, *Ingenious Yankees*; Meyer, *Networked Machinists*; Pursell, *Early Stationary Steam Engines*; M. Smith, *Harpers Ferry Armory*; Stapleton, *Transfer*.
[113]Stapleton, *Transfer*, 114.

set up his experimental production system for manufacturing breech-loading rifles with interchangeable parts, he had to obtain some components for his production machines from outside forges and foundries. Two of them were as far away as Baltimore. He therefore sent "patterns and drawings to the founders, who cast the respective parts and then shipped them to Harpers Ferry where they were finished and assembled into complete machines."[114] This small piece of evidence, confirming the use of drawings at Harpers Ferry, casts some doubt on Booker's blanket judgment, "that mass production could be, and was, effected without any call upon drawings, for the simple reason that, at that time, drawings were inadequate for this purpose."[115]

Engineering drawing, as Latrobe found, was essential to the installation of steam engines and heavy machinery. Mechanical drawing became indispensable in the first decades of the nineteenth century with the spread first of canals and steamboats and then the advent of railways. Robert Moore, by 1852 joint owner of the Harkness, Moore & Co. locomotive builders of Cincinnati, OH, recalled joining Anthony Harkness in 1828 to make "patterns and drawings" for steamboat machinery and sugar mills.[116]

Mechanical engineers, responsible for building or maintaining locomotives, the largest, most complex, and most expensive mechanisms yet seen, wanted accurate and detailed drawings to curb waste of materials and reduce other costs. Examples of the drawings' use are found in the mid-1830s recollections of George Escol Sellers. Drawings were submitted when competing for orders from railway operating companies. More drawings were made when modifications were required. If subcontractors (like boiler-makers) supplied components, then drawings specified the components. Drawings helped to prove originality and priority of design.[117]

The widest range of industrial drawings in the United States prior to the mid-1830s was made because they were legally obligatory. These were patent drawings, a requirement in submitting an application for a patent of invention at that time. In December 1836 fire completely destroyed the Patent Office building and its contents in Washington, DC. Early patents were lost and had either to be reconstructed or reissued.[118] Consequently, scholars have used these to understand the early development of industrial technology. One impression is that before 1830 for patent drawings recording the designs of textile inventors, almost as many displayed 3D-perspective views as 2D elevations and plans to scale (see Table 3.4, p. 73). Their purpose, presumably, was to reveal the form of the patented device.

Before 1840 drawings clearly became important to machine makers, both for design and construction purposes. As well as reducing waste, drawings helped in monitoring production progress; assisted in settling disputes among contracting parties; were used in cataloguing wooden patterns for castings; and guided assembly, erection, maintenance, repair, modification, and replacement. However, as John Brown has shown, they were not uniformly used across manufacturing industry in the United States.[119] Nor do we know a great deal about the drawing-board conventions that emerged before the 1830s. After 1850 machine drawings became even more important, not least as instruments of managerial control over the shop floor.

---

[114]M.R. Smith, *Harpers Ferry Armory*, 228.
[115]Booker, *History of Engineering Drawing*, 187.
[116]J.H. White, *Cincinnati Locomotive Builders*, 34.
[117]Ferguson (ed.), *Early Engineering Remiscences*, 163–171.
[118]An estimated 7,000 models, 9,000 drawings, 230 books, and all records of patent applications and grants were lost. US Department of Commerce, *Story of the United States Patent Office*, 7. Some pre-1836 patent specifications had been reported in technical magazines and so have survived. Others, figuring in patent litigation, survive in the early records of the district courts. Yet others are sometimes preserved in the private papers of patentees or their associates.
[119]J.K. Brown, "Design Plans."

It was evident in the 1830s, both in the United States and in Britain, that drawings were one of the three languages (with mathematics and physical science) of the mechanical engineer. As observed in Boston in 1839 by Timothy Claxton, an English immigrant whitesmith-turned inventor with European experience:

> A man who is well versed in arithmetic and geometry, together with the laws or properties of simple machines, (improperly called mechanical powers), with their effects, when combined, as in compound machines,—I say, a man that understands these, and the laws of attraction, of simple and compound forces, the various properties of aeriform [*sic*] and watery fluids, with some knowledge of chemistry,—such a man, if he sees a plan on paper, hears it described, or reads of it, especially if he is a practical man, will be able to judge of its merits more correctly, in less time, and consequently at less expense, than if he were merely a workman. In the latter case, he must construct his model. If this is done in a bungling manner, it will not perform, but he may conclude it would if it had been well made. This is all guesswork. ...Now if our mechanics understood the principles of science, what time and expense would be saved![120]

In these new, more accurate "machine plans," by the 1820s the perspective filter (to coin a phrase) was being relegated to a subordinate role as a drawing convention in both Britain and in America.

VISUAL ASPECTS OF THE GENERATION OF NEW KNOWLEDGE IN EARLY-ANGLO-AMERICAN INDUSTRIAL ENGINEERING

Markham's drawings and his industrial experience draw attention to three aspects of the generation of new knowledge (invention) during the early nineteenth century. The nature of invention and the impact of technology diffusion on inventive capabilities is the first. A second is design and dreaming. Third, there is the institutional dimension.

Markham evidently originated several improvements to imported British textile technology, technology he received via his Middlebury, Vermont, Waltham, Massachusetts, and Dover, New Hampshire, training and experience. One was a ground-to-rollers arrangement to feed cotton into a carding machine (see Fig. 2.2, p. 27). Others were fly-wheel spinning (see Fig. 2.10, p. 33 and Figs. 3.27–3.32), possibly a spooling machine (Figs. 3.33 and 3.34), and a dresser (Fig. 3.35, see also Figs. 2.6–2.9, pp. 31–32). One mechanism repeatedly appears in the fragments of his notebooks, in his Sketchbook, and in his drawings. This device was the double or twin-cone pulley. It appears in Markham's lap winder (Fig. 3.36), his spinning machine (see Fig. 3.27), his dresser (see Figs. 2.3 and 2.4, pp. 28, 29; Table A3.2), his warping frame (Fig. 3.37), possibly in his power loom, and in other unspecified mechanical recombinations (see Fig. 2.5, p. 30, and Figs. 3.38–3.40). This device appears so often in Markham's work, it is almost as common as his signature.

This raises two questions. What did he intend in using the double-cone pulley? What indeed did he understand about the process of discovering applications for that pulley, that is, the process of invention? Although it is not directly known what he might have understood by the process of invention, and he in all likelihood worked intuitively, one of the sources that was available publicized ideas about the essence of invention.

Articles in Rees's *Cyclopaedia* offered criteria in machine building and stories, albeit fragmentary ones, about inventive effort. Because the *Cyclopaedia* was reprinted in America and widely circulated, these

[120]Claxton, *Memoir*, 146.

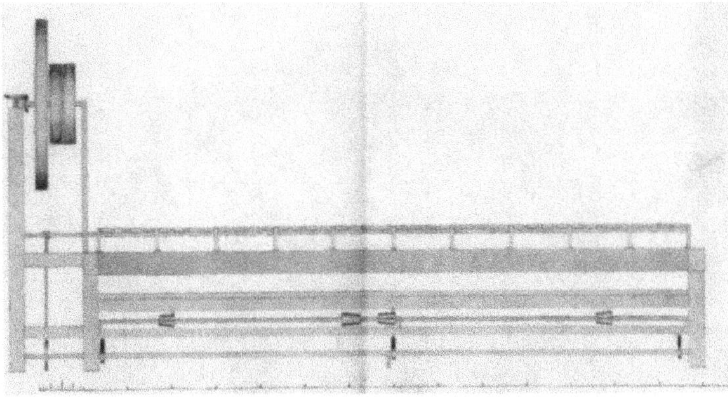

FIGURE 3.27 Spinning machine, front (?) elevation. See also Figure 6.66 (p. 274) for further analysis.

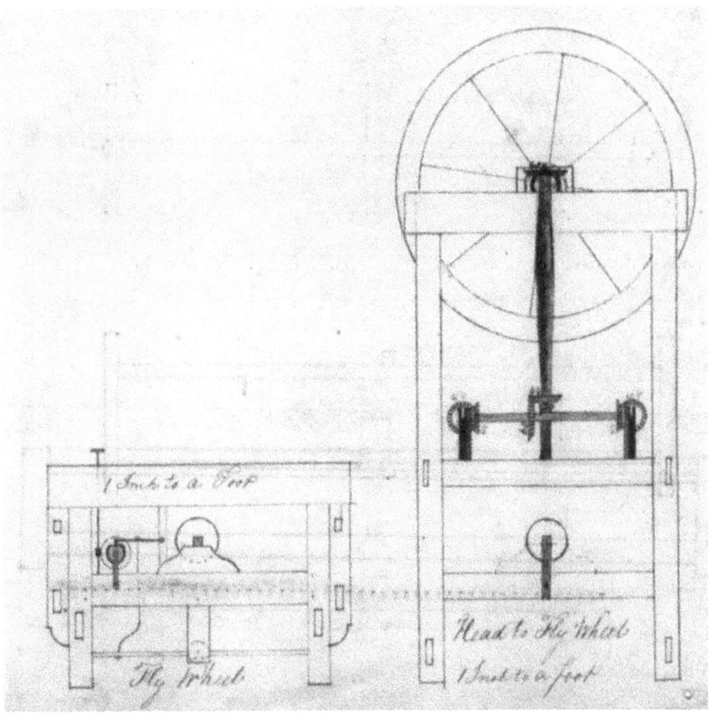

FIGURE 3.28 Flywheel spinning frame, side elevations. See also Figure 6.67 (p. 276) for further analysis.

ideas would have moved along networks of mechanics and machine builders. It is almost certain, from circumstantial evidence (see Appendix 2), that Markham would have known of them.

In his article for Rees on "Machinery," the brilliant young polymath engineer John Farey, Jr. (see Appendix 2), observed, "The grand object of all mechanism, or machinery, is to convey and modify the motion of the first mover of the machine, and communicate it in a proper manner to the subject to be operated upon." Expanding a bit, he continued, "In contriving any machinery, the engineer should always

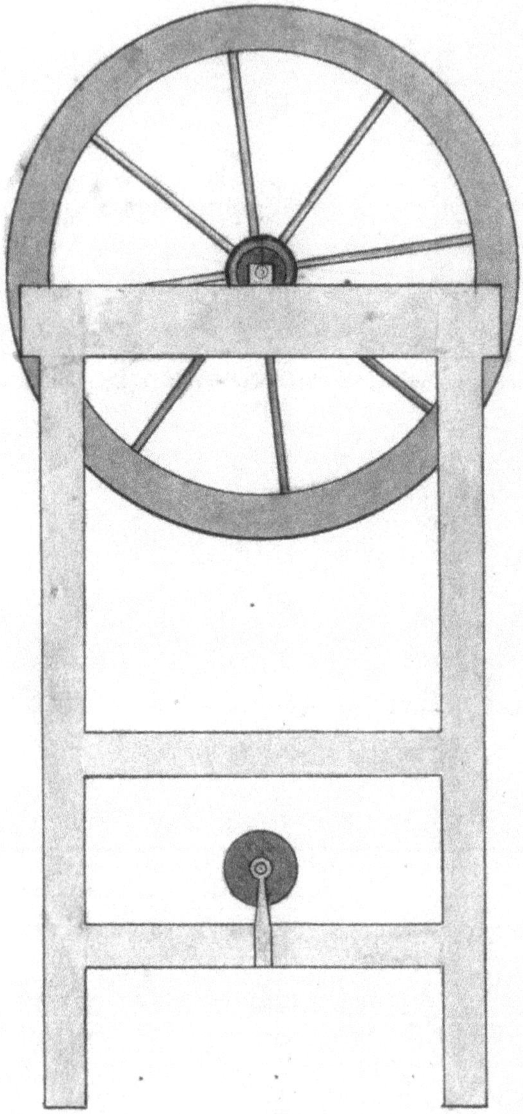

FIGURE 3.29 Flywheel spinning frame, side elevation. See also Figure 6.68 (p. 278) for further analysis.

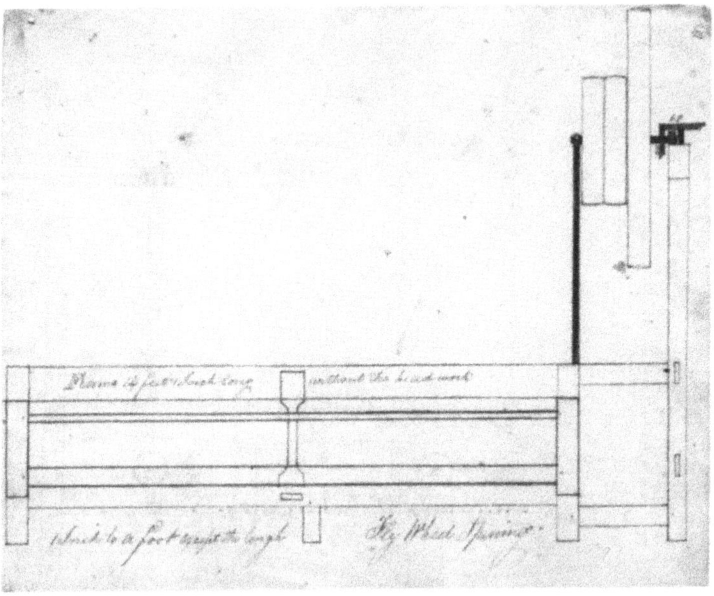

GURE 3.30 "Fly Wheel Spinning," rear elevation. See also Figure 6.69 (p. 280) for further analysis.

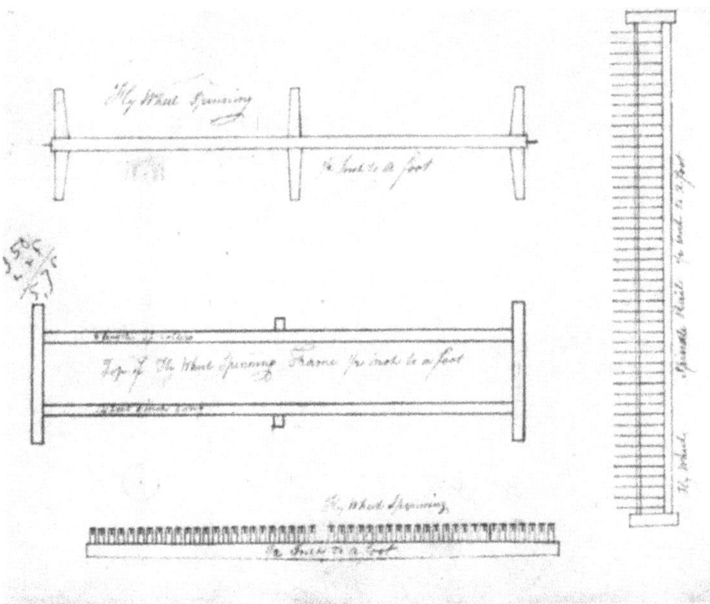

GURE 3.31 Flywheel spinning frame, components. See also Figure 6.70 (p. 282) for further analysis.

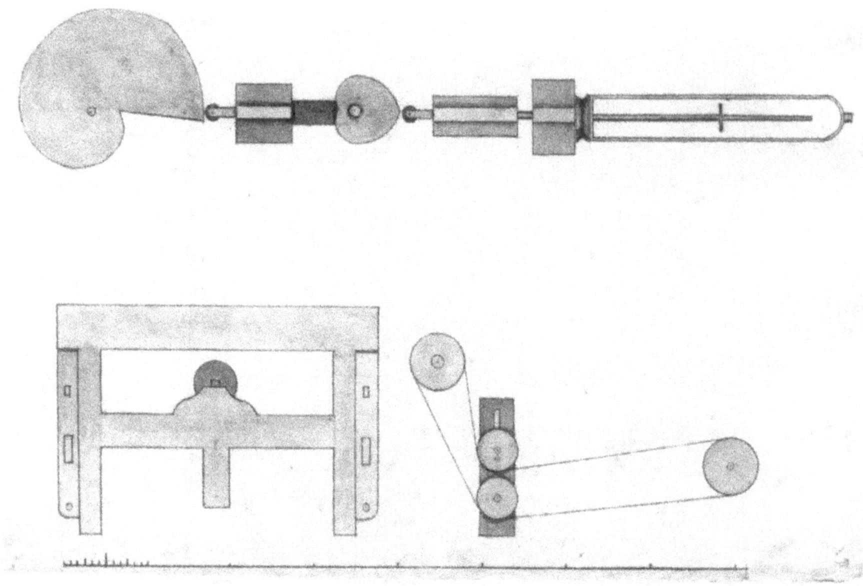

FIGURE 3.32 Flyer drives. See also Figure 6.71 (p. 284) for further analysis.

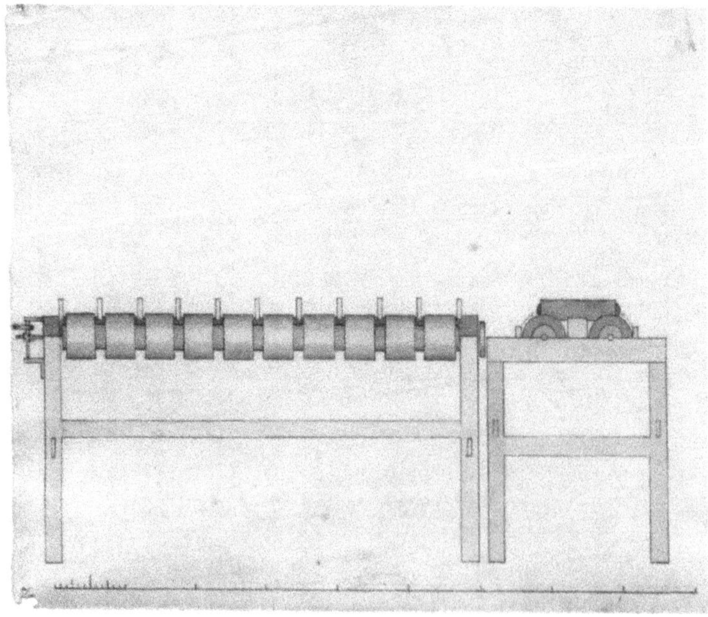

FIGURE 3.33 Spooling machine, front and side elevations. See also Figure 6.72 (p. 288) for further analysis.

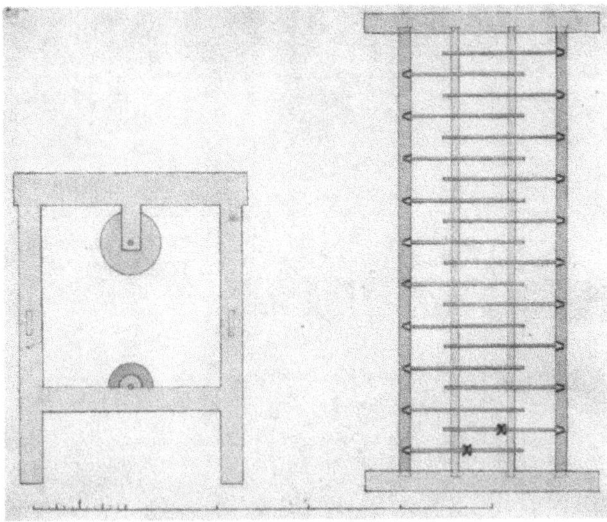

IGURE 3.34 Spooling machine (?), side elevation and plan. See also Figure 6.73 (p. 290) for further analysis.

IGURE 3.35 Dresser, side elevation. See also Figure 6.75 (p. 294) for further analysis.

remember that nothing contributes more to the perfection of a machine, especially if it is massive and ponderous, than great uniformity of motion."[121] That article appeared in London in July 1812 and in Philadelphia early the following year. In applying his double-cone mechanism, Markham took the advice to heart. Double-cone pulleys were used in the English double speeder roving frame for the purpose of effecting change in rotational speed smoothly. With that effect, and presumably with that intent, Markham used the mechanism in his dresser, listed it in his technical notes and in his sketchbook, and tinkered with it for other applications (see Table A3.2, p. 368). Similarly, his preoccupation with fly-wheel spinning (see Figs. 3.28–3.31) suggested he was concerned with achieving "uniformity of motion" and an increased smoothness of operation.

Likewise pertinent were the invention stories recounted by Thomson in his article for Rees on "Cotton Manufacture" (1808). James Hargreaves got the idea for his spinning jenny when witnessing a trifling domestic accident with a family spinning wheel knocked over by children playing in his house: "The

[121]Farey, "Machinery," cols. 4–5.

FIGURE 3.36 Loose sheets. See also Figure 6.19 (p. 166) for further analysis.

thread still remained in the hand of the spinner, and as the arms and periphery of the wheel were prevented by the framing from any contact with the floor, the velocity it had acquired still gave motion to the spindle, which continue to revolve as before." Hargreaves was riveted. "[He] surveyed this with mingled curiosity and attention. He expressed his surprise in exclamations which are still remembered, and continued again and again to turn round the wheel as it lay on the floor, with an interest which was at that time mistaken for mere indolence."[122] Having already tried to spin three or four threads at once on a single spinning wheel, Hargreaves was already thinking about the problem. He proceeded to use two bars of wood to act as a clasp for several threads, thereby substituting for the spinner's left hand. In this fashion, according to the story, the jenny was conceived.

In a similar way, Thomson credited Richard Arkwright with the idea of his drafting rollers (in progressively faster pairs, drawn between which the roving was stretched into yarn):

> The first hint for effecting this improvement, he accidentally received from seeing a red hot iron bar elongated, by being passed through iron rollers. Between this operation and that of elongating a thread, as now practised in

---

[122]J. Thomson, "Cotton Manufacture" (1808) col. 7.

spinning, there is no mechanical analogy; yet this hint being pursued, has produced an invention, which, in its consequences, has been a source of national and individual wealth unparalleled in the annals of the world.[123]

With apparent perversity, Thomson underplayed Samuel Crompton's achievement in inventing the spinning mule. In his 1808 Rees article he referred to Crompton's "ingenuity" but said nothing about the process of invention. It was "a compound of the two machines of Arkwright and Hargreaves."[124] In Thomson's 1812 article he was positively dismissive: "This machine was introduced by a Mr Crumpton, who lately received a reward of £5,000 from Parliament for the invention, which, as before mentioned, consisted only in the combination of Hargreaves' spinning jenny with Sir Richard Arkwright's drawing rollers."[125]

What did these invention stories imply? Their message would have been well known to British and American mechanics in the 1790s and 1800s. In short, it was that the essence of invention lay in the recombination of familiar mechanical elements to form a new mechanism. In 1797 the engineer Joseph Bramah, when contesting James Watt's steam-engine patent, expressed the idea, but he was by no means the first: "Invention [consists in] those efforts of the mind and understanding which are calculated to produce new effects from the varied applications of the same cause, and the endless changes producible by different combinations and proportions."[126] The universality of the recombinatory principle was affirmed a year later by Richard Lovell Edgeworth, jointly authoring a text with his daughter Maria Edgeworth, both educators: "Invention is nothing more than the new combination of materials."[127] Recombination derived from the association principle of the philosopher David Hartley (1705–1757).[128]

The dominating feature of all the invention stories in Thomson's article in Rees is that they required the inventor first to spot fresh potential in components of real but disparate mechanisms, and then to reassemble them in a new combination. Exercises in recombining mechanical components, the Edgeworths thought, should be devised for children and incorporated in children's education. They regretted that Benjamin Donne's promised solid geometrical shapes were not available for children's use.[129] Physical vision, of course, was aided by memory and imagination. In his chapter on "Memory and Invention," Richard Edgeworth observed, "Memory is chiefly useful as it furnishes materials for invention."[130]

How could Crompton's downgrading be explained? Thomson may have harbored some animosity toward Crompton, envied the success of his mule, or resented his parliamentary premium (he received his award of £5,000 in 1812.[131] More likely, as Crompton, a weaver, gained no royalties or capital investments from his invention, Thomson had in mind another aspect of the theory of invention current in the 1790s. Richard Edgeworth expressed it: "We must judge in general of the ease or difficulty of any invention, either by the number of ideas necessary to be combined, or by the dissimilarity or analogy of

[123]J. Thomson, "Cotton Manufacture" (1808), col. 10.

[124]J. Thomson, "Cotton Manufacture" (1808), cols. 15–16.

[125]J. Thomson, "Manufacture of Cotton" (1812), col. 37.

[126]Bramah, *A Letter to the Lord Chief Justice* (1797), 77, 83; quoted in MacLeod, *Inventing*, 220.

[127]Edgeworth, *Practical Education*, 592.

[128]Hartley's ideas raised questions for twentieth-century neuroscientists. See Boden, *Mind as Machine*, chapter 12 and *passim*. Possibly the concept of recombination owed something to mathematical principles: Emerson, *Doctrine of Combinations*. In the mid-twentieth century the combinatory principle was famously invoked by the economist Joseph Schumpeter to explain the nature of entrepreneurship: Schumpeter, *Theory of Economic Development*, 74–94. In the 1990s, evolutionary psychologists affirmed a universal capacity for combinatorial thinking as the essence of creative activity (of which invention is one manifestation): "that is what our combinatorial adapted minds are for"; Pinker, *How the Mind Works*, 362.

[129]Edgeworth, *Practical Education*, 448; Donn, *Essay on Mechanical Geometry*, x.

[130]Edgeworth, *Practical Education*, 587.

[131]Farnie, "Samuel Crompton."

those ideas." In its simplest form, the mule seemingly involved two ideas, both related to spinning. Thomson may well have downgraded Crompton's achievement because he recalled these criteria. Whatever, all these views of the inventive process would have been familiar to the American middle classes, and very likely to Markham, because not only was Rees's *Cyclopaedia* published in the United States, as seen, but also the Edgeworths's *Practical Education* was thrice published in Markham's lifetime in New York (1801), Providence, RI (1815), and Boston (1823).[132]

Many years later when American textile engineers looked back on their activities, as a number did in old age, one or two dredged up recollections of the sources of their inventive ideas. Aza Arnold, one of these textile engineers, then living in Washington, DC, at the outbreak of the Civil War, remembered how it happened for him. The germ in Arnold's mind was a story his father had told him in 1803 about a wagon with two driving wheels fixed on a single shaft. A crank in the middle allowed a person to power the shaft but not to turn the wagon right or left. Over the years Arnold thought and talked to fellow mechanics about this and eventually invented the "compound motion (so-called) as early as 1818, and while repairing a [William] Hines speeder in 1820, it occurred to me, that by using it, I could simplify the speeder, and produce a much more usefull [*sic*] machine." He consulted the leading mechanics in his Rhode Island networks—David Wilkinson, Larned Pitcher, and Ira Gay—all of whom thought his ideas impracticable. Against their advice, Arnold went ahead, put his speeder into operation in 1822, and patented his mechanism, the differential gear, in January 1823. This important invention introduced the differential gear to the textile industry and to the wider world. Arnold, because of some confusion about the wagon story, suppressed the details of the source of his invention when his recollections were published in 1861. So it appears that by the pursuit of the recombinatory principle, the differential gear of the textile industry had its origins in a road vehicle, exactly where it has been most widely used in the twentieth century.[133]

In Britain there was some ambivalence about certain aspects of the invention phenomenon. If recombination was the essence of invention, who practiced it? Bramah asserted that God had made it every man's duty to exploit nature.[134] Did all have similar inventive capacities? Before the 1750s it was believed that Providence released new inventions of major significance to the talented elite, leaving artisans merely to improve the "big" inventions. Mechanization in the second half of the century forced economic theorists to change this view. The role of Providence was limited to supplying raw materials; people were responsible for making the best use of those resources, primarily via labor-saving inventions.[135] And "invention was still regarded as a largely anonymous achievement."[136]

In Markham's lifetime, this human-centered, more egalitarian understanding of inventiveness was modified, but not submerged, by the emerging concept of creative genius. [137] An elitist theory re-emerged,

---

[132]Maria and Richard L. Edgeworth, *Practical Education* (2 vols., New York: G.F. Hopkins, Brown & Stansbury, 1801; 2 vols., Providence, RI: J.F. Lippitt, 1815; 1 vol., Boston: S.H. Parker, 1823). For other evidence on early-nineteenth-century understandings of the invention process, see MacLeod, *Heroes of Invention*, 161–180.

[133]Rhode Island Historical Society, Zachariah Allen Papers, Miscellaneous Papers 1809–1862, Miscellaneous Papers Box, Aza Arnold to Zachariah Allen, 4 September 1861 and 12 June 1862. *Transactions of the Rhode Island Society for the Encouragement of Domestic Industry* 1861, 80–85. Jeremy, "Technological Diffusion," reprinted in *Artisans*.

[134]Quoted MacLeod, *Inventing*, 220.

[135]MacLeod, *Inventing*, 201–222.

[136]MacLeod, *Inventing*, 220–221.

[137]MacLeod, *Heroes of Invention*, 161–180. The myth of the heroic inventor-industrialist reached full flower in the hands of Samuel Smiles whose *Self Help* (1859) and other best-selling industrial hagiographies made the myth unassailable for decades. In the twentieth century the notion that inventors are uniquely endowed was rejected by historians of technology. Usher, *History of Mechanical Inventions*, 56–83; Rosenberg and Vincenti, *Britannia Bridge*, 75–79.

not least because the likes of Arkwright and Watt gathered posthumous reputations over which their successors hotly disputed.[138]

In the United States the heroic theory ran counter to the Republic's foundational democratic principles. In the American textile industry, the myth of the heroic industrialist-inventor emerged long after the death of Samuel Slater in 1835. In his lifetime Markham faced no restraints from an elitist concept of inventiveness.[139] Markham's career and his papers almost self-evidently establish the link between the diffusion of technical knowledge and invention. The array of mechanisms progressively revealed by the publication of visual and related evidence, partly in the *Repertory of Arts* (which published 20 percent of British textile patents between 1794 and 1812; 30 percent cent between 1813–1824[140]), partly in the multiplying volumes of Rees's *Cyclopaedia*, simply added to the number of mechanisms available for inventive mechanics to recombine in new ways to meet changing market perceptions of the quantity and quality of goods demanded under the social constraints imposed by the new Republic.[141]

The second visual knowledge theme prompted by the Markham papers is that of design and "dreaming."[142] Allowing visual recollections to swim together in the mind, whether in daydreams or vivid flashbacks, could, kaleidoscope-like, suggest new recombinations from old ones. Externalizing these dreams on paper could be highly creative. "The effect of concentrating the geometric aspects of manufacture in a drawing is to give the designer a much greater 'perceptual span' than the craftsmen had."[143] Markham could think almost without a limit on his drawing board. Yet envisaging successful inventions required the newly conceived technical device to be firmly related to factor cost and savings and market opportunities. Unfortunately, very little of his perceived economic opportunities has been preserved in Markham's papers, apart from a comparison between his spinning frame spindle costs and the Boston Manufacturing Co.'s comparable spindle prices, which can be deduced from other sources (Figs. 3.41A and 3.41B; also see Appendix 5).

However, his technological dreams are partly betrayed in his Sketchbook (see chapter 6, p. 186). Of the hundred or so miniature sketches, some appear imaginative and fanciful, such as the top sketch on Figure 2.5 (p. 30). Numerous others (e.g., Fig. 3.42) echo known equipment. Do they depict machines Markham conceived at Waltham, MA, or Dover, NH, or Middlebury? Whatever their purpose, these choice miniatures, freshly and sharply sketched, are informed by the eye of an artist in their execution and their balanced composition on each page. The question remains: Do these sketches record the early-nineteenth-century New England mechanic as observer or as dreamer? In view of the manner in which early-nineteenth-century American engineers like Latrobe organized their design and production operations, it is likely that this was a preliminary design document, a record of imaginative ideas.

There remains the institutional aspect of the drawings, the puzzle of Markham's neglect of patent protection. The only evidence that any of his inventions had commercial value, and therefore were worth

---

[138]For the controversy over Arkwright's reputation, see MacLeod, *Heroes of Invention*, 193–198; Jeremy, "British and American Entrepreneurial Values." Compared to Arkwright, James Watt, that other titan of early industrial technology, fared much better. The champions of Watt's claims to pre-eminence in steam-engine technology quickly emerged triumphant as regards his posthumous reputation. MacLeod, *Heroes of Invention*, *passim*; MacLeod, "James Watt"; Torrens, "Jonathan Hornblower."

[139]For Slater's reputation, which reached heroic proportions between 1890 and the 1960s, see Conrad, "'Drive that Branch,"' 1–12 especially. Jeremy, "British and American Entrepreneurial Values," argues that his first biographer, the Reverend George Savage White (1836), scarcely knew Slater and succeeded in burying him under long excerpts from a mixture of contemporary sources about the British and American cotton industries but left him with the epitaph of "The Father of American Manufactures" (title page of White, *Slater*).

[140]Jeremy, *Transatlantic Industrial Revolution*, 47.

[141]Primarily the agrarian ideal promoted by Jefferson before the War of 1812 and the urban-industrial vision of the Hamiltonians.

[142]Morrison, "Some Notes on Visions" in *From Know-How to Nowhere*; Hindle, 'The Contriving Mind' in *Emulation and Invention*; Ferguson, *Engineering and the Mind's Eye*, 44–51.

[143]J. Christopher Jones quoted in Baynes and Pugh, *Art of the Engineer*, 11.

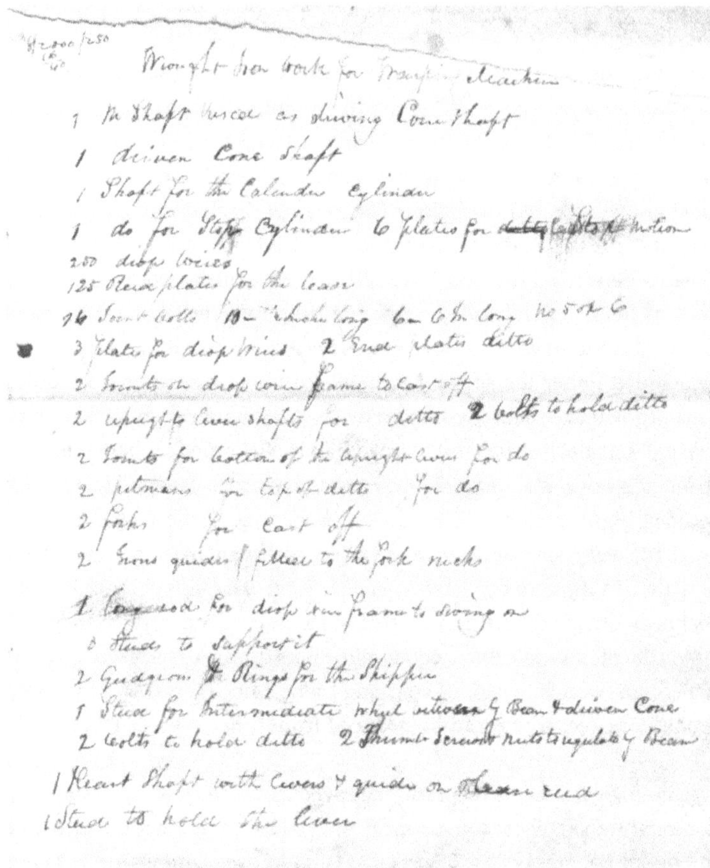

FIGURE 3.37 Loose sheets. See also Figure 6.16 (p. 158) for further analysis.

patenting, comes from an inventory of the Dover Cotton Factory at Dover, NH, dated 29 June 1820.[144] The firm's equipment then included "2 warping machines with stop motions with 2 dressing of Adams and Markhams plan" all valued at $1,000. The listing does not make it clear whether Adams and Markham's design was confined to the dressing machines or included the warping machines as well. Only one possible warping-machine design survives among Markham's workshop drawings (Fig. 3.43). At any rate, at least one of Markham's inventions reached the marketplace. Why did he fail to file any patents, to protect his inventor's rights from invasion by freeloading rivals?

Securing an American patent was neither expensive nor impossibly inconvenient for most American citizens, even those living in Vermont far away from the government Patent Office in Washington, DC. The cost of a patent in Markham's lifetime was $30, equivalent to three weeks' wages of a cotton mule spinner in the 1820s. The legal formalities (at some cost no doubt) could have been completed through Daniel Chipman, the Middlebury attorney and fellow freemason; possibly even through Markham's brother David, a trainee lawyer.

[144]Jeremy, "Innovation," 66.

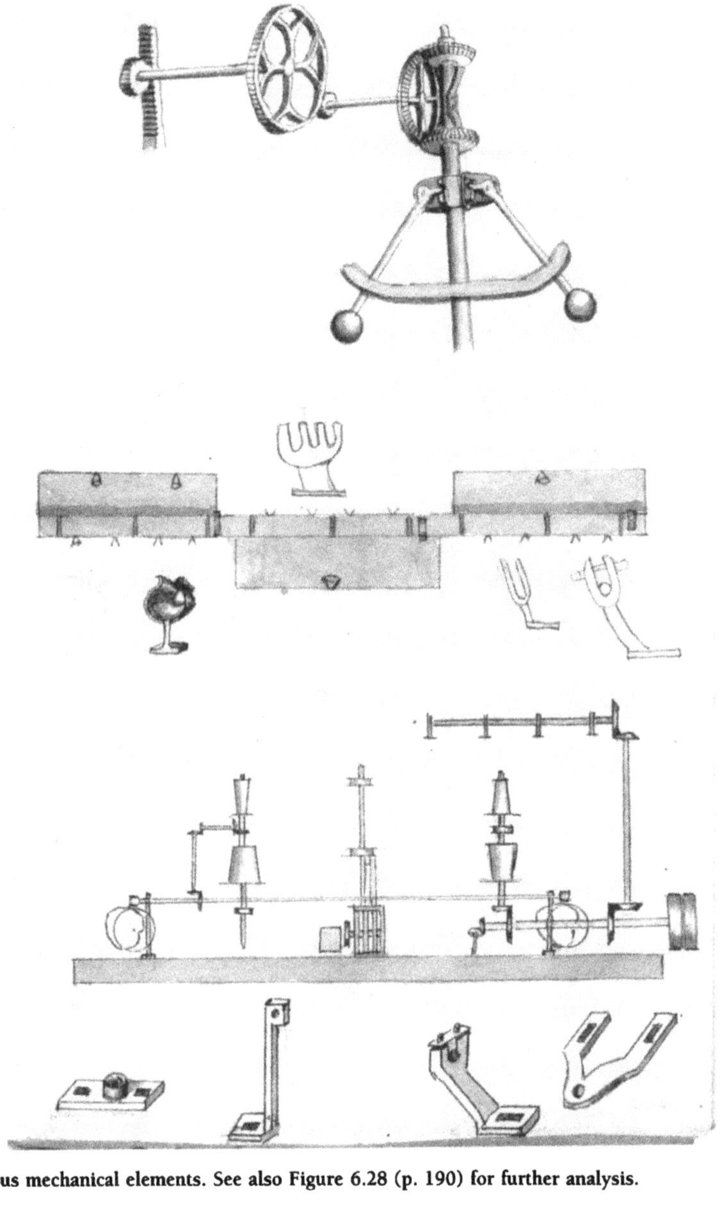

GURE 3.38 Miscellaneous mechanical elements. See also Figure 6.28 (p. 190) for further analysis.

A variety of reasons may have deterred Markham from patenting. He may have regarded his devices as insignificant, not worth patenting. This was unlikely because he carefully preserved drawings of his dressing machines and because they commanded as much as $200 or $300 each.

Another possibility is that he did not consider patent protection an economically acceptable strategy. Many inventions were devised in Britain but never patented. Their inventors thought that the inventions were before their time, before returns would cover capital invested in development and the patent process.

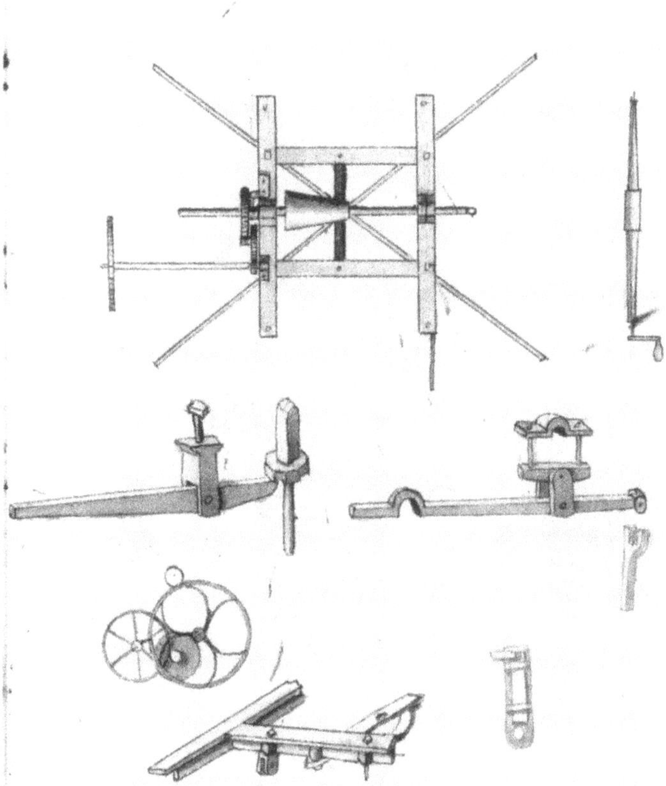

FIGURE 3.39 6 Miscellaneous mechanical elements. See also Figure 6.31 (p. 196) for further analysis.

Instead they chose the alternative of secrecy or of simple suppression. John Farey, a widely experienced engineer, in 1829 reported that "There are many secret inventions of value, which it is not the interest of the inventors to disclose under the present [patent] system."[145] Astonishingly, Richard Roberts, one of the leading engineers of his day, told the 1851 Lords Committee on Patents that he had contrived

[145]John Farey in "Report [on]...the Law of Patents," *Parliamentary Papers* 1829 (332) III, 134.

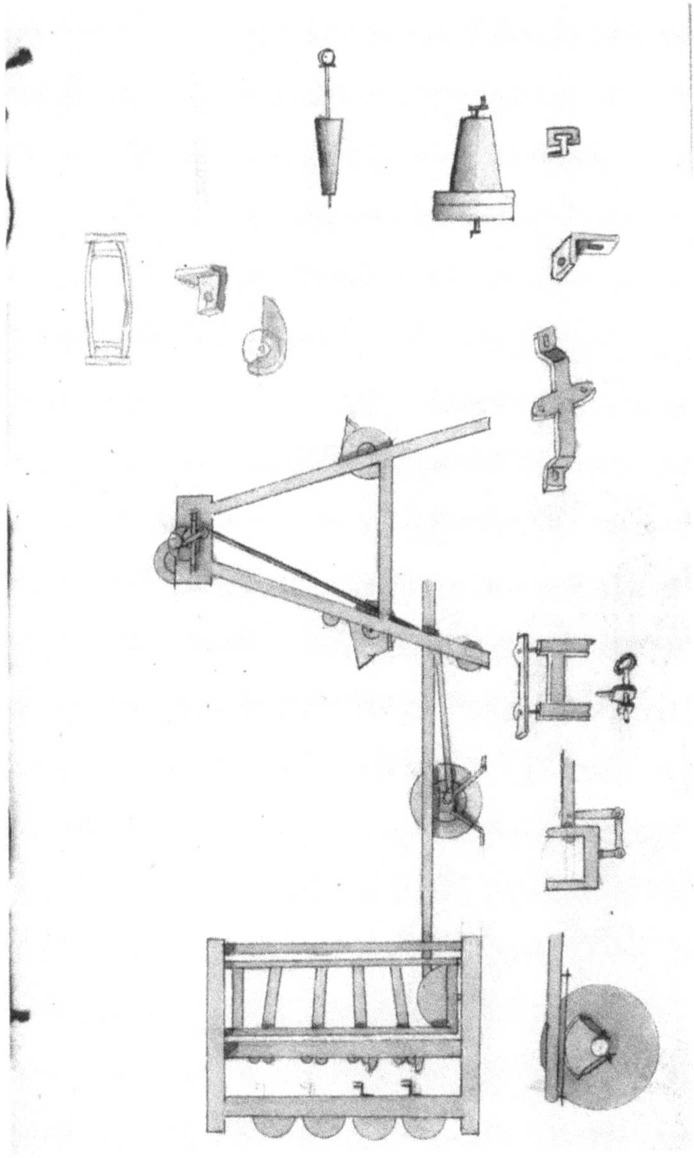

FIGURE 3.40 Dresser mechanisms. See also Figure 6.33 (p. 200) for further analysis.

nearly a hundred inventions, all with written descriptions and many with drawings, but that he chose to patent only 13 [15 eventually] of them, leaving the rest locked up, so great had been the costs of developing and then securing and defending patents.[146] The high cost of British patents was usually justified on grounds that it discouraged the patenting of frivolous inventions. Given that American patent

[146]Richard Roberts, evidence in "Report [on]...the Law [on] Patent for Inventions," *Parliamentary Papers* 1851 (486) XVIII, 187.

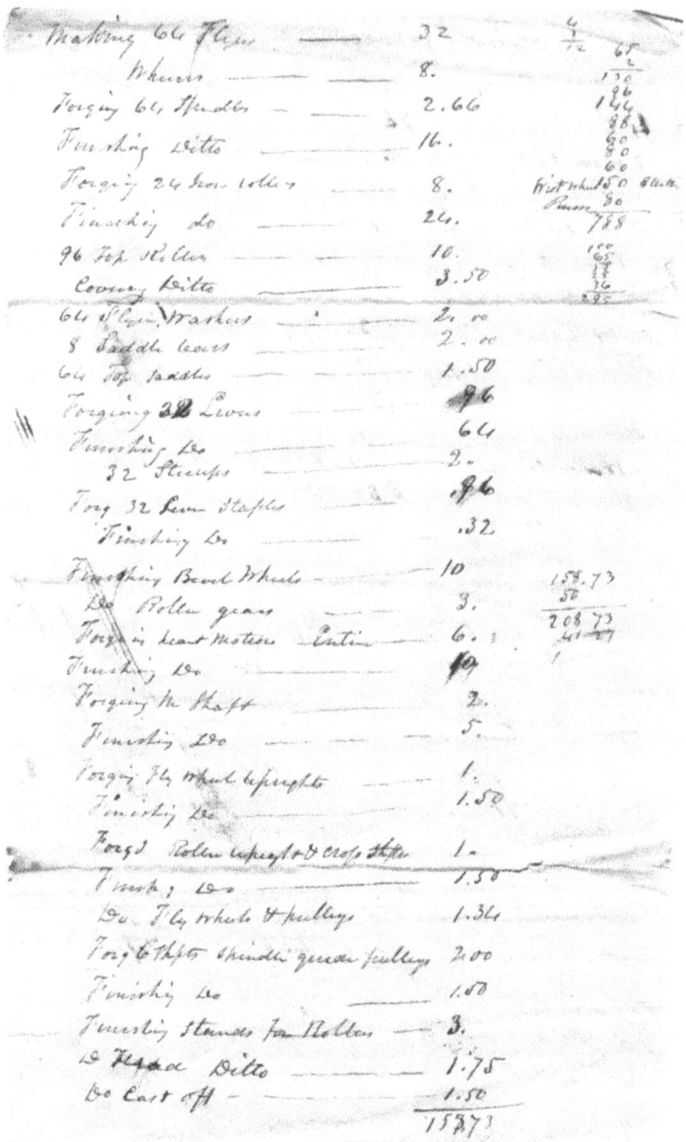

Making 64 Flyers — 32
Wheels — 8.
Forging 64 Spindles — 2.66
Finishing Ditto — 16.
Forging 24 Iron rollers — 8.
Finishing do — 24.
96 Top Rollers — 10.
Covering Ditto — 3.50
64 Flyer Washers — 2.00
8 Saddle levers — 2.00
64 Top Saddles — 1.50
Forging 32 Levers — .96
Finishing Do — 64
32 Stirrups — 2.
Forg 32 Lever Staples — .96
Finishing Do — .32
Finishing Bevel Wheels — 10
Do Roller gears — 3.
Forging heart motion Entire — 6.
Finishing Do — 10
Forging M Shaft — 2.
Finishing Do — 5.
Forging Fly wheel uprights — 1.
Finishing Do — 1.50
Forged Roller uprights & Cross Shafts — 1.
Finishing Do — 1.50
Do Fly wheels & pulleys — 1.36
Forg 6 Shafts spindle guides pulleys 2.00
Finishing Do — 1.50
Finishing Stands for Rollers — 3.
Do Head Ditto — 1.75
Do Cast off — 1.50
155.73

FIGURE 3.41A Loose sheet. See also Figure 6.23A (p. 176) for further analysis.

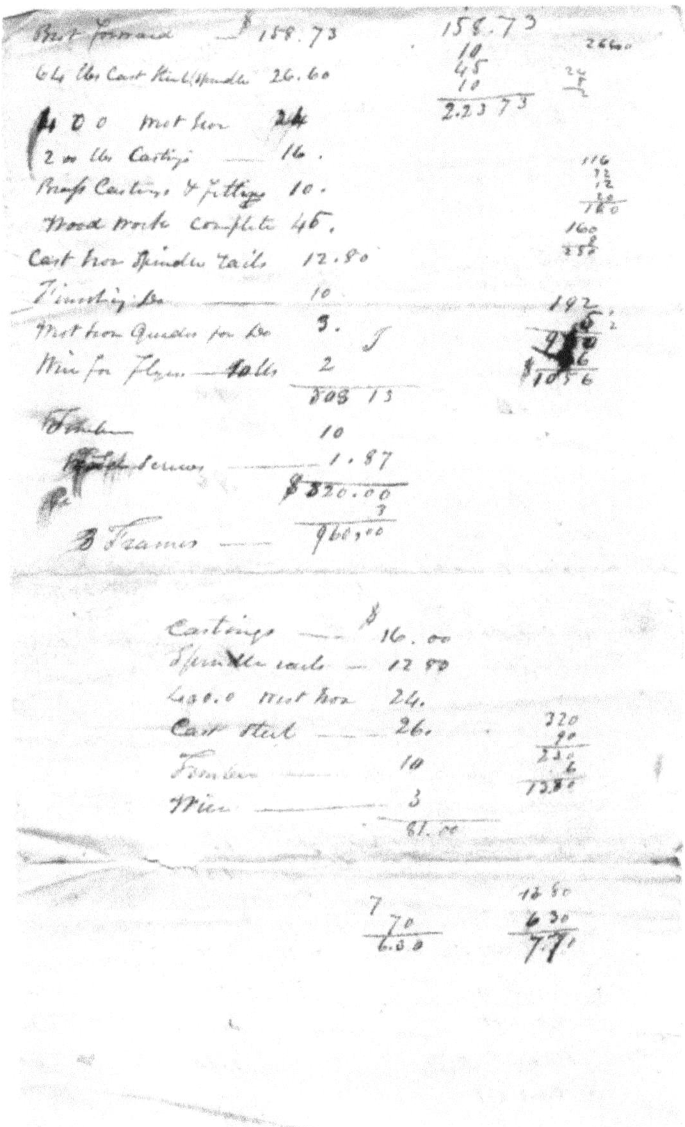

FIGURE. 3.41B Loose sheet. See also Figure 6.23B (p. 178) for further analysis.

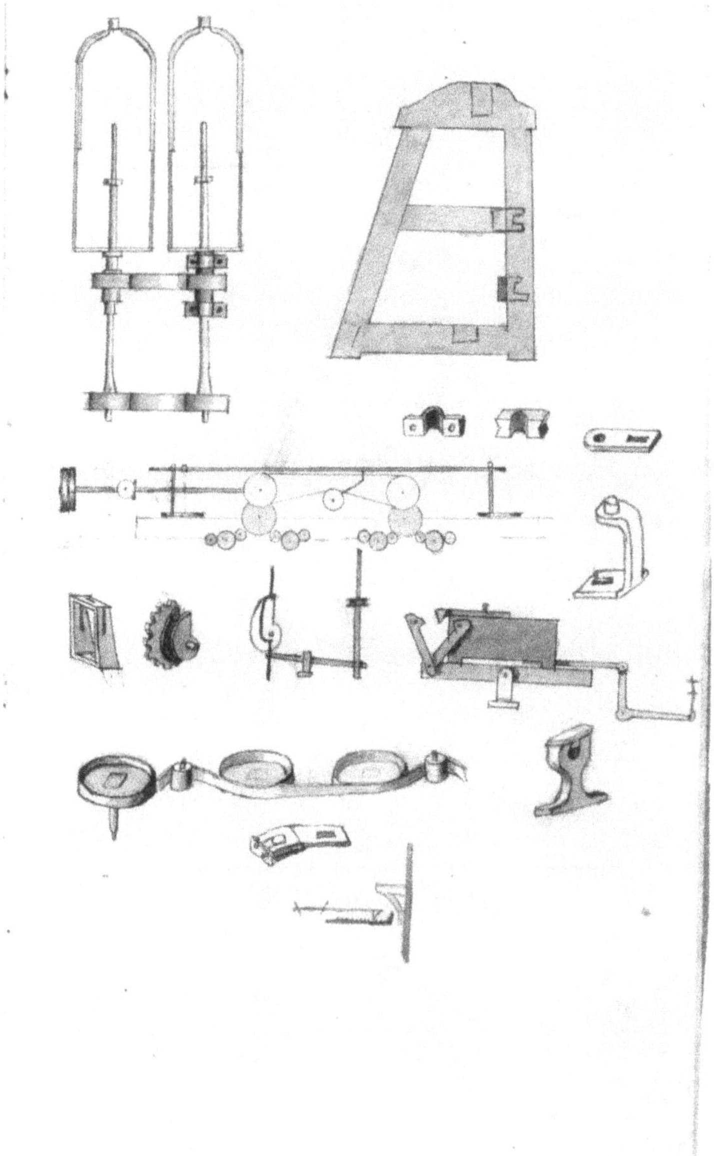

FIGURE. 3.42 Roving-frame sketches. See also Figure 6.29 (p. 192) for further analysis.

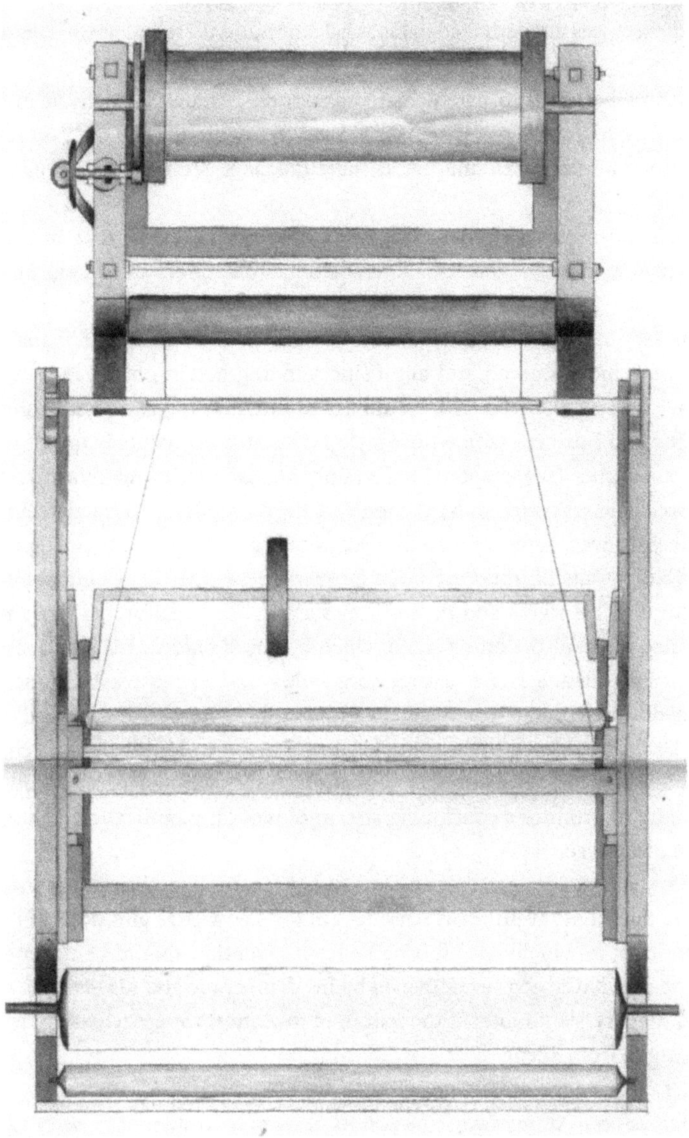

FIGURE. 3.43 Intermediate warping frame (?), perspective sketch (?). See also Figure 6.74 (p. 292) for further analysis.

protection was much cheaper than British patents and that Markham was able to sell his dresser, it is unlikely that the dresser was not perfected to his satisfaction and therefore needed temporary postponement of patent protection.

Not all Americans bothered to patent their inventions. Samuel Slater lived in the United States for thirty-five years before he took out a patent. It is hard to believe that he did not devise modifications to the British inventions he imported and could therefore have secured the benefits of American patent monopoly.[147]

Was Markham's dresser quickly copied by rivals, before he could take out a patent? This is one possibility. Plagiarism was not uncommon. Both small manufacturers and a large business like the Boston Manufacturing Co. were quite willing to steal profitable inventions.[148]

Markham's dresser may have been too similar to the Waltham dresser. The main difference seems to have been Markham's more smooth and automatic winding action on the loom warp beam (see Table A3.2, p. 368). This, he may have thought, would thrust him into litigation with a much larger and better-resourced New England business with whose style he became acquainted, and of which he disapproved in 1820–1821, if not earlier (see chapter 1). Certainly Massachusetts' manufacturers were most ready to defend their patent rights in court in the second and third decades of the nineteenth century.[149] Perhaps he felt outside the networks, some defined by their willingness to patent their improvements, as recently identified by Meyer.[150] Outside these or other primary networks, he would minimize his chances of learning of invasions of his rights and perhaps miss assistance in enforcing those rights.

If Markham filed a patent the only way in which he could enforce his monopoly rights was through litigation, either in or outside court, a time-consuming and expensive business. Knowledge of legal hazards would readily have come from Daniel Chipman, his capitalist backer and Masonic brother, and his own brother David Markham, both in the legal profession in Middlebury. Such a daunting prospect may have persuaded him to sell his dresser to manufacturers without patent protection and to depend on his own ingenuity to produce a continuing stream of modest modifications that enabled him to retain some competitive advantage.

Supposing Markham did believe that any one of his recorded modifications was novel to himself, he may have regarded the act of invention as a product of the knowledge and skills of the whole community, past and contemporary; in which case all benefits from invention should be returned to the community. In these matters he may have been strengthened by his democratic and his Masonic convictions. As noted earlier, elitist and democratic notions of the nature of invention were circulating in England at this time. But all this is to argue from silence.

If, as appears, he shared the dresser design with an individual called Adams, any dresser modifications may not have been entirely of his own conception. Adams then may have persuaded Markham against patent protection. The last possibility explaining Markham's failure to patent is the least likely: that he was planning to file a patent but was overtaken by his final fatal illness. This is most unlikely in the case of the dresser, the machine on which he lavished most effort, apparently, because he was selling dressers some time before 1820 and did not die until five years later.

---

[147]The US Patent Act of 1793 required applicants for a patent to allege that their improvement was "not known or used before the application." A popular introduction to American patent law, written by the Philadelphia lawyer Peter Browne, citing *Evans v. Eaton*, reported Justice (Bushrod) Washington's charge to the jury that "The discovery must be not only useful, but *new*. It must not have been *known*, or *used, before, in any part of the world.*" *Franklin Journal* 2 (1826) 350. This was the position before 1836.

[148]Jeremy, *Transatlantic Industrial Revolution*, 177–178, 185–186, 193–194, 216.

[149]Khan, *Democratization of Invention*, 69–78.

[150]Meyer, *Networked Machinists*, 66–72.

Perhaps the most likely explanation for Markham's failure to patent is that he did not regard his inventions as sufficiently important and distinctive as to warrant patent protection and the consequent trouble of monitoring and defending his monopoly rights through litigation.

MARKHAM'S VISUAL ENGINEERING KNOWLEDGE

Markham's machine drawings suggest that he possessed more than ordinary artistic skills—one tiny hint of his ability is indicated by the fact that on occasion he could not resist the flick of his brush to indicate shadow at the feet of a machine frame (see Fig. 2.9, p. 32). His cognitive interests sharpened his visual sense. When he was at Waltham, MA, between 1819 and 1821 Markham made a pencil sketch, partly completed in watercolor, of the Boston Manufacturing Co.'s mills (Fig. 2.1, p. 23).[151] It reveals a keen sense of perspective and solidity, and a preference for slightly vivid color. But it tells even more about Markham's mentality.

Whether he took his paper and pencils, watercolors, and brushes with him when he moved among mills, or whether he purchased some or all of them in Waltham, or nearby Boston, it suggests that he was (a) interested in two-dimensional representation of his visual experience, (b) curious about aspects of new industrial forms outside the machine shop, (c) alert to opportunities of visual registration and visual recording, (d) willing to use spare moments in long working days and weeks to go apart and sketch and think, (e) eager to record memories of what he might ruminate upon when up north in Middlebury, (f) hopeful that he could show his sketches elsewhere and thereby diffuse Waltham's industrial forms. Was it a commissioned sketch? With his artistic talent, was he invited either by his Waltham employers or his Middlebury contacts like Chipman to make this sort of visual evidence? At present it is impossible to know. More generally, the sketch demonstrates that neither his natural skill nor his professional interest was confined to machine drawing. As his biography (see chapter 1) indicates, for Markham, interest in life included but certainly went beyond the machine shop.

However, it is the machine drawings that have survived and which tell us yet more about Markham's mentality and visual knowledge. To detect these, the drawings must be placed in his world of industrial machines and technical engineering drawing of his day, as is attempted here. That context was distinguished by transition and flux, emanating from the flows of new machines from Europe; American determination to secure new technology; and technical, economic, and social pressures on and within factory and workshop.

Visual knowledge in the workshop, it is argued here, was shaped for Markham by four broad influences. First, he would have absorbed the shift away from perspective drawing to elevations and plans, as suggested by the work of Oliver Evans, if no one else. Second, Markham's outlook was shaped by the skilled migrants, like Gordon, and the publications, like the installments of Rees, that transferred new technical knowledge from Europe to the infant American Republic. Third, his workshop horizons owed something to the activities of American mechanics, chiefly in the northern New England network pivoted on Waltham, MA, in modifying imported machines to meet local economic and social circumstances. Fourth, European engineering and educational ideas about the nature of machine building and invention would have reached him via his transatlantic community of migrant mechanics directly or via middle-class backers.

---

[151]First reported by Richard Candee, "Architecture and Corporate Power in the Early Waltham System."

The revolution in machine drawing had begun before Markham was born but was still underway in his lifetime. It required the individual machine maker to perceive and then represent the 3D on 2D paper in a nonperspective way. This was not easy to do, particularly when perspective was the norm in many post-Renaissance forms of illustration. Nor was it easy to "read" elevations. The orthogonal projection would have been even more difficult but, as it was, in Markham's lifetime, Monge's orthogonal projections had no impact on either British or American nonmilitary machine making. Comparison with the survey of drawing techniques in twenty-seven British books for mechanics (see Table 3.1, p. 44), and those in Oliver Evans's *Young Mill-Wright's Guide* (Table 3.2, see p. 70), shows that Markham (see Table A5.1, p. 379) was not left behind in the shift from the perspective convention toward the more unambiguous forms of representation (for mechanics), particularly the elevation and the plan.

In depiction and in content, the single most informative set of visual information reaching the United States in Markham's lifetime was Rees's *Cyclopaedia.* Because the drawings were accompanied by lengthy textual explanations, the work constituted a remarkable window onto Britain's new industrial world. As such it was a triumph for free trade in human knowledge. For cotton-manufacturing technology, Rees offered elevations; for manufacturing technology of woolens, the older perspective view prevailed. The cotton-machine drawings could be regarded as exemplars of the new technique of machine drawing, apart from the fact that they left out the key datum of scale. The Rees volumes were also important in revealing to the public at large the form or design of the new textile machines for the first time, as well as many other kinds of mechanism. Drawings in the Rees plates gave American proprietors and mechanics definitive benchmark information against which the claims and competencies of migrant strangers might be checked.

Circumstantial evidence suggests that Markham would have consulted the articles and plates in Rees. Comparisons between his drawings and those in Rees suggest some fairly close correspondence with carding, drawing, and roving-frame designs, and with the Arkwright-derived spinning frame. Comparing his drawings of the spinning mule to those of the Rees plate, it is apparent that Markham was drafting more simple designs. This may be illusory if some of his drawings of component or final layout elevations and plans were lost. Markham's designs for power looms were more advanced than those shown in the Rees plates, possibly because of contributions from Joseph Gordon, possibly because, at Waltham, Markham would have met the different type of power loom, pioneered by the Boston Manufacturing Co. engineers.

However, with the international diffusion of new techniques during the early industrial period, the significance of drawings must not be overstated. Designs in the Rees plates required a cluster of skills to convert them into operating machines. Markham's drawings needed Markham to interpret them and to mobilize and coordinate machine-making resources. At the very least he would have required access to appropriately skilled artisans with metalworking facilities for making iron castings, forging wrought-iron, casting steel, and making brass castings; carpentry work for wooden framing would have been most readily available locally.

Migrant artisans played a more important part than drawings or models, partly because tacit knowledge could not be expressed by means of "inscriptions," partly because a complete record of explicit knowledge was not published in conjunction with drawings before the 1830s. Markham, and the American cotton-manufacturing industry, became indebted to immigrant mechanics like William Gilmour in Rhode Island and Joseph Gordon in Vermont, for bringing tacit and explicit knowledge of the power looms that were currently transforming Britain's textile districts in the first two decades of the nineteenth century. The roles of Gilmour and Gordon were in some respects similar to that of Slater, who successfully introduced Arkwright's waterpowered spinning technology to the United States in the 1790s. However, neither

Gilmour nor Gordon became major capitalists controlling several factories and acting as nodes in mechanic networks, as did Slater.

Machine drawings published in Britain, whether in Rees's *Cyclopaedia* or in the *Repertory of Arts* (which published 16 percent of textile patent drawings before 1812 and 25 percent from 1813–1824[152]), presented an increasing range of mechanical movements and combinations from which American inventors could select new combinations and thereby experiment to see which inventions best suited American circumstances. Artisans, immigrant or indigenous, would add to knowledge about European alternatives. Machine makers' networks confirmed or modified that knowledge. Unfortunately too little is known about Markham's machinist networks: their structure, their membership, and how they linked to those centered on Waltham and elsewhere in northern New England. Nor is it known how Markham's modifications, like his use of double-cone pulleys for smoother speed change, or his dresser, were received in machine makers' networks.

Behind the drawings were attitudes, widely shared in the transatlantic world of migrant mechanics. Foundational was the preparedness to test, accept, and modify imported engineering techniques and ideas: in sharp contrast to other cultures, like those of Japan and China, in Markham's day. Beyond this was the belief that invention was no more than "the new combination of materials" and that this act of recombining was a universal human propensity. Such an ideology harmonized richly with the democratic principles of the young American Republic and New World demands for labor-saving solutions to the technical problems of sparse populations striving to civilize the wilderness. The quest for American economic and industrial independence from Britain, complementing its political independence, imaginably sharpened the young Markham's determination to master machine making, and gave a special pride to the visual record of his own future achievements as he inscribed the cover of his portfolio, "I. E. Markham's Property Middlebury Vermont 8th Nov 1814."

Drawings as knowledge tend to be neglected by historians outside the fine arts. In essence drawings possess two characteristics: their content and their representational conventions. Content was *what* was being depicted. Convention was *how* content was depicted. In Markham's day both the content and the conventions of machinery drawings were undergoing major transitions, with conventions being adjusted to keep up with changes in content.

Obviously, machine drawings communicate a great deal of content. As such their advantages for manufacturers and machine makers were numerous. A published set of engravings like those in Rees's *Cyclopaedia*, providing a representative view of the technology of a whole economy, offered a definitive catalogue of the forms of the first generation of industrial machines. For Americans, they became a standard by which to identify lags behind Europe, to test claims to novelty, and to measure the aggregate extent of American improvements.

Workshop drawings confined to one firm were more specific in function and advantages. In the chain of value-creating activities of any firm, whether a machine maker or textile manufacturer, technology is, in Porter's terms, one of several crucial support activities in creating competitive advantage.[153] Although little is known of Markham's value chain, it is certain that the design of technology, both for products and for the means of making those products, was critical. His drawings testify to an undeclared significance, which, it is argued here, emerges by placing them in their long-term and transatlantic contexts.

---

[152]Jeremy, *Transatlantic Industrial Revolution*, 47, 264–267.
[153]Porter, *Competitive Advantage*, 33–61.

Clearly, by comparing Markham's machine drawings to those published on both sides of the Atlantic over the previous decades, it is evident that he was part of a new trend. Under the impulse of the widening spread of machine technology, the old monopoly, or tyranny, of perspective was dented, or demolished. A new set of conventions for 2D representations of 3D machine objects was being forged. The advantages of a new way of seeing were manifold as drawing conventions came to serve content more effectively.

The new (to engineering) conventions of elevation and plan, taking each machine or machine component in turn, allowed greatly improved technical accuracy, compared to the perspective filter. An accurate 2D representation of a machine was important because of its implications up and down the value chain. Drawings in which ambiguity was minimized reduced misunderstanding and errors in ordering parts from suppliers. In Markham's case this would have related to patterns for forgings and instructions for wrought ironwork at the local forge. Possibly other work was contracted out, unless more components than the evidence indicates were fashioned in-house. Likewise, customers would benefit if they could see Markham's drawings. Drawings made it much easier for Markham to communicate his ideas when discussing them with the owner of a distant mill site or factory. With drawings in front of him a prospective customer could form a clearer idea of how far one machine or modification was more fit-for-purpose than another in, say, matching workers' skills, or lack of them. Again, drawings had a vital marketing potential. Furthermore, drawings bolstered the machine maker's reputation, for new ideas and clever modifications yielding useful labor-savings, increased output, and so forth.

In-house, mechanical drawings had other advantages. Within his own machine shop, if Markham could teach his employees to "read" his drawings, they could be left to get on with the job of making components and, by realizing a degree of uniformity, minimize the filing and fitting that were necessary in assembling a machine built of wood and iron. The drawings, once made, served as masters. Provided they were securely held, and their long survival after Markham's demise suggests that they would have been, they acted as a form of insurance should the wooden patterns at the forge or in his own machine shop be destroyed.

Over the decades following the death of Markham in 1825 many of the engineering drawing conventions that his and others' workshops followed were modified at the margin and widely shared. Drawing rules of several machine shops merged to become the conventions of a wider network or region, eventually of a whole country. When, in the second half of the nineteenth century, if not earlier, mechanical drawing rules and conventions attained wide currency they became the engineer's language, as Nasmyth observed. In North's terms, society's systems of rules (institutions) reduced the costs of economic transactions.[154] In ways already elaborated, mechanical drawing, so formalized, reduced the costs of machine maker and machine user: the more unambiguous the drawing conventions, the lower the costs of communication.

But there was also content to consider. Drawings were precious to Markham not only because they became instruments controlling how to build his machines. Interlocked with this merit was their function in serving as a way of collecting ideas about machine designs, about what to build. Such ideas began in the mind of the engineer. Markham, dreaming between times in the workshop, seized the opportunity to imagine and scribble ideas in rough, which, sifted later, might be carefully sketched in a more permanent record, perhaps like his Sketchbook (see section 2, "Sketchbook," chapter 6). Detailed drawings of machine components and something akin to a whole layout would be pored over and discussed in detail before costly investments in construction began. In his northern New England network of machinists, Markham could likewise discuss his ideas on the basis of sketches showing where he was thinking of going (or had been) in his design work, much as Aza Arnold did.

---

[154] North, *Institutions*.

Long before the drawing office became an instrument of managerial control of the shop-floor, machine drawings (in preserving specific content that could be decoded by means of widely shared conventions) reduced costs by extending knowledge of a machine or machine technology beyond the immediate involvement and life of the machine maker or designer, beyond the orbit of the originating machine shop, and beyond the depreciated life of a single machine. In time, and shaped by more conventions than Markham knew, mechanical drawing would become the universal and distinguishing language of engineers. Always, however, machine drawings, as a medium of creativity, have held the potential for incubating and ordering inchoate ideas as well as fostering low-cost experimentation in recombination; and the generation of new technological arrangements that gave prime movers competitive advantage.

Chapter 4

The Significance of Isaac Markham and His Workshop Papers

☆ ☆ ☆ ☆ ☆ ☆ ☆

MARKHAM'S CAREER AND WORK WERE RELATIVELY STRAIGHTFORWARD and not untypical of machinists of his day. He was trained in a machine shop; he tramped among other machine shops in the country; he returned to his birthplace, where he managed not only the machine shop but also the textile factory; tragically he died at the age of thirty, leaving much of his early promise unfulfilled. In two matters, however, he was very atypical. In his lifetime he was especially fortunate to be working in a machine shop to which an immigrant Scottish mechanic brought designs of some of the first cotton power looms to be used in the United States. Second, after his death, his drawings were preserved. Nothing so complete and comparable for the textile industry has survived from America's antebellum decades. As a result we have a much clearer view of the work of an early industrial New England textile machinery maker. With the aid of insights from other academic disciplines, Markham's career emerges with new significance.

A modern knowledge revolution, informed by many disciplines from philosophy and psychology to computing and management, has awakened interest in all forms of knowledge. Machine drawings are a form of knowledge. They can act as a store of knowledge, a vehicle for communicating knowledge, a canvas on which new combinations of visual ideas can be represented and tested, they can serve as a cost-saving step in the stages of machine building, and much else.

Markham's drawings can be seen as evidence of the transition from preindustrial to industrial visual forms of technical knowledge and of a much wider knowledge revolution. The shift from perspective views of machines, predominant in Europe between the fifteenth and eighteenth centuries, to more accurate plans, elevations, and cross-sections, and eventually to First Angle Projection (pioneered in France but not adopted for civilian engineering in the United States until after the mid-nineteenth century) profoundly changed engineering on both sides of the Atlantic. Machinists had to learn to see, represent, and interpret visual depictions of machinery in an entirely new way. Those who were part of the transition did not always find it easy. Once the skill was mastered, creativity, experiment, costing, and construction were vastly facilitated. The new skill of machine drawing, however, set its practitioners apart from those relying on traditional handicraft ways of machine making.

The machine-drawing shift was part of the knowledge revolution sweeping the United States in the first half of the nineteenth century. Visual knowledge became increasingly important within and beyond the technological domain. Patent drawings, diagrams in encyclopaedias or handbooks, illustrations in books and newspapers, all contributed to the wider knowledge revolution. Prime among these during Markham's lifetime was the American republication of Rees's *Cyclopaedia*, a vast 39-million-word treasure house of knowledge, not least about the processes and machines then powering and sustaining the international leadership of the British economy. In the accompanying plates, Rees's draftsmen presented both perspective and elevation-and-plan views. Although the general public needed the help of perspective views, the alternative technique of plans and elevations pointed forward to a new world of specialized machine builders and machine shops. Americans, including Markham, proved quick to learn.

Another dimension of Markham's drawings has further significance. Organizational studies by sociologists, anthropologists, economists, and analysts of international business, have emphasized the importance of networks in organizations.[1] From an historical angle, America's early industrial period has been re-evaluated in terms of networks of machinists. David Meyer has traced their growth and persistence radiating out from centers or nests such as those found in (for textiles) Providence, Rhode Island; Philadelphia; and Waltham, Massachusetts. Linking these and other centers and mill machine shops were itinerant mechanics, like Markham.[2] Ross Thomson has layered this networked model with a dense web of knowledge-diffusing institutions and shown how intra-industry networks fed into complex, inter-industry networks of machinists. Instrumental in this process were social organizations like the Patent Office, scientific journals, professional associations, and public colleges.[3] Markham stood at the dawning of these developments. Before his death in 1825, his experience and the records he made reveal a number of important aspects of mechanic networks of knowledge.

First they demonstrate that networked technical knowledge was being reduced to visual forms. Contextualized, Markham's drawings were made during, and were a part of, the quiet but revolutionary shift from the perspective to the projective in representing machines on paper. Being workshop drawings, only 10 percent of Markham's drawings use perspective views (see Table A5.1, p. 379). In contrast, during Markham's working life drawings in British texts for mechanics used the convention of perspective in 48 percent of their printed drawings, which was much lower than drawing practices before 1812, when that proportion was 68 percent (see Table 3.1, p. 44). In this he was probably influenced by the pioneering work of millwright Oliver Evans (see Table 3.2, p. 70), whose perspective views in his *Young Mill-Wright's Guide* were down to 36 percent of all its drawings; possibly too he was influenced by the imported English conventions of the eminent immigrant architect and engineer, Benjamin Henry Latrobe. For the individual mechanic, improved machine drawings promised creativity without limit, experimentation with curbed costs, construction with minimized ambiguity, and marketing with the least misunderstanding.

For the machinist network, machine drawings had other important implications. Assuming that the smallest size of the network was two mechanics, drawings reduced what otherwise would be tacit knowledge and allowed the more experienced to pass along a certain amount of technical knowledge, that which was made explicit by the drawing, into the network. Presumably Markham was able to learn about power looms more readily from his immigrant instructor, Joseph Gordon, by visualizing, sketching, and then drawing Gordon's information. Of course, if Gordon had been as adept as Markham, the interchange of visualized ideas would have been mutual.

Most obviously, Markham's drawings demonstrate the ubiquity of inventiveness even in the remotest of early machine shops, at the extremity of a mechanic network. In the hands of an ingenious machinist, modest technical changes (like Markham's various applications of twin-cone pulleys) could be introduced; with knowledge of a major technological breakthrough, radical developments like the power loom could be transplanted from abroad (as Markham's power-loom drawings illustrate). His batten motion (see Fig.

[1]Van Wijk et al., "Knowledge and Networks."
[2]Meyer, *Networked Machinists.*
[3]R. Thomson, *Structures of Change.*

2.11, p. 34; see also Appendix 4) was, similarly, an American attempt to find a mechanical simplicity consistent with production for basic styles of cloth in rural markets. How many more contemporary mechanics were likewise searching for their own trade-offs between mechanical simplicity and complexity to meet demands in localized clothing styles, machine construction, maintenance and operation costs, and the quantity and quality of labor supplies? As shown earlier (see chapter 3), visual means materially contributed to the generation and diffusion of new technical knowledge, and in these activities networks were crucial.

Markham's experience of New England networks revealed that they could relay technical knowledge extremely quickly. In part this was a result of the expansion of printing and publishing houses. One of their biggest projects was to reprint Rees's *Cyclopaedia*, the multipart compendium of knowledge that took eighteen years to complete in England and fourteen years to be reprinted in the United States. By the time the letter "C" (for Cotton &c) was reached, comprehensive drawings and textual explanations of Britain's factory-based textile technology were circulating within twelve months or so of availability in London. Rees's *Cyclopaedia's* appearance in a remote part of New England in turn depended on a network of publishers (see Table 3.5, p. 77 and Table 3.6, p. 77).

The *Cyclopaedia* moved new knowledge along mechanic networks by inanimate means. In contrast, Lancashire's favored power loom, patented in 1813, was introduced to New England in 1816 and to Middlebury, Vermont, in 1817 by immigrant Scots machinists. This looked very fast in comparison to diffusion rates in the late eighteenth century. Speed of knowledge along networks does not of course necessarily equate with effectiveness of knowledge transfer. In these cases it did.

As has long been recognized, transatlantic networks established or fed America's internal networks of machinists. One set of machinists was comprised of immigrants from Britain or Europe: Samuel Slater, the Scholfield family, the Du Pont family, and William Gilmour, to name the more prominent in the textile industry. Another group consisted of American industrialists who visited Europe with the intention (premeditated or opportunistic) of obtaining new knowledge. Francis Cabot Lowell found out about power looms. Zachariah Allen went to spy on Britain's woolen districts. Charles Richmond took knowledge of a differential gear for a roving frame and used it to bargain for information and machine parts; on the same mission he recruited an experienced calico printworks superintendent. Daniel Treadwell, Rumford Professor at Harvard, went on a wide-ranging knowledge-collecting mission and was instrumental in recruiting a very big fish for the American cotton industry, James Montgomery. A third group of machinists transferring knowledge along transatlantic networks was comprised of Americans who moved permanently to Europe. Best known of these was Joseph Chesseborough Dyer who conveyed to England American improvements in the card-clothing machine.[4] Markham's instructor, Joseph Gordon—the immigrant Scots mechanic, is a hitherto unsung figure in the gallery of networked mechanics. He introduced to northern New England a power loom that differed in some respects (after he and Markham had finished with it) from those developed at Waltham or built along English lines in Rhode Island (see Appendix 4). Immigrant networks were a powerful vehicle of knowledge transfer.

Markham's experience in accessing new technical knowledge disclosed how multilayered and reinforcing networks of knowledge transfer could be. For a sight of Rees's *Cyclopaedia*, Markham was dependent on one or another of the only two Middlebury citizens who could afford to subscribe to such an expensive work. Because one of them did not move to Middlebury until 1818, this left Daniel Chipman, an influential lawyer. Young Markham belonged to a least three networks that connected him with Chipman. The latter

[4]Jeremy, *Transatlantic Industrial Revolution*, *passim*; Jeremy, *Technology and Power*, 13–28; Dyer, "Notes."

would have regularly met Markham's father because both were Freemasons and their local lodge, the Union Lodge, met in Markham's father's tavern when Markham was growing up. After Markham's father died in 1813, Masonic obligations of mutual protection required Chipman to keep a close eye on the late Captain Markham's two sons, who had not then reached their majority. Markham's older brother trained in Chipman's law office. In addition, Chipman had an economic interest in Page's cotton factory, where Markham started work at the age of seventeen in 1812. In due course, Isaac Markham himself joined the Union Lodge. So it can safely be surmised that Masonic, family, and business networks gave Markham privileged access to the wide-ranging view of textile technology that was available in Rees's *Cyclopaedia*.

Markham earned a strong reputation as a mechanic, which in turn would have enhanced any machinist network with which he associated. His first employer, David Page, Jr., supplied him with a glowing testimonial in 1819 (see chapter 1). At this time, Markham, twenty-four years old, was draftsman, machinist, and superintendent of the mill's machine shop. This was probably a small shop with no more than a handful of employees. Even so, Markham's supervisory responsibilities, his skill as a worker in iron and brass, and his machine-drafting skills set him apart. No other power looms, in Page's opinion, worked as well as Markham's. Page would earlier have encountered the Waltham loom.[5] He knew what he was looking at. In contrast to machinists in a large machine shop like that at Waltham, where there was considerable subdivision of labor, Markham accumulated a wide range of skills. Most of all, he was familiar with the designs of spinning and weaving machinery, how the parts were made, how they were assembled together, and how they operated. In proof of this there were his drawings and, as Page said, "they speake for themselves."

After his death, Markham's reputation for mechanical ingenuity and industriousness was acknowledged in his obituary notice. Allowing for the customary tendency toward hyperbole (which Markham's obituarist claimed to have avoided), the author wrote of Markham's " mind of singular felicity of invention, originality and strength of conception, and of rich and abundant resource." Through his effort he had "carefully cultivated his uncommon powers until he had become one of the first mechanicians in this section of our country" (see chapter 1). The exceptionally large crowds that gathered for his funeral on 20 September 1825 in the Congregational Church in Middlebury, swelled by the support of his own Masonic lodge and of other lodges in the area, was additional independent proof of Markham's high reputation and status in Middlebury and northern New England. All of which is some demonstration that machinists, and by implication their networks, were accumulating esteem and status in New England.

Machine drawings and machine-drawing skills had another important role in machinists' networks. With the shift from perspective to projective conventions, together with science and mechanics, they raised the learning barriers to entry for newcomers to the network. In this way they contributed to enhancing the professional status of engineering. Markham never lived to see these developments come to fruition.

Although mechanic networks were primarily cemented by the technical knowledge of the workshop, with all its practices and concerns, other social bonds played into machinists' workshop networks. In Markham's case his entry into the elite of machinist ranks would have been denied without family and Masonic connections. Equally important, through the social networks of family, church, and Masonic

[5]He came from Jaffrey, New Hampshire, the birthplace of Samuel Batchelder, who was known to the Boston Manufacturing Co. directors at Waltham; and Batchelder was a partner of Jesse Holton who built spinning machinery for Page after 1812.

lodge, the attitudes and practices of the machine shop were underpinned by core life values: honesty, brotherly concern, loyalty, pride in one's work.

In sum, machine drawings were becoming crucial to early industrial machinists' networks. Drawings and the skills of drawing had the potential for creating and storing new knowledge. If, as in the case of Markham and Gordon, the new knowledge was held within a network, drawings enabled some tacit knowledge to be made explicit. In turn they had the potential for releasing explicit knowledge to other members of the network, or to other persons new to the network. Indeed they might be the means of endowing entry to the network. Drawings could simultaneously diffuse new knowledge and extend the machinist network. Equally, if held in secret, drawings had the potential to inhibit diffusion and growth of knowledge. The appearance of the machines on which drawings were based of course eroded secrecy. It is hard to avoid the conclusion that the role of drawings and drawing skills in creating and diffusing knowledge and in the formation of machinists' networks has been much underestimated and largely because, being often bulky and not easy to understand by the layperson, they have less readily been chosen for preservation.

Chapter 5

Editorial Method

☆ ☆ ☆ ☆ ☆ ☆ ☆

THE MARKHAM PAPERS ARE HELD IN THE SHELDON MUSEUM at Middlebury, Vermont, USA. However, a significant number of the drawings are unidentified; and no introductory text or annotations, such as are provided here, accompany them.

IDENTIFICATION OF THE DRAWINGS

As related in Appendix 1, where their provenance is reported, Markham's drawings were discovered to be of considerable significance in the 1980s. The sequence in which they were then placed was based on preliminary identification of the function of the pictured machines. Consequently, the primary author made a second sift (again following the textile processing sequence) through the drawings, a sift that was further refined for final editing. The three attempts to identify the drawings have resulted in the sequence that is presented in this edition.

EDITORIAL APPARATUS

To explore all the aspects of these drawings, each is accompanied by up to five elements of editorial apparatus: title and catalogue number, type of workshop drawing, document description, commentary, and comparators.

Drawing Title and Catalogue Number

Titles are sometimes listed on the drawings themselves; otherwise they have been supplied by the primary author. The catalogue numbers were assigned by the Sheldon Museum based on the physical characteristics of the documents and drawings, not the content.

Workshop Drawing Type

The five main types are: scale drawing, drawing without a scale, perspective drawing, sketch, doodle.

Document Description

Paper

Kind, for example, sheet, notebook, fragment, and so on. Dimensions, in centimeters (height x width). Watermark, if any. Drawing medium: pencil, ink, etc. Condition: ink faded, paper foxed, stained, tears, holes, pinholes, and so on.

Commentary

Identification of the equipment illustrated. Explanation of the functions of the machinery.

Comparators

Drawings: published or unpublished

Artifacts: British or American

Allusions to earlier (sometimes, when comparators are lacking, successor) machine designs and reference numbers to those major comparators illustrated in chapter 3 ("Seeing in a Different Way" and "The Diffusion of Visual Engineering Knowledge in the Early American Republic"), in Appendix 3, and in Appendix 4. About a quarter of the 300 or so British textile patents issued in the period 1790–1830 were published. These have been consulted and a number of key ones are presented in chapter 3, in Appendix 3, and in Appendix 4. A small number of museum examples of machines surviving from the late eighteenth and early nineteenth century have been cited in the notes to the drawings.

CROSS-REFERENCING

Cross-referencing is organized as follows:

References to Part I (Introduction)

> Editorial text: chapters 1 through 4
> Figures: Fig. 1.1, Fig. 1.2, 2.1, 3.1, and so on
> Tables: Table 3.1, 3.2, etc.

References to Part II (Drawings and Papers)

> Editorial text: Chapter 6 sections, 1, 2, 3, 4, and so on.
>
> Documents: Documents and drawings use the same numbering system, cited sequentially in the text: Fig. 6.1, 6.2, etc.

References to Part III (Appendices)

> Editorial text: Appendix 1, Appendix 2, Appendix 3, and so on.
> Figures: Fig. A3.1, Fig. A3.2, Fig. A3.3, etc.
> Tables: Table A2.1, Table A2.2, Table A2.3, etc.

IDENTIFICATION OF MECHANISMS

For the identification of mechanisms, the primary author has consulted the appropriate volumes in the list of sources; the engineering specialists named in the acknowledgments; Barber's *Engineer's Sketchbook of Mechanical Movements*; and fallen back on his own knowledge of early textile machinery, accumulated while a Visiting Research Associate in the Smithsonian Institution's National Museum of History and Technology (now the National Museum of American History), Washington, DC, 1969–1970, and Curator of the Merrimack Valley Textile Museum at North Andover, MA (now the American Textile History Museum at Lowell, MA), 1970–1973, and since. Other specialist assistance is gratefully recorded in the Acknowledgments.

☆ ☆ ☆ ☆ ☆ ☆ ☆

Part II

THE PAPERS AND DRAWINGS OF ISAAC MARKHAM

☆ ☆ ☆ ☆ ☆ ☆ ☆

Chapter 6

Markham's Drawings and Papers

☆ ☆ ☆ ☆ ☆ ☆ ☆

1. NOTEBOOKS AND LOOSE SHEETS OF WRITING

Notebooks

These two small notebooks (notebook 1 runs from Figs. 6.1 to 6.6, notebook 2 from Figs. 6.7 to 6.14), compiled ca.1821, record waterwheel data, information on mule gear and pulley sizes, and, mostly, information and calculations that Markham needed to establish correct drafting distances between rollers or between cylinders in several different machines. The latter were carding machines, drawing frame, stretcher, mule, but here evidently the water frame was omitted. The common feature of these machines is that all involved drafting (stretching and therefore making smaller in diameter) the "rope" of cotton initially produced by the carding machine. Clearly this was critical information needed by Markham both as machine builder and manager of a cotton yarn production line.

His recording of data from Scottish mills (in the Paisley area) is interesting. He seems to accept a Scottish manufacturing rule for power requirements: 828 spindles (mule and water frame, undistinguished) per horsepower (Fig. 6.11). Such an estimate was subject to numerous variables. Montgomery, in 1841, cited an 1837 estimate that one horsepower was needed to drive 500 mule spindles or 100 throstle spindles, with their preparation equipment. But this was the subject of great controversy when the relative merits of water and steam power as energy sources were discussed.[1] The scribbles and doodles (Figs. 6.24B and 6.25 [pp. 182 and 184]; also Fig. 6.45A [p. 226] and 6.45B [p. 227]) reveal Markham in his lighter moments.

[1]Jeremy, *Technology and Power*, 294, 338 n31, *passim*.

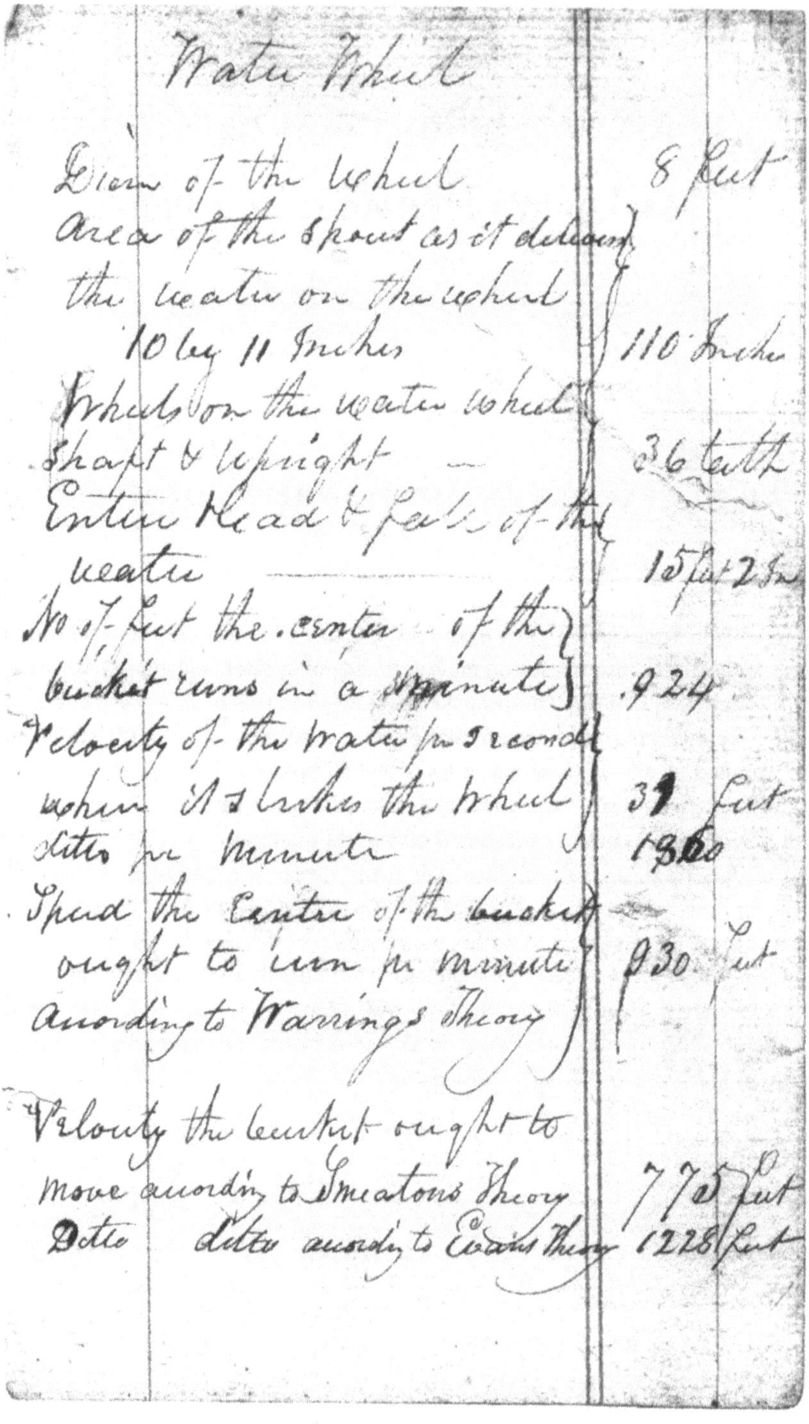
Water Wheel

Diam of the wheel	8 feet
Area of the spout as it delivers the water on the wheel 10 by 11 Inches	110 Inches
Wheels on the water wheel Shaft & upright	36 teeth
Entire Head & fall of the water	15 feet 2 In
No of feet the center of the bucket runs in a minute	924
Velocity of the water per second when it strikes the wheel	31 feet
ditto per minute	1860
Speed the Centre of the bucket ought to run per minute according to Warrings Theory	930 feet
Velocity the bucket ought to move according to Smeatons Theory	775 feet
Ditto ditto according to Evans Theory	1228 feet

FIGURE 6.1 Notebook page.

Catalogue number: IEM 34.6

Document Description

Measurement: 16.2 x 9.8 cm

Medium: Ink, ruled with red ink

Condition: tear near top of page

Transcription

Water wheel

Diam of the wheel	8 feet
Area of the spout as it delivers the water on the wheel 10 by 11 inches	110 Inches
Wheels on the water wheel, shaft and upright	36 teeth
Entire head and fall of the water	15 feet 2 in
No of feet the center of the bucket runs in a minute	924
Velocity of the water per second when it strikes the wheel	31 feet
ditto per minute	1860
Speed the center of the bucket ought to run per minute according to Warring's Theory[2]	930 feet
Velocity the bucket ought to move according to Smeatons Theory[3]	775 feet
Ditto ditto according to Evans Theory[4]	1228 feet

---

[2]William Waring, teacher in the Friends Academy, Philadelphia, found an error in the hydraulics calculations relating to waterwheels published in English works at much the same time as did Oliver Evans. Whereas English science lecturers believed that an undershot waterwheel delivered its maximum quantity of work when the wheel rim moved a third as fast as the impinging stream of water, Waring and Evans concluded that maximum work would be achieved at one half of stream velocity. Waring reported his conclusions in the *Transactions of the American Philosophical Society*. Evans called the American findings the "New Theory" but later modified them. See Evans, *Young Mill-Wright*, 59–63 and Ferguson's introduction to the 1990 reprint.

[3]John Smeaton, *An Experimental Enquiry Concerning the Natural Powers of Water and Wind to Turn Mills and Other Machines Depending on a Circular Motion* (London: reprinted from the *Philosophical Transactions of the Royal Society*, 1760). John Smeaton, *An Experimental Examination of the Quantity and Proportion of Mechanic Power Necessary to Be Employed in Giving Different Degrees of Velocity to Heavy Bodies from a State of Rest* (London: I and J Taylor, 1794). John Smeaton, *An Experimental Enquiry Concerning the Natural Powers of Water and Wind to Turn Mills and Other Machines Depending on a Circular Motion. And An Experimental Examination of the Quantity and Proportion of Mechanic Power Necessary to Be Employed in Giving Different Degrees of Velocity to Heavy Bodies from a State of Rest* (London: I and J Taylor, 1796).

[4]Oliver Evans, *The Young Millwright and Miller's Guide* (Philadelphia: the author, 1795; reprinted 1807).

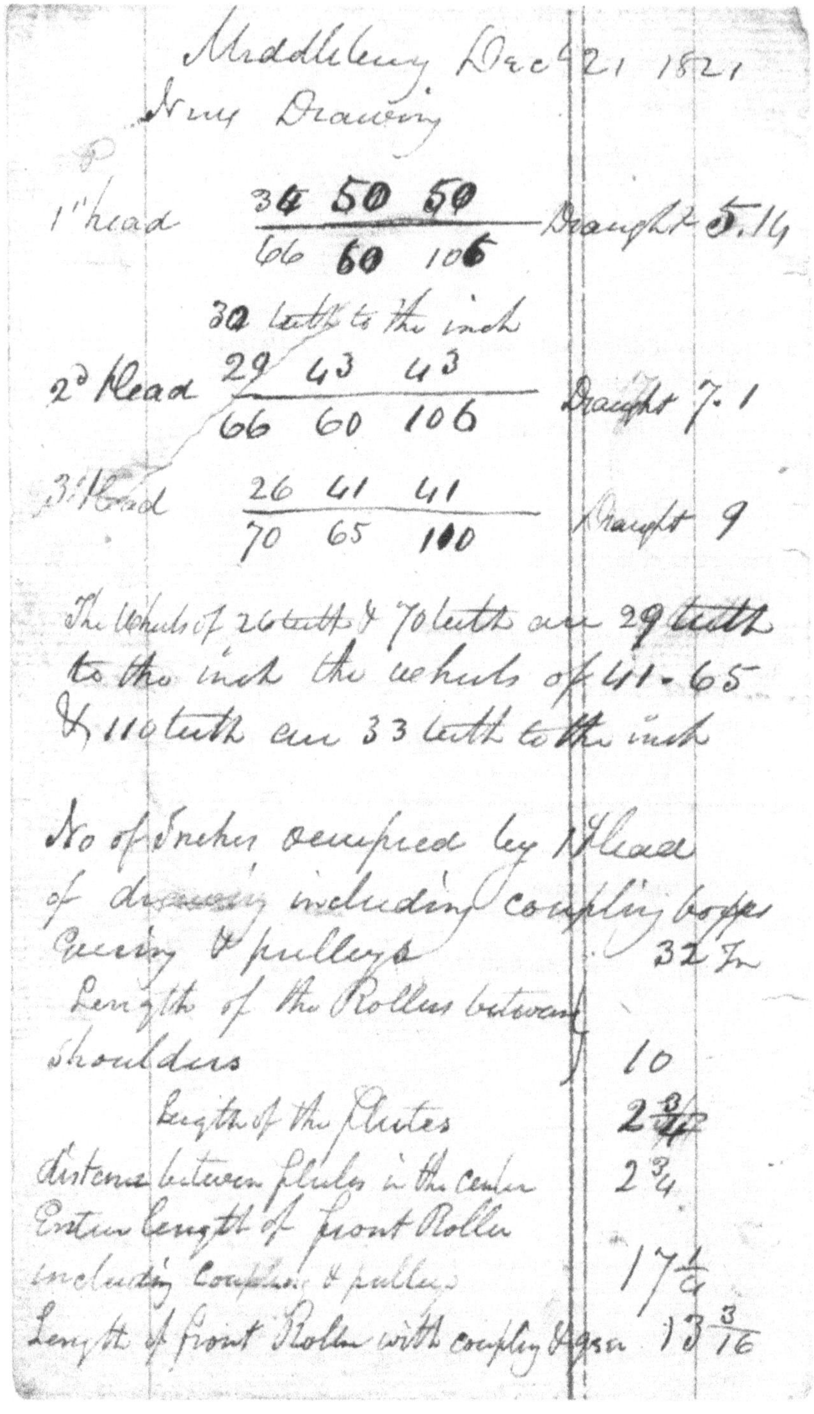

FIGURE 6.2 Notebook page.

Catalogue number: IEM 34.5

Document Description

Measurement: 16.2 x 9.8

Medium: Ink, ruled with red ink

Condition: tear near top of page

Transcription

Middlebury Dec 21 1821
New (?) Drawing

1st head	34 50 50 / 66 60 106	Draught 5.14
	30 teeth to the inch	
2d Head	29 43 43 / 66 60 106	Draught 7.1
3d Head	26 41 41 / 70 65 110	Draught 9

The wheels of 26 teeth and 70 teeth are 29 teeth to the inch the wheels of 41, 65 and 110 teeth are 33 teeth to the inch

No of inches occupied by 1 Head of drawing including coupling boxes

Geering & pulleys	32 In
Length of the Rollers between shoulders	10
Length of the flutes	2 3/4
Distance between flutes in the centre	2 3/4
Entire length of front Roller including coupling & pulley	17 1/4
Length of front Roller with coupling & geer	13 3/16

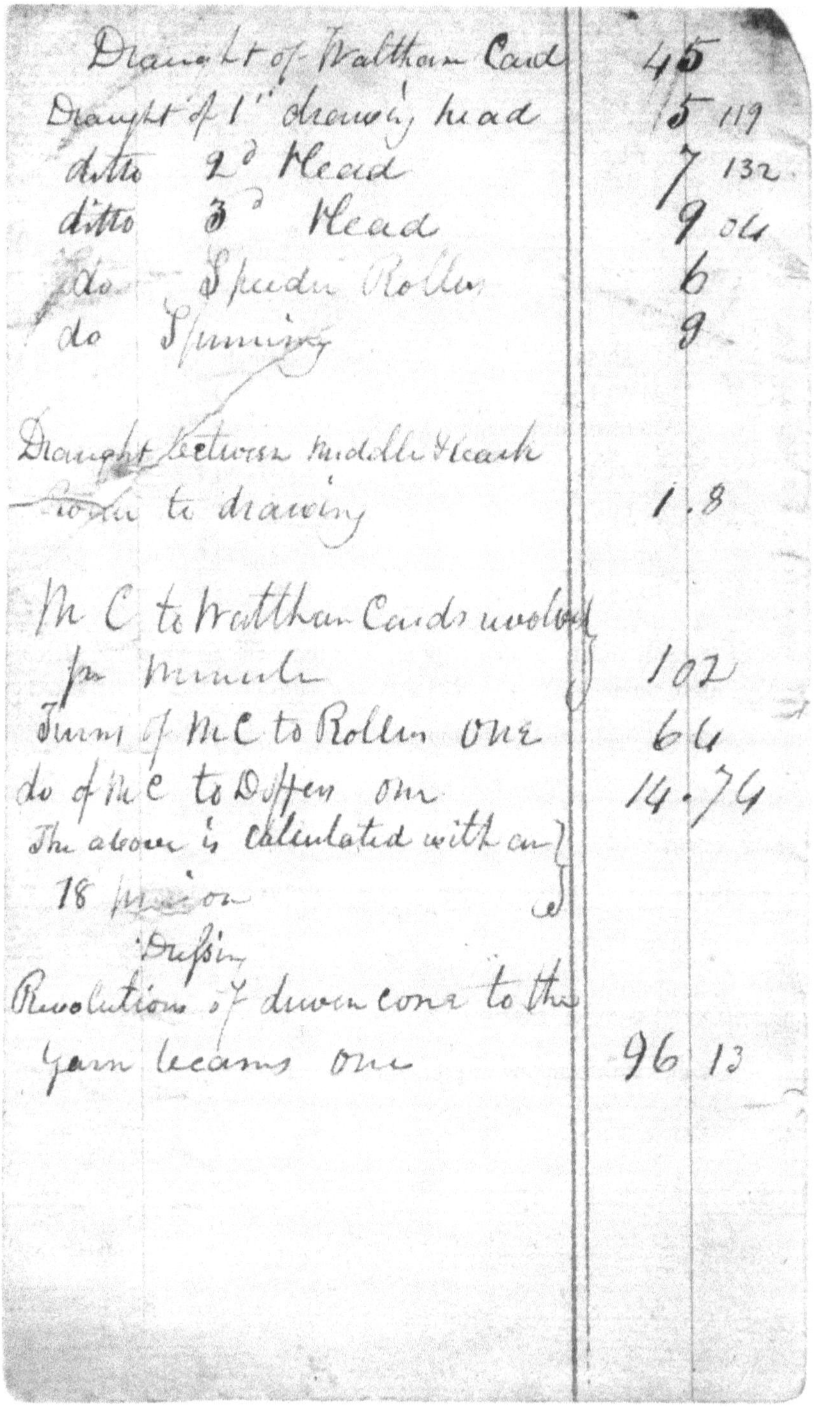

FIGURE 6.3 Notebook page.

Catalogue number: IEM 34.1

Document Description

Measurement: 16.2 x 9.8 cm

Medium: Ink, ruled with red ink

Condition: tear near top of page

Transcription

Draught of Waltham Card			45
Draught of 1st drawing head		5	119
ditto 2d Head		7	132
ditto 3d Head		9	04
do Speeder Roller		6	
do Spinning		9	
Draught between middle & back Roller to drawing		1.8	
MC [main cylinder] to Waltham cards revolves per minute		102	
Turns of MC to Roller	one	64	
do of MC to Doffer	one	14.74	
The above is calculated with an 18 pinion Dressing			
Revolutions of driven cones to the yarn beams one		96 13	

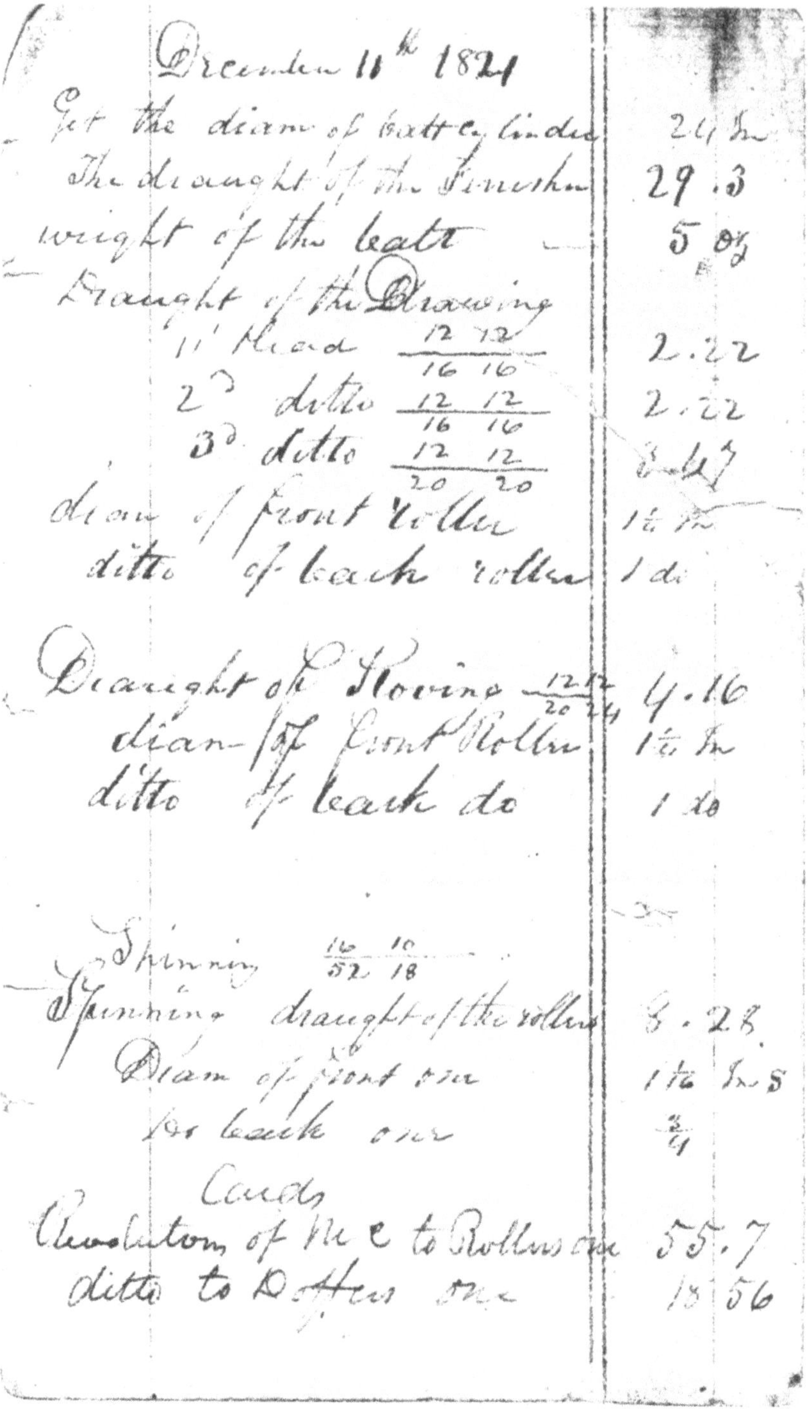

FIGURE 6.4 Notebook page.

Catalogue number: IEM 34.2

Document Description

Measurement: 16.2 x 9.8 cm

Medium: Ink, ruled with red ink

Condition: tear near top of page

Transcription

December 11th, 1821

Get the diam of batt cylinder		24 In
The draught of the Finisher		29.3
weight of the batt		5 oz
Draught of the Drawing		
1st Head	$\frac{12\ 12}{16\ 16}$	2.22
2d ditto	$\frac{12\ 12}{16\ 16}$	2.22
3d ditto	$\frac{12\ 12}{20\ 20}$	3.67
diam of front roller		1 In
ditto of back roller		1 do
Draught of roving	$\frac{12\ 12}{20\ 24}$	4.16
diam of front roller		$1\frac{1}{4}$ In
ditto of back do		1 do
Spinning	$\frac{16\ 10}{52\ 18}$	
Spinning draught of the rollers		8.28
Diam of front one		$1\frac{1}{16}$ Ins
Dr back one		$\frac{3}{4}$
Cards		
Revolution of MC to Rollers one		55.7
ditto to Doffer one		18.56

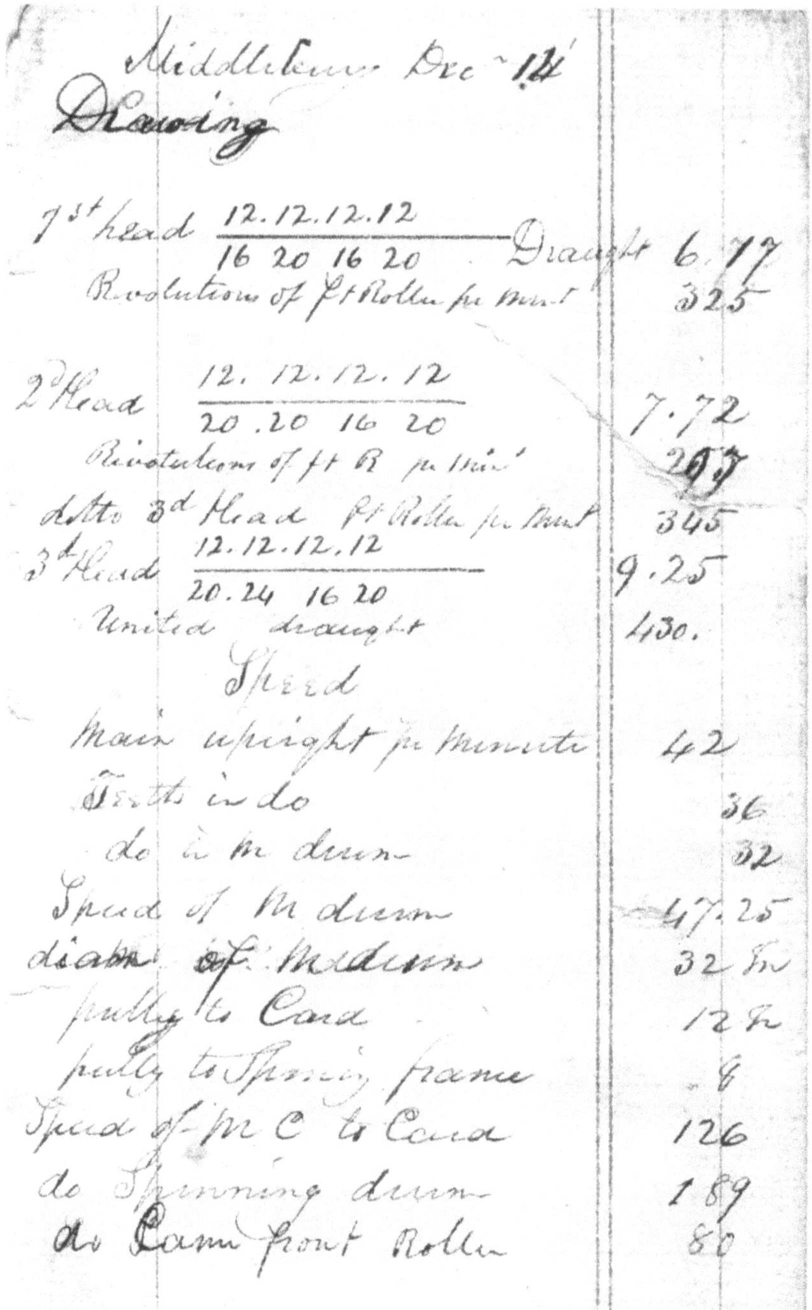

Middleton Dec 14
Drawing
1st head 12.12.12.12 / 16 20 16 20 Draught 6.77
Revolutions of Ft Roller per minute 325
2d Head 12.12.12.12 / 20.20 16 20 7.72
Revolutions of ft R per min [illegible]
ditto 3d Head Ft Roller per minute 345
3d Head 12.12.12.12 / 20.24 16 20 9.25
United draught 430.
Speed
Main upright per minute 42
Teeth in do 36
do in M drum 32
Speed of M drum 47.25
diam of M drum 32 In
pulley to Card 12 In
pulley to Spinning frame 8
Speed of M C to Card 126
do Spinning drum 189
do Card front Roller 80

FIGURE 6.5 Notebook page.

Catalogue number: IEM 34.4

Document Description

Measurement: 16.2 x 9.8 cm

Medium: Ink, ruled with red ink

Condition: Tear near top of page

Transcription

Middlebury Dec 14
Drawing

1st head	12. 12. 12. 12 / 16 20 16 20	Draught	6.17
Revolution of ft [front] Roller pr min't			325
2d Head	12. 12. 12. 12 / 20 20 16 20	Draught	7.72
Revolution of ft [front] Roller pr min't			205 (207?)
ditto 3d Head ft Roller pr min't			345
3d Head	12. 12. 12. 12 / 20 24 16 20		9.25
United draught			430
Speed			
Main upright pr minute			42
Teeth in do			36
do in M(ain) drum			32
Speed of M drum			47.25
diam of M drum			32 In
pully to card			12 In
pully to spinning frame			8
Speed of MC to card			126
do spinning drum			189
do same front roller			80

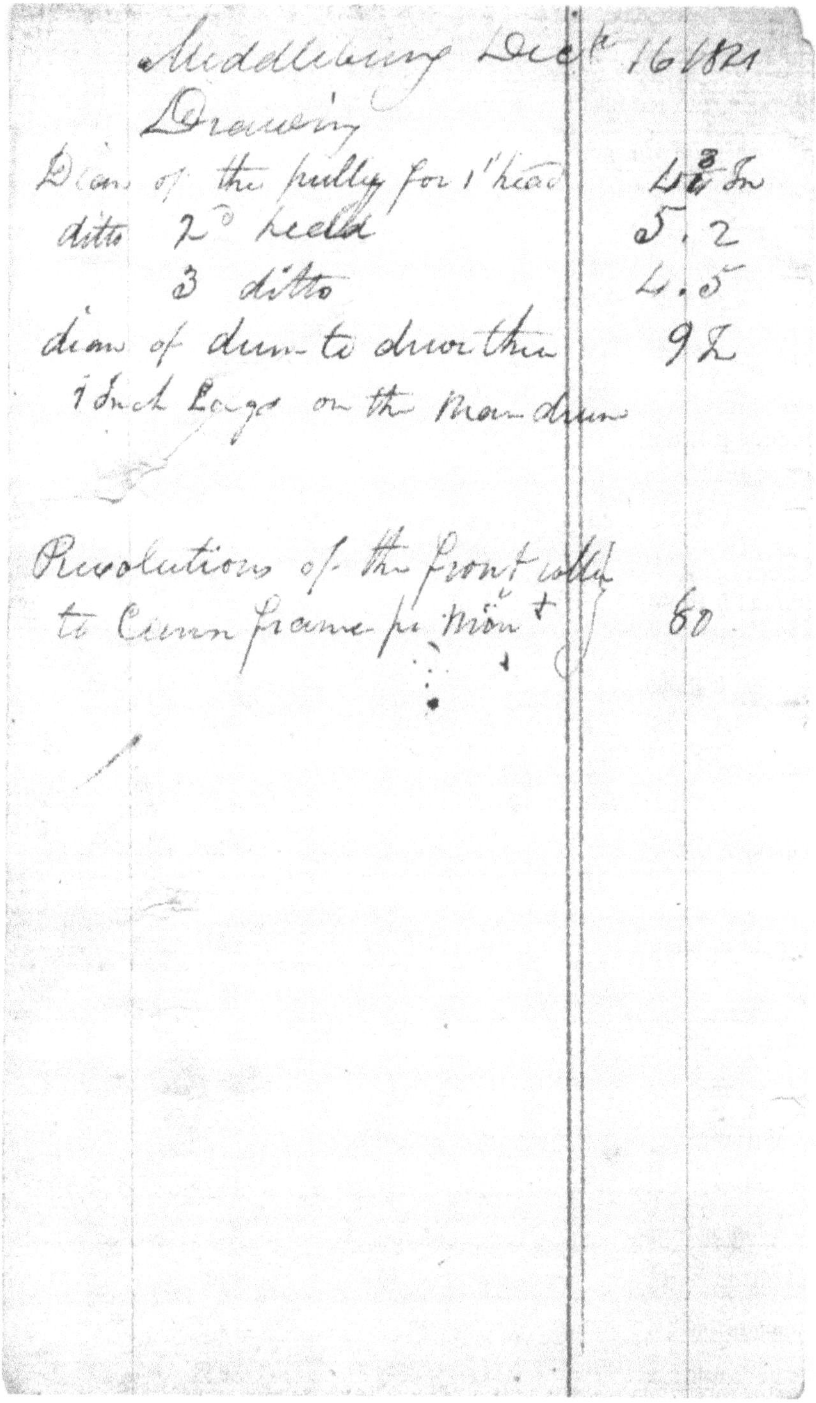

Middlebury Decr 16 1821
Drawing
Diam of the pulley for 1st head 4 3/4 In
ditto 2d heads 5.2
3 ditto 4.5
diam of drum to drive them 9 2
1 Inch Lags on the Main drum

Revolutions of the front rolls
to Cumn frame per Minut 80

FIGURE 6.6 Notebook page.

Catalogue number: IEM 34.3

Document Description

Measurement: 16.2 x 9.8 cm

Medium: Ink, ruled with red ink

Condition: tear near top of page

Transcription

Middlebury Decr 16, 1821

Drawing

Diam of the pulley for 1st head	$4^3/_4$ In
ditto 2nd head	5.2
3 ditto	4.5
diam of drum to drive them	9 In
1 inch Lags on the main drum	
Revolutions of the front roller to can frame pr min't	80

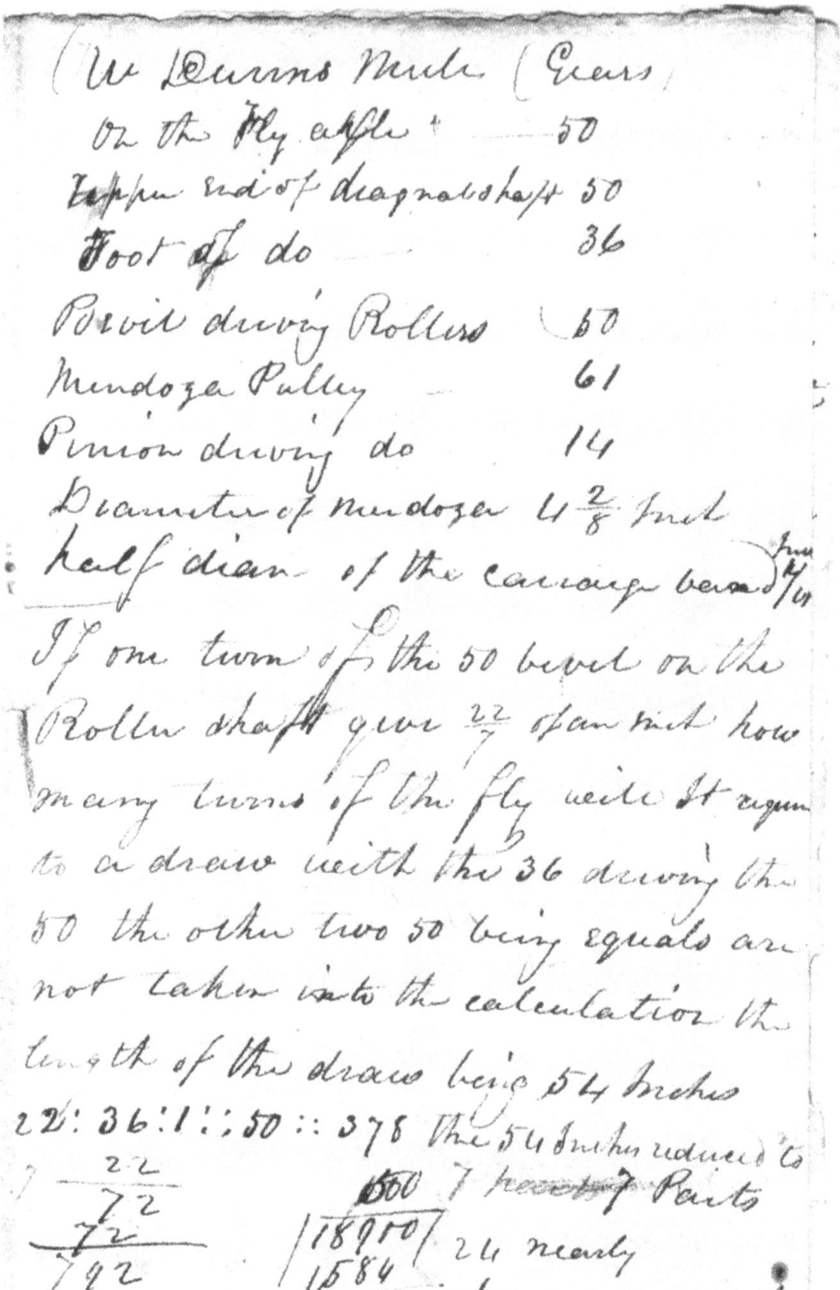

(W Dunns Mule (Gears
On the Fly axle 50
Upper end of diagonal shaft 50
Foot of do 36
Bevil driving Rollers 50
Mendoza Pulley 61
Pinion driving do 14
Diameter of mendoza 4 2/8 Inch
half diam of the carriage beam 1/4 Inch
If one turn of the 50 bevel on the Roller shaft give 22/7 of an inch how many turns of the fly will it require to a draw with the 36 driving the 50 the other two 50 being equals are not taken into the calculation the length of the draw being 54 Inches
22 : 36 : 1 :: 50 :: 378 the 54 Inches reduced to 7 parts
22 / 72 / 72 / 792
1500 / 18900 / 1584 / 3060 / 3168
24 nearly
23 1 taken out for the slack draught

FIGURE 6.7 Notebook page.

Catalogue number: IEM 33.1

Document Description

Measurement: Full sheet 20.3 x 17 cm; sheet folded in half twice and torn to make eight pages (see Figs. 6.7–6.14)

Medium: Ink

Condition: Minimal tears

Transcription

W Durrns [or Dunns]
mule[5]

	Gears
on the Fly axle	50
Upper end of diagonal shaft	50
Foot of do	36
Bevil driving Rollers	50
Mendoza Pulley[6]	61
Pinion driving do	14
Diameter of mendoza	$4^2/_8$ Inch
half diam of the carriage band	7 Inch

If one turn of the 50 bevel on the Roller shaft give $^{22}/_7$ of an inch how many turns of the fly will it require to a draw with the 36 driving the 50 the other two 50 being equals are not taken into the calculation the length of the draw being 54 Inches

22: 36: 1: :50: : 378 the 54 Inches reduced to 7 Parts

22 / 72	560 / 18,900	
72 / 792	1584 / 3060	24 nearly
	3168	1 / 23 1 taken out for the slub draught

---

[5]Unidentified.

[6]The first (late 19th century) and second (1989) editions of the *Oxford English Dictionary* correctly tracked the use of the word "mendoza" to the geared wheel in the spinning mule, which effected the movement of the spindle carriage away from the roller beam (thereby stretching the yarn to a predetermined fineness) to John Wood's British patent of 14 June 1803 (2711), but then attributed it to "[? The (Spanish) surname of the inventor]." This attribution is wrong. No such inventor has ever been identified. Much more likely, "mendoza" derived from Daniel Mendoza (1763–1836), the name of the Jewish bare-knuckle boxer who was Champion of England, 1792–1796, a decade when use of the spinning mule was spreading in Britain.

The mendoza wheel rotated on a pin at the end of a pivoted arm that was knocked down and into a meshing position with another gear (by a rope, wire, and lever sequence) causing the outward movement of the spindle carriage: a violent form of clutch. It was not surprising therefore that the action reminded mule spinners of the knockout blows Mendoza delivered: hence mendoza wheel.

J.Thomson, "Manufacture of Cotton"; Montgomery, *Carding and Spinning Master's Assistant*, 153; Catling, *Spinning Mule*, 87–88; *Jewish Encyclopaedia*; *Encyclopaedia Judaica*. Drawings of Wood's patent are shown in Fig. 3.19, p. 67.

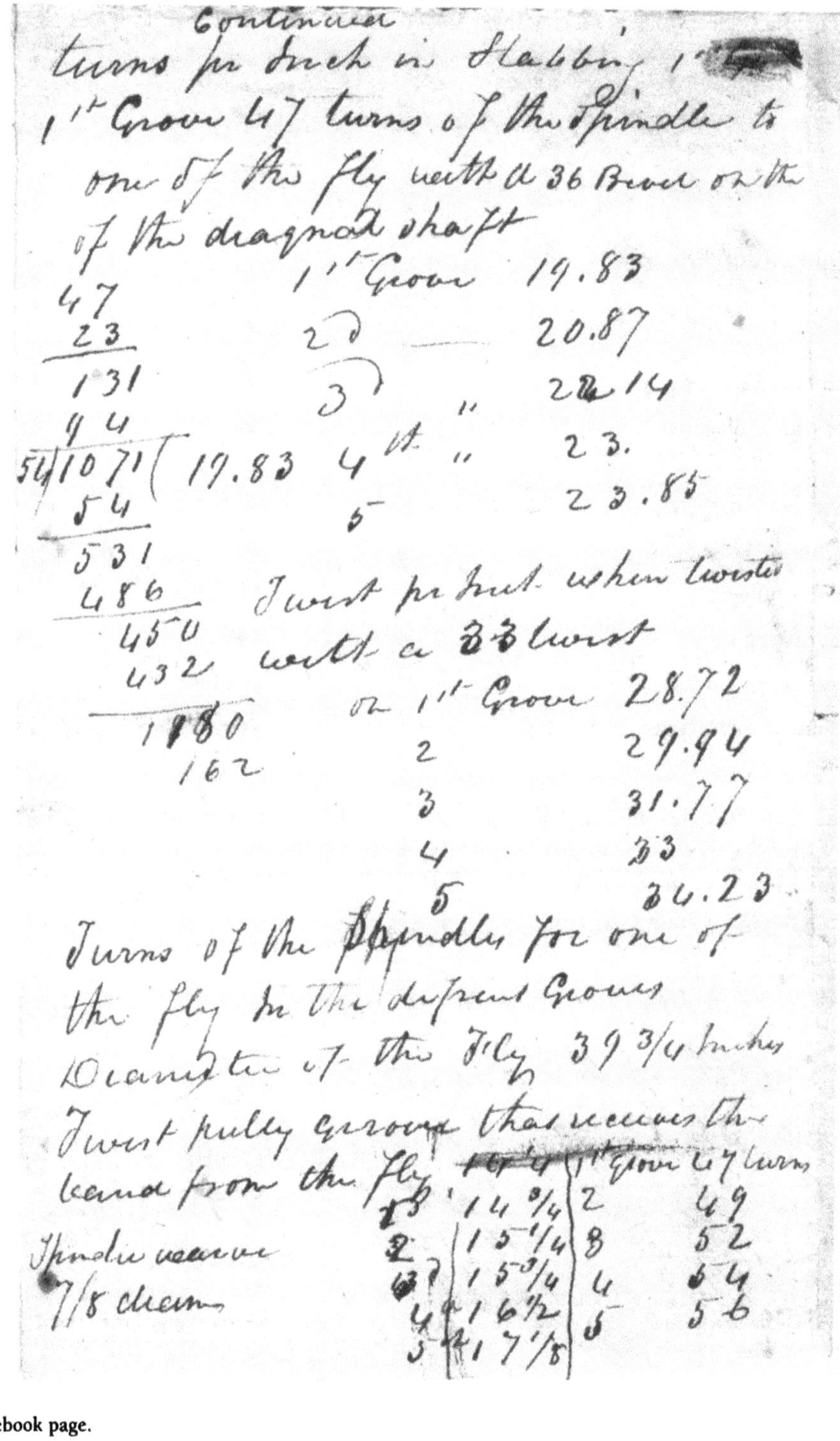

FIGURE 6.8 Notebook page.

Catalogue number: IEM 33.2

Document Description

Measurement: Full sheet 20.3 x 17 cm; sheet folded in half twice and torn to make eight pages (see Figs. 6.7–6.14)

Medium: Ink

Condition: Minimal tears

Transcription

Continued
turns per Inch in Slubbing

1st Grove[7] [groove] 47 turns of the spindle to one of the fly with a 36 bevel on the [missing] of the diagonal shaft

47	1st Grove	19.83
23	2d	20.87
131	3d	24.14
94	4th	23
54/1071 (19.83	5	23.85
54		
531		
486		
450		
432		
1180		
162		

Twist pr Inch when twisted with a 33 twist

On	1st Grove	28.72
2	29.94	
3	31.77	
4	33	
5	34.23	

Turns of the spindles for one of the fly in the different Groves

Diameter of the Fly 39 3/4 Inches

Twisted pully Grrove that receives the band from the fly

Spindle warve 7/8 diam

		1st Grove	47 turns
1st	14 3/4	2	49
2	15 1/4	3	52
3d	15 3/4	4	54
4d	16 1/2	5	56
5th	17 1/8		

[7]Groove.

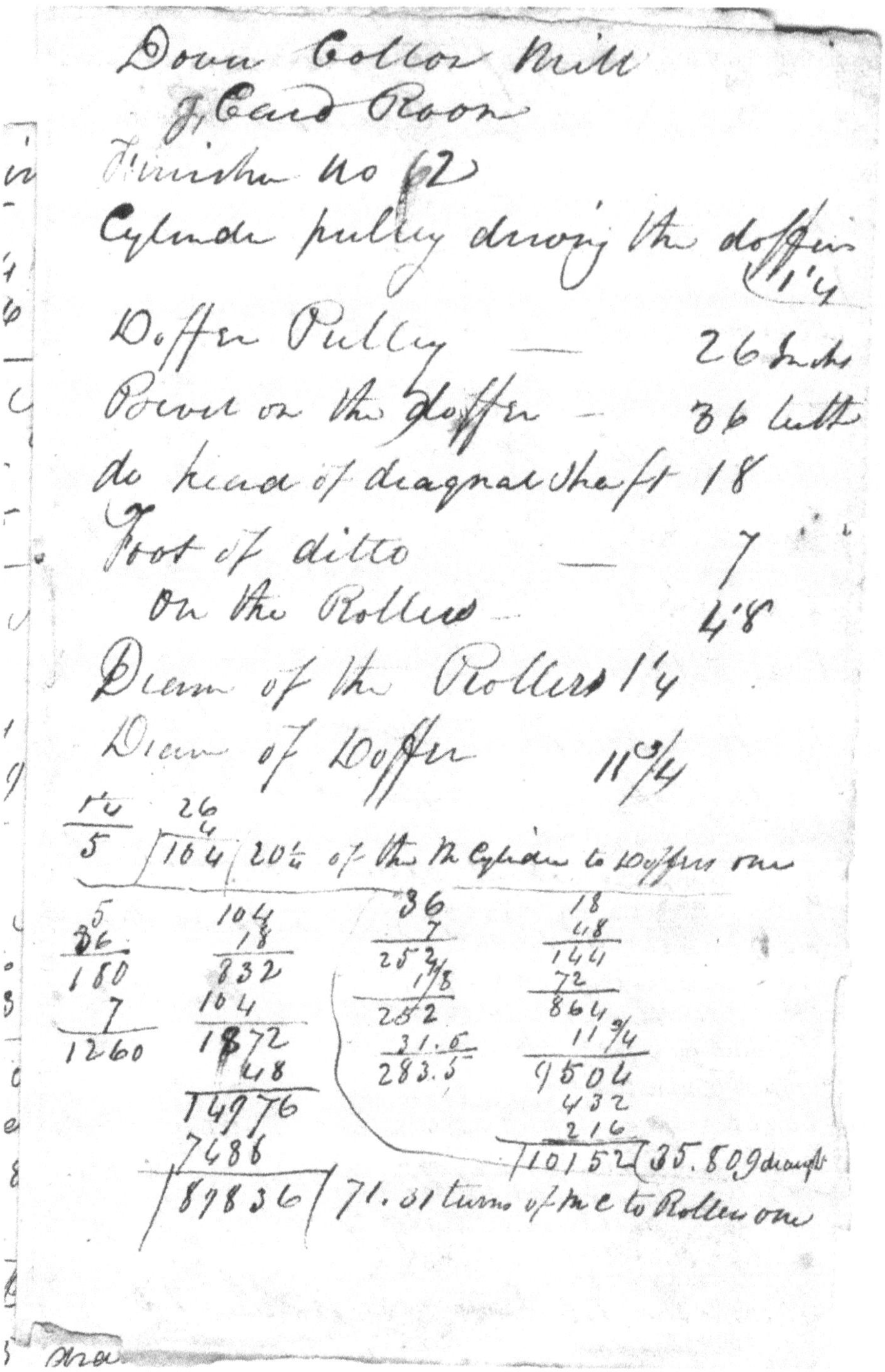

FIGURE 6.9 Notebook page.

Catalogue number: IEM 33.3

Document Description

Measurement: Full sheet 20.3 x 17 cm; sheet folded in half twice and torn to make eight pages (see Figs. 6.7–6.14

Medium: Ink

Condition: Minimal tears

Transcription

Dover Cotton Mill[8]
F Card Room

Finisher no 2	
Cylinder pulley driving the doffer	1¼
Doffer Pully	26 Inches
Bevel on the doffer	36 teeth
do head of diagonal shaft	18
Foot of ditto	7
On the Rollers	4' 8
Diam of the Rollers	1¼
¼ Diam of Doffer	11¾

1¼	26	
5	4	20¼ of the cylinder to Doffer's one
	104	

[ARITHMETICAL CALCULATIONS NOT TRANSCRIBED]

71.31 turns of MC to Roller's one

[8]The Cotton Factory at Dover, NH, ran 2,000 spindles and 50 power looms, employing 15 men, 5 boys, and 105 women and girls in 1820. This was before a mid-1820s expansion into calico printing. See US National Archives, RG 29, MS returns of the 1820 Census of Manufactures, New Hampshire, document 16; Jeremy, *Transatlantic Industrial Revolution*, 105–117.

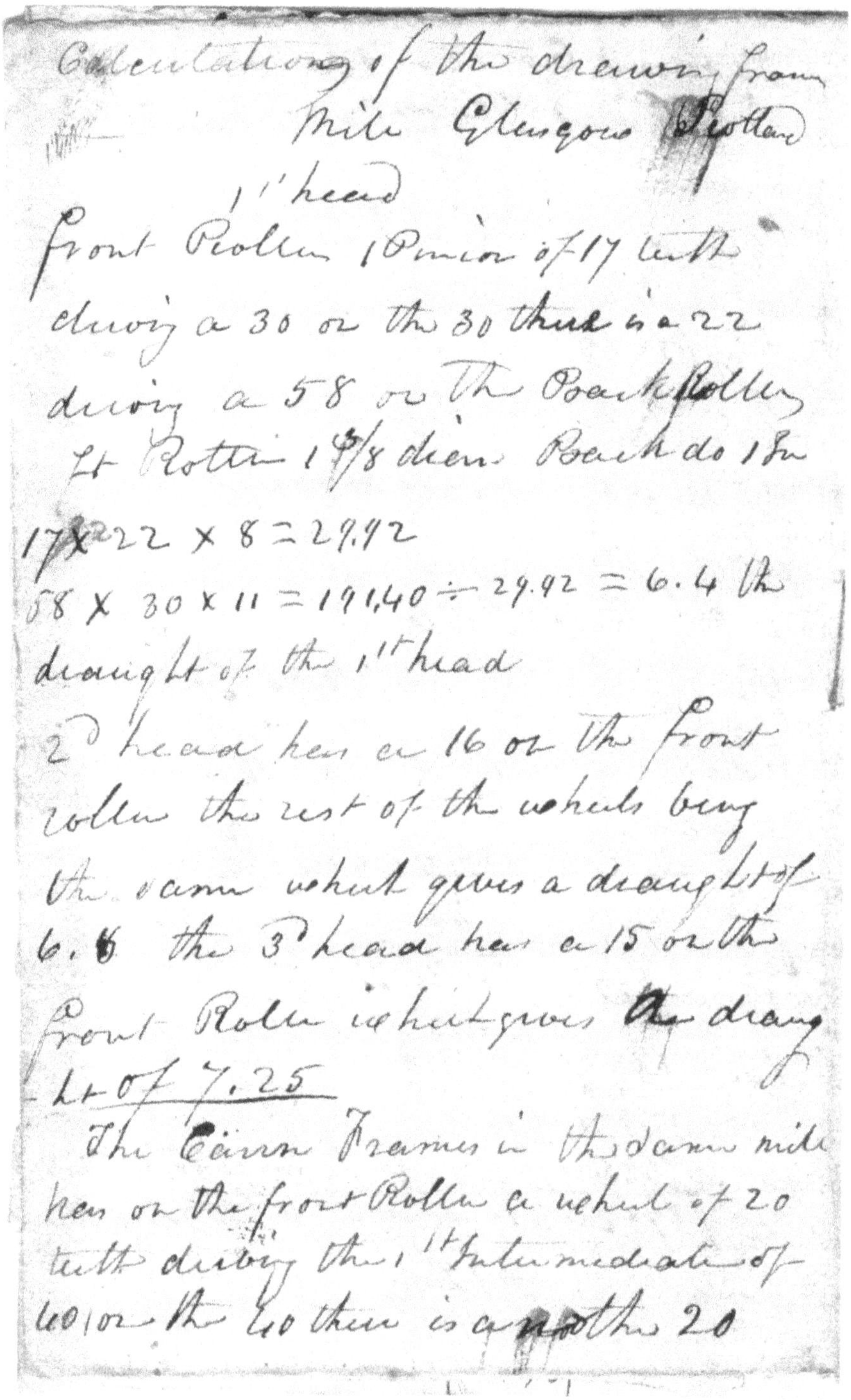

Calculations of the drawing from
Mile Glasgow Scotland
1st head
front Roller 1 Pinion of 17 teeth
driving a 30 on the 30 there is a 22
driving a 58 on the Back Roller
Ft Roller 1 3/8 diam. Back do 1 3/4
17 x 22 x 8 = 2992
58 x 30 x 11 = 19140 ÷ 2992 = 6.4 the
draught of the 1st head
2d head has a 16 on the front
roller the rest of the wheels being
the same which gives a draught of
6.8 the 3d head has a 15 on the
front Roller which gives the draug
ht of 7.25
The Cairn Frames in the same mill
has on the front Roller a wheel of 20
teeth driving the 1st Intermediate of
60 on the 60 there is another 20

FIGURE 6.10 Notebook page.

Catalogue number: IEM 33.4

Document Description

Measurement: Full sheet 20.3 x 17 cm; sheet folded in half twice and torn to make eight pages (see Figs. 6.7–6.14)

Medium: Ink

Condition: Minimal tears

Transcription

Calculation of the drawing frame—Mill Glasgow Scotland

1st head

front Roller 1 Pinion of 17 teeth

driving a 30 on the 30 there is a 22

driving a 58 on the Back Roller

ft Roller 1 4/8 [?] diam Back do 1 In

17 x 22 x 8 = 29.92

58 x 30 x 11 = 191.40 / 29.92 = 6.4 the

draught of the 1st head

2d head has a 16 on the front roller the rest of the wheels being the same which gives a draught of 6.8 the 3d head has a 15 on the front roller which gives a draught of 7.25

The Cann Frames in the same mill has on the front Roller a wheel of 20 teeth driving the 1st Intermediate of 40, on the 40 there is another 20

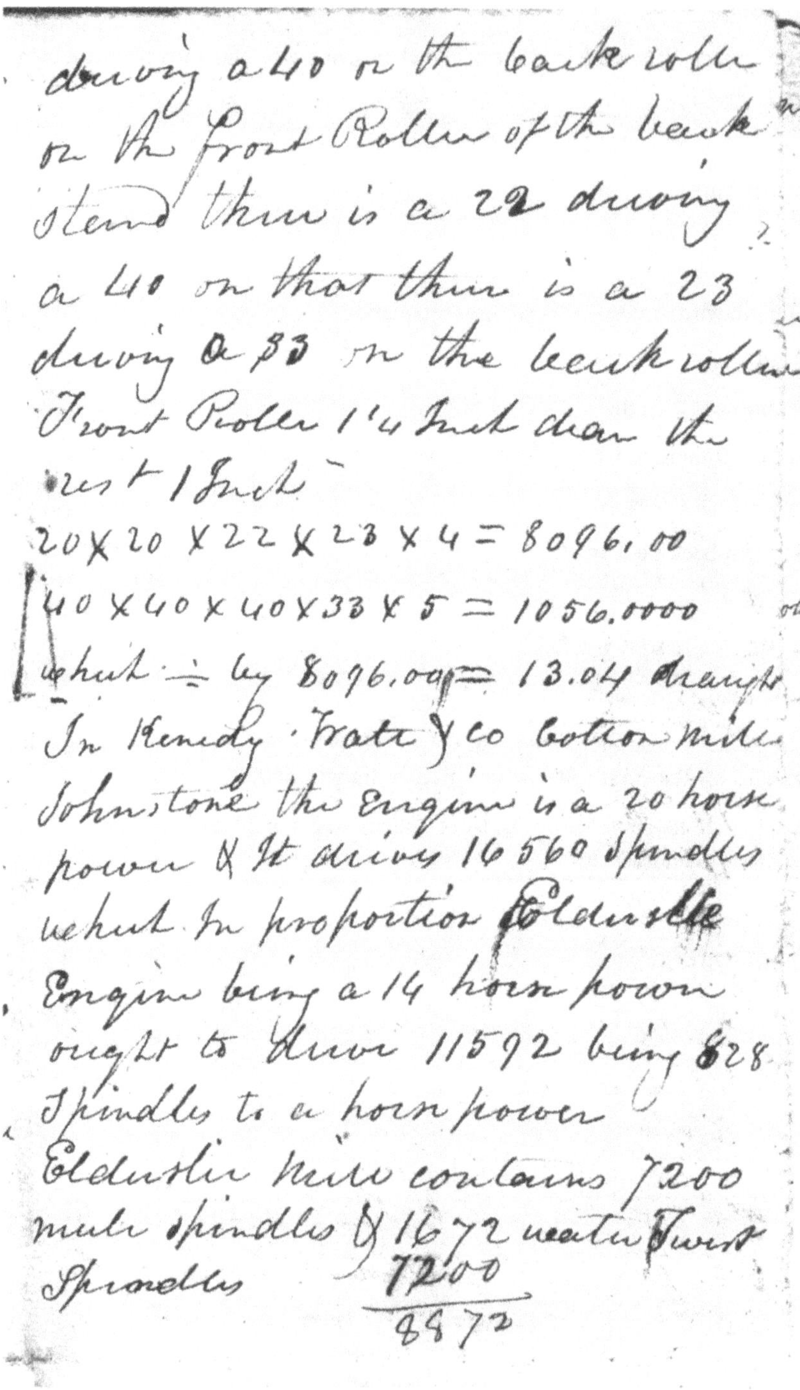
driving a 40 on the back roller
on the Front Roller of the back
stand there is a 22 driving
a 40 on that there is a 23
driving a 33 on the back roller
Front Roller 1¼ Inch diam the
rest 1 Inch
20 X 20 X 22 X 23 X 4 = 8096.00
40 X 40 X 40 X 33 X 5 = 1056.0000
which ÷ by 8096.00 = 13.04 draughts
In Kenedy Watt & Co Cotton mills
Johnstone the Engine is a 20 horse
power & It drives 16560 Spindles
which In proportion Elderslie
Engine being a 14 horse power
ought to drive 11592 being 828
Spindles to a horse power
Elderslie Mill contains 7200
mule spindles & 1672 water Twist
Spindles 7200
8872

FIGURE 6.11 Notebook page.

Catalogue number: IEM 33.5

Document Description

Measurement: Full sheet 20.3 x 17 cm; sheet folded in half twice and torn to make eight pages (see Figs. 6.7–6.14)

Medium: Ink

Condition: Minimal tears

Transcription

driving a 40 on the back roller
on the back front Roller of the back stand there is a 22 driving a 40 on that to there is a 23 driving a 33 on the back roller

Front Roller $1^1/_4$ Inch diam

the rest 1 Inch

20 x 20 x 22 x 23 x 4 = 8096.300

40 x 40 x 40 x 33 x 5 = 1056.0000

which / by 809600 = 13.04 draught

In Kennedy Watt & Co cotton mill Johnstone[9] the engine is a 20 horse power & it drives 16,560 spindles which in proportion Elderslie Engine being a 14 horse power ought to drive 11,592 being 828 spindles to a horse power. Elderslie Mill[10] contains 7200 mule spindles & 1672 water twist spindles

$$\frac{7200}{8872}$$

[9]In 1828 there were 20 cotton mills, totaling over 151,000 spindles, at Johnstone and neighborhood, near Paisley in Renfrewshire, Scotland, according to a local directory of that date. Although none belonged to Kennedy, Watt & Co., a Robert Watt is listed in the local directory as a cotton spinner. For this local directory information, we are grateful to Jean Mclean of the Local Studies Library, Central Library, Paisley, Scotland.
[10]The Elderslie Mill (and Old Mill) near Johnstone, belonged to Browns Malloch & Co., cotton spinners, in 1828 (source as in previous footnote).

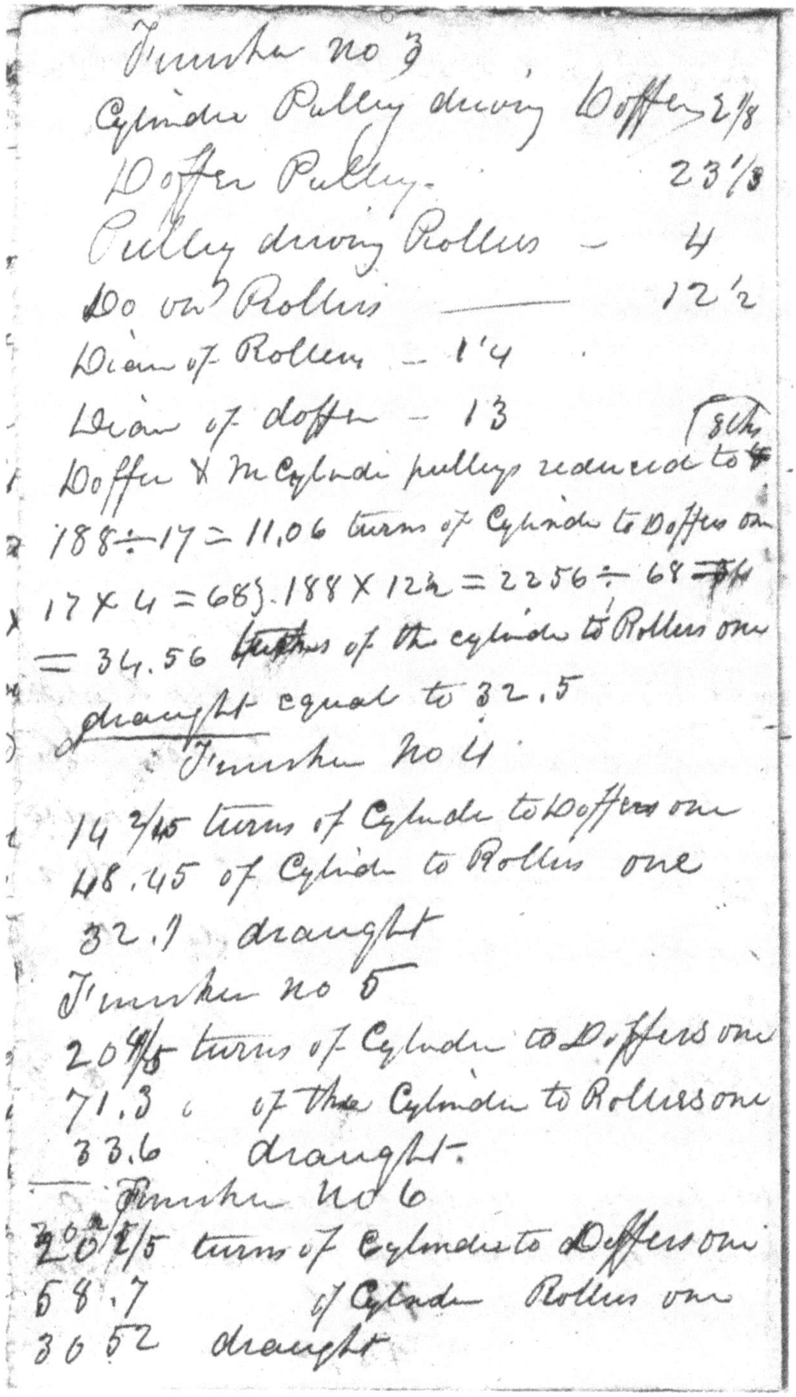

Finisher no 3
Cylinder Pulley driving Doffer 2 1/8
Doffer Pulley 23 1/8
Pulley driving Rollers — 4
Do on Rollers ——— 12 1/2
Diam of Rollers — 1 1/4
Diam of doffer — 13
Doffer & m Cylinder pulleys reduced to 8th
188 ÷ 17 = 11.06 turns of Cylinder to Doffers one
17 × 4 = 68}. 188 × 12 1/2 = 2256 ÷ 68 =
= 34.56 turns of the cylinder to Rollers one
draught equal to 32.5
Finisher No 4
14 2/15 turns of Cylinder to Doffers one
48.45 of Cylinder to Rollers one
32.1 draught
Finisher no 5
20 4/5 turns of Cylinder to Doffers one
71.3 " of the Cylinder to Rollers one
33.6 draught.
Finisher No 6
20 2/5 turns of Cylinder to Doffers one
68.7 of Cylinder Rollers one
30 52 draught

FIGURE 6.12 Notebook page.

Catalogue number: IEM 33.6

Document Description

Measurement: Full sheet 20.3 x 17 cm; sheet folded in half twice and torn to make eight pages (see Figs. 6.7–6.14)

Medium: Ink

Condition: Minimal tears

Transcription

Finisher No 3

Cylinder Pulley driving Doffer		$2^1/_8$
Doffer Pulley		$23^1/_8$
Pulley driving Rollers		4
Do on Rollers		$12^1/_2$
Diam of Rollers	$1^1/_4$	
Diam of doffer	13	

Doffer & M Cylinder pulleys reduced to 8ths

188 ÷ 17 = 11.06 turns of cylinder to Doffer's one

17 x 4 = 68 188 x 12 = 2256 ÷ 68

= 34.56 turns of the cylinder to Rollers one

draught equal to 32.5

Finisher no 4

$14^2/_{45}$ turns of Cylinder to Doffer's one

48.45 of Cylinder to Rollers one

32.9 draught

Finisher no 5

$20^4/_5$ turns of Cylinder to Doffer's one

71.3 of Cylinder to Rollers one

33.6 draught

Finisher no 6

$20^2/_5$ turns of Cylinder to Doffer's one

58.7 of Cylinder Rollers one

30.52 draught

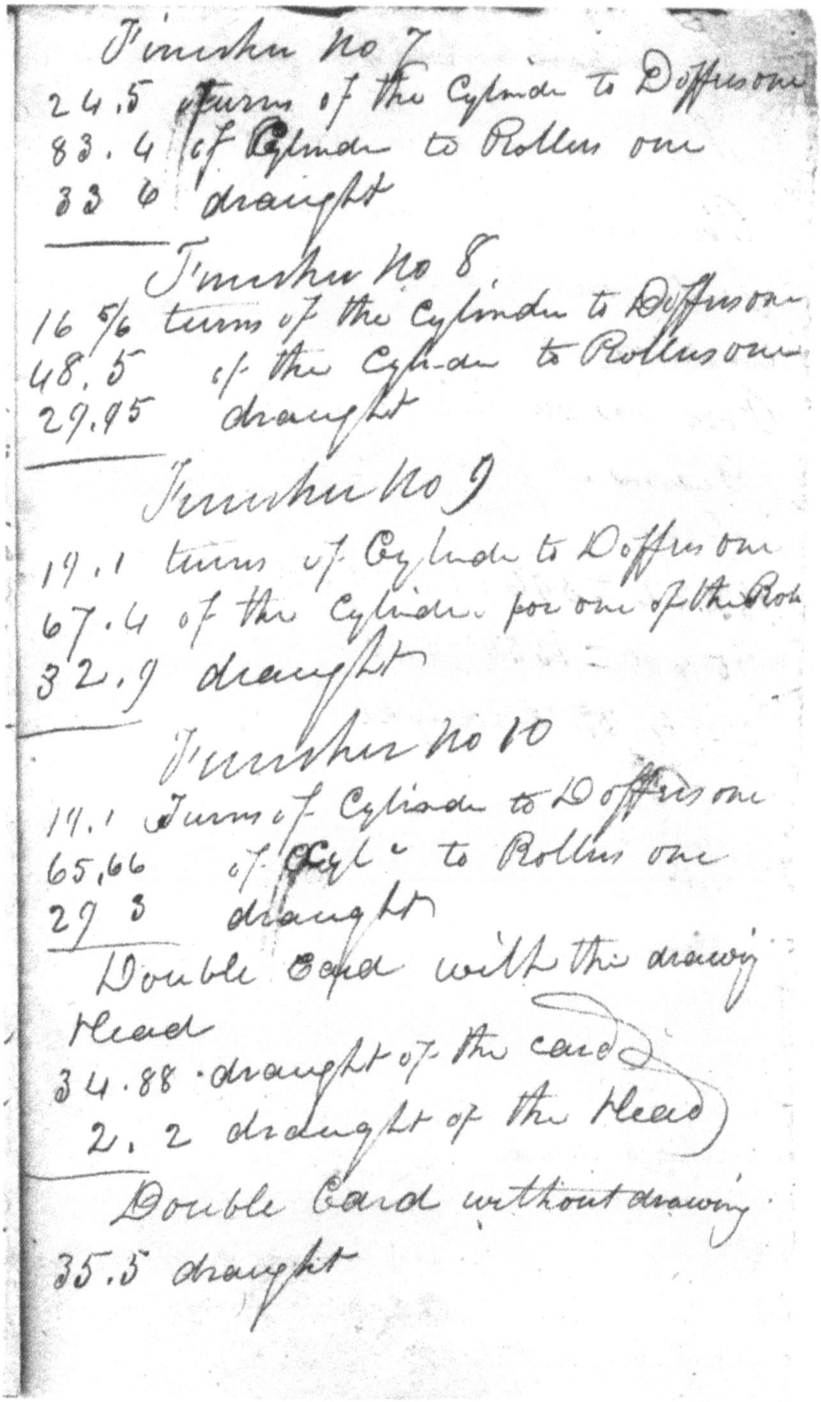

Finisher No 7
24.5 turns of the Cylinder to Doffer one
83.4 of Cylinder to Rollers one
33.6 draught

Finisher No 8
16 5/6 turns of the Cylinder to Doffer one
48.5 of the Cylinder to Rollers one
29.15 draught

Finisher No 9
19.1 turns of Cylinder to Doffer one
67.4 of the Cylinder for one of the Rol
32.9 draught

Finisher No 10
19.1 Turns of Cylinder to Doffer one
65.66 of Cyl. to Rollers one
27.3 draught

Double Card with the drawing Head
34.88 draught of the card
2.2 draught of the Head

Double Card without drawing
35.5 draught

FIGURE 6.13 Notebook page.

Catalogue number: IEM 33.7

Document Description

Measurement: Full sheet 20.3 x 17 cm; sheet folded in half twice and torn to make eight pages (see Figs 6.7–6.14)

Medium: Ink

Condition: Minimal tears

Transcription

Finisher no 7

24.5 turns of the Cylinder to Doffer's one

83.4 of Cylinder Rollers one

33.6 draught

Finisher no 8

16 5/6 turns of the Cylinder to Doffer's one
48.5 of Cylinder Rollers one

29.95 draught

Finisher no 9

19.1 turns of Cylinder to Doffer's one

67.4 of the Cylinder for one of the Rollers

32.9 draught

Finisher no 10

19.1 Turns of Cylinder to Doffer's one

65.66 of Cyl to Rollers one

29.3 draught

Double Card with the drawing Head

34.88 draught of the card

2.2 draught the Head

Double Card without drawing
35.5 draught

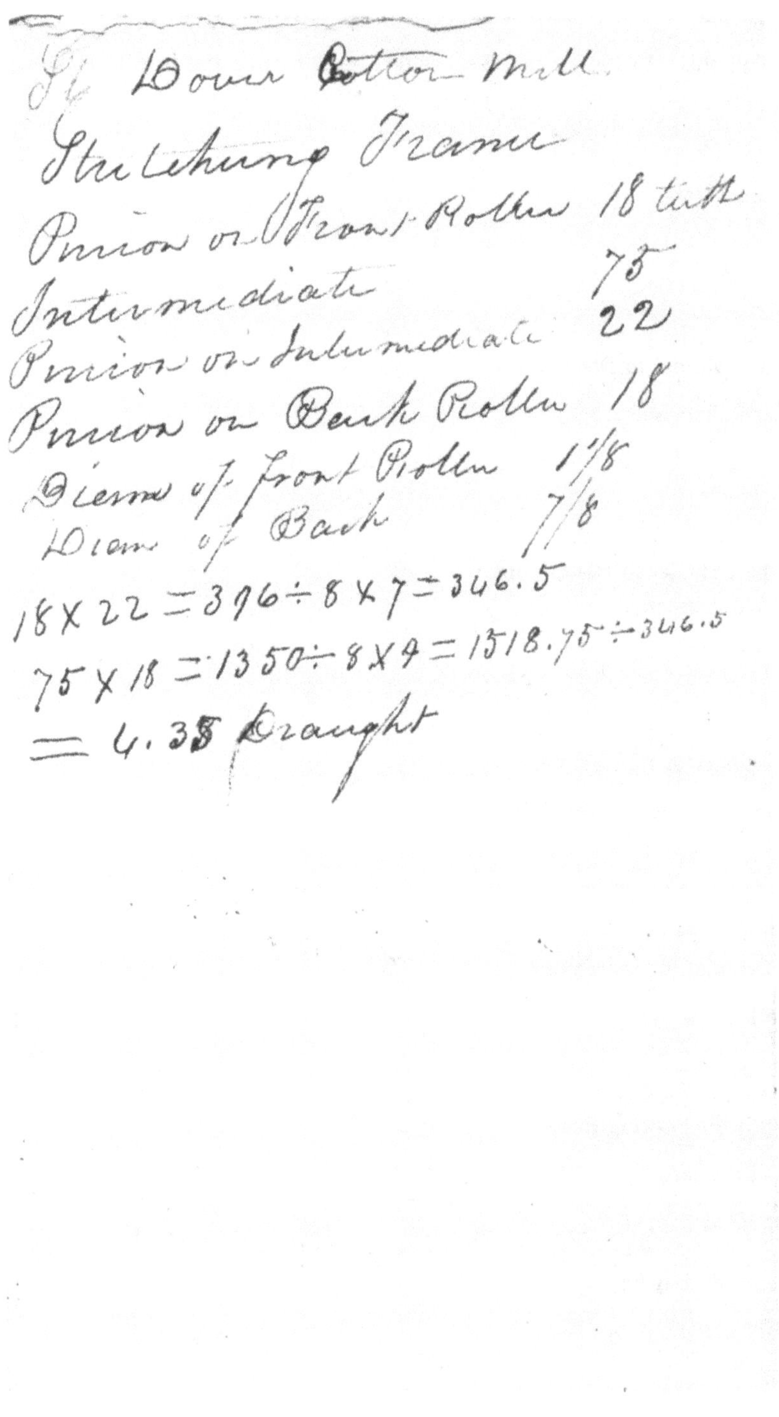

Of Dover Cotton Mill
Stretching Frame
Pinion on Front Roller 18 teeth
Intermediate 75
Pinion on Intermediate 22
Pinion on Back Roller 18
Diam of Front Roller 1 1/8
Diam of Back 7/8
18 X 22 = 396 ÷ 8 X 7 = 346.5
75 X 18 = 1350 ÷ 8 X 9 = 1518.75 ÷ 346.5
= 4.38 Draught

FIGURE 6.14 Notebook page.

Catalogue number: IEM 33.8

Document Description

Measurement: Full sheet 20.3 x 17 cm; sheet folded in half twice and torn to make eight pages

Medium: Ink

Condition: Minimal tears

Transcription

Dover Cotton Mill[11]

Stretching Frame[12]

Pinion on Front Roller	18 teeth
Intermediate	75
Pinion on Intermediate	22
Pinion on Back Roller	18
Diam of front Roller	1 1/8
Diam Back	7/8

18 x 22 = 396 ÷ 8 x 7 = 346.5

75 x 18 = 1350 ÷ 8 x 9 = 1518.75 ÷ 346.5 = 4.35 Draught

[11]See note 8.

[12]Stretching frame: a spinning mule, used to reduce the fineness of roving, displaced by the fly frame in the late 1820s. Montgomery, *Carding and Spinning Master's Assistant*, 132–135.

Loose Sheets

These stray sheets are both surprising and revealing. They are surprising because the recording of such vital information for Markham, both as machine maker and as mill manager, seems so unsystematized and vulnerable to accidental loss. Perhaps he kept them in a special folder. The folds, creases, tears, blots, and torn away parts seem to belie this possibility. Perhaps he kept them on his person. Perhaps they were superseded in the last three or four years of his life.

At any rate, the technical and economic knowledge preserved on the sheets was as vital an accompaniment to the drawings as the information held in the Notebook. They include an index to the wooden patterns required as moulds to make castings for a speeder and a double speeder, both types of roving frame (see Fig. 6.15A, 6.15B). Without such an index a good deal of time would be lost hunting through the bank of patterns kept either at the workshop or at the forge. Another sheet listed the parts needed to build a warping machine, involving 25 shapes of wrought iron (see Fig. 6.16). One sheet specified components of a dresser, including a mixture of 20 kinds of metal parts, including brass reeds, five kinds of iron casting, as many as 22 wooden parts, three kinds of pulleys, and five kinds of bolts (see Fig. 6.17A, 6.17B). And some of that document is missing.

Five sheets give machine dimensions for a carding machine, a double speeder, and a dressing frame. Particularly interesting is the first of these (see Fig. 6.18), which details ratios of card-delivery rollers in relation to a drawing frame's feed rollers. In other words, this records speed ratios between two different but sequential factory machines—one step in the important exercise of establishing a balance along a production line giving a predictable output from a given amount and quality of cotton.

There are two sheets of notes detailing economic information about machine construction. One lists the many parts and their costs for building three 64-spindle spinning frames. It begins by enumerating some 32 steps associated with making 64 flyers and accompanying iron rollers, saddles, gears, flywheels, and pulleys, totaling an expense of $158.73 (see Fig. 6.23A). The reverse of the sheet (Fig. 6.23B) shows almost as much expended on raw materials ($150.40). This gives an overall figure for a 64-spindle spinning frame of $320. Presumably neither this nor the price for three similar machines ($960) included profit. The $81 for raw material seems to be an attempt to isolate raw material from labor and other costs.

What is not clear is how much of all this work was done in-house and how much had to be secured from outside Markham's machine shop. Given the detail of the costs, the size of the whole business (8 men, 34 women, 3 children, running 840 spindles, 24 power looms and all related equipment, as well as a machine shop in 1820), and given the other artisan businesses in the town and vicinity of Middlebury, it is possible that all except the iron castings and some of the forgings were made in-house. But it is impossible to tell. The key functions of design, component construction, assembly, fitting, operation, maintenance, and repair would have kept the machinists busy without engaging in an excessive amount of casting and forging such as a much larger machine shop like that at Waltham, with its 40 mechanics at the 1820 census date, would have undertaken.[13]

[13]US Census of Manufactures, 1820, National Archives film M-279, roll 3, VT doc 278, MA doc 33.

The last loose sheet of data (see Fig. 6.24A) records machine output for a speeder (roving frame): information that is critical for a machine maker trying to sell a speeder to a customer, like a mill manager, or in operating the machine.

Inventory of Patterns for Machine Components

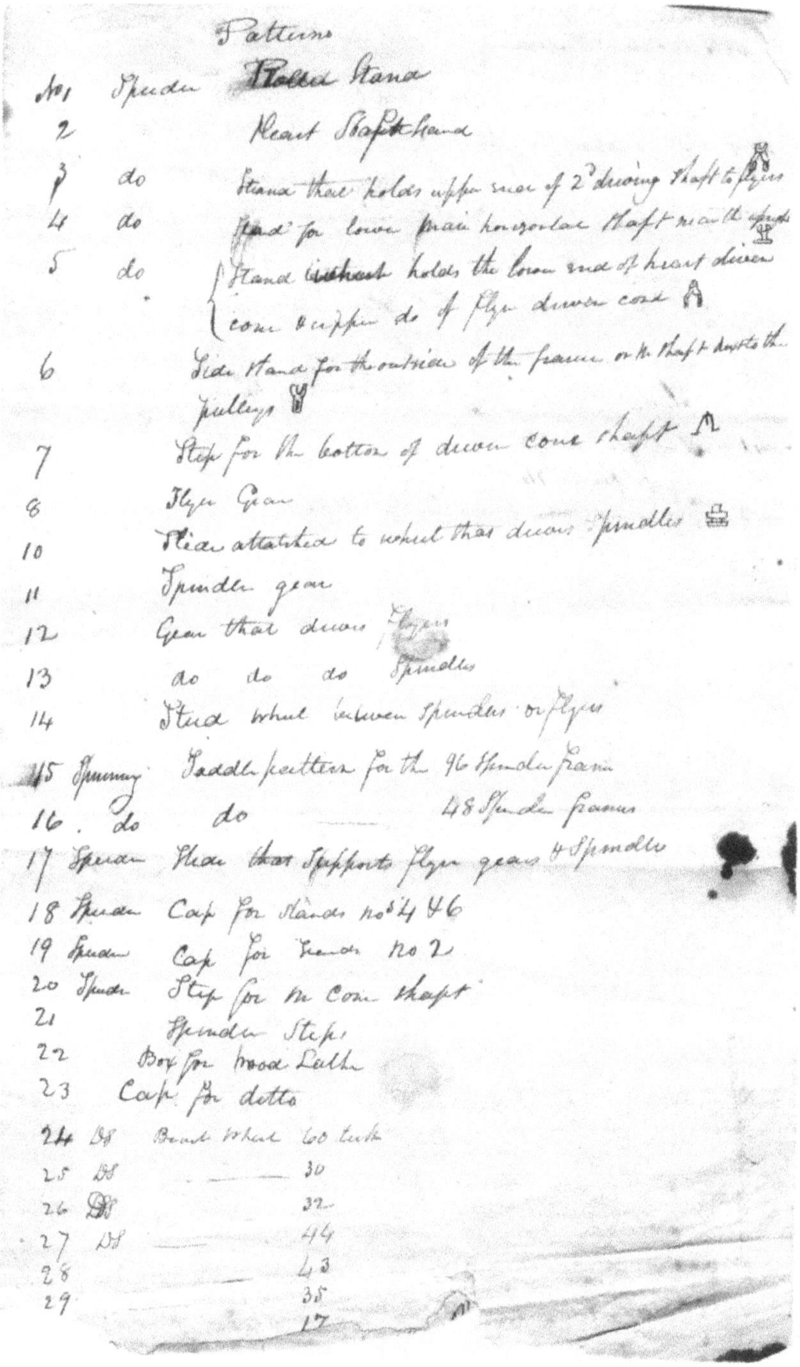

Patterns

No1 Speeder ~~Roller~~ Stand
2 Heart Shaft Stand
3 do Stand that holds upper end of 2d driving shaft to flyers
4 do Stand for lower main horizontal shaft in all speeders
5 do Stand which holds the lower end of heart driven cone & upper do of flyer driven cone
6 Side Stand for the outside of the frame on the shaft & drives the pulleys
7 Step for the bottom of driven cone shaft
8 Flyer Gear
10 Plate attached to wheel that drives spindles
11 Spindle gear
12 Gear that drives flyers
13 do do do Spindles
14 Stud wheel between Spindles or flyers
15 Spinning Saddle pattern for the 96 Spindle frame
16 do do 48 Spindle frames
17 Speeder Plate that supports flyer gears & Spindles
18 Speeder Cap for Stands nos 4 & 6
19 Speeder Cap for Stands no 2
20 Speeder Step for the Cone shaft
21 Spindle Steps
22 Box for Wood Lathe
23 Cap for ditto
24 DS Bevel wheel 60 teeth
25 DS 30
26 DS 32
27 DS 44
28 43
29 35
17

FIGURE 6.15A Loose sheet. Speeder.

Catalogue number: IEM 35.1a

Document Description

Measurement: 32.5 x 20.2 cm

Medium: Ink

Condition: Small stains, tears on bottom of sheet; on reverse of Figure 6.15B

Transcription

Patterns

1	Speeder	Roller stand
2		Heart shaft stand
3	ditto	stand that holds upper end of second driving shaft to flyers
4	ditto	stand for lower main horizontal shaft near the upright
5	ditto	stand which holds lower end of heart driven cone & upper ditto of flyer driven cone
6.		side stand for the outside of the frame or the shaft next to the pulleys
7		step for the bottom of the driven cone shaft
8		Flyer gear
10		slide attached to wheel that drives spindles
11		spindle gear
12		gear that drives flyers
13		ditto ditto ditto spindles
14		Stud wheel between spindles flyers
15	Spinning	saddle pattern for the 96 spindle frame
16		ditto 48 spindle frame
17	speeder	slider that supports flyer gears and spindles
18	speeder	cap for stands Nos 4 & 6
19	speeder	cap for stands No 2
20	speeder	step for the cone shaft
21		spindle steps
22		box for wood lathe
23		cap for ditto
24	DS [Double speeder]	bevel wheel 60 teeth
25	DS	30
26	DS	32
27	DS	44
28		43
29		35
30	DS	17

31 DS Cyclar Frame
32 DS Cap to ditto
33 DS Stand for worm shaft to heart
34 DS Support for top of main upright to rollers
35 DS Stand to hold the large Stud wheels 2
36 DS Stand to hold the stationary end of the levers that regulate the bands on the cones
37 & 38 Are the feet of this Stand made right & left to change
39 Support for the end of the Iron front roller gear end
40 Studs to hold small Stud wheels between Spindles & between flyers
41 Cap for no 33
42 Cap for no 34
43 Spinning ~~roller~~ Stand for the rollers

FIGURE 6.15B Loose sheet. Speeder.

Catalogue number: IEM 35.1b

Document Description

Measurement: 32.5 x 20.2 cm

Medium: Ink

Condition: Small stains, tears on bottom of sheet; on reverse of Figure 6.15A

Transcription

31	DS	cycloid frame
32	DS	cap to ditto
33	DS	stand for worm shaft to heart
34	DS	support for top of main (?) upright to rollers 
35	DS	stick to hold the large stud wheels
36	DS	stand to hold the stationary end of the levers that regulate the bands on the cones
37 & 38		are the feet of the stand made right and left to change
39		support for the end of the iron front roller geer end
40		stick to hold small stud wheels between spindles & between flyers
41		cap for no 33
42		cap for no 34
43		spinning stand for the rollers

Endorsement

...days after date I promise to pay Arthur Latham C[o?][14]

[14]This could have been Arthur Latham, first postmaster of Lyme, NH, or his son Arthur Jr., members of a family who kept a general store at Lyme in the nineteenth century. Their papers are in the Library of Dartmouth College, Hanover, NH.

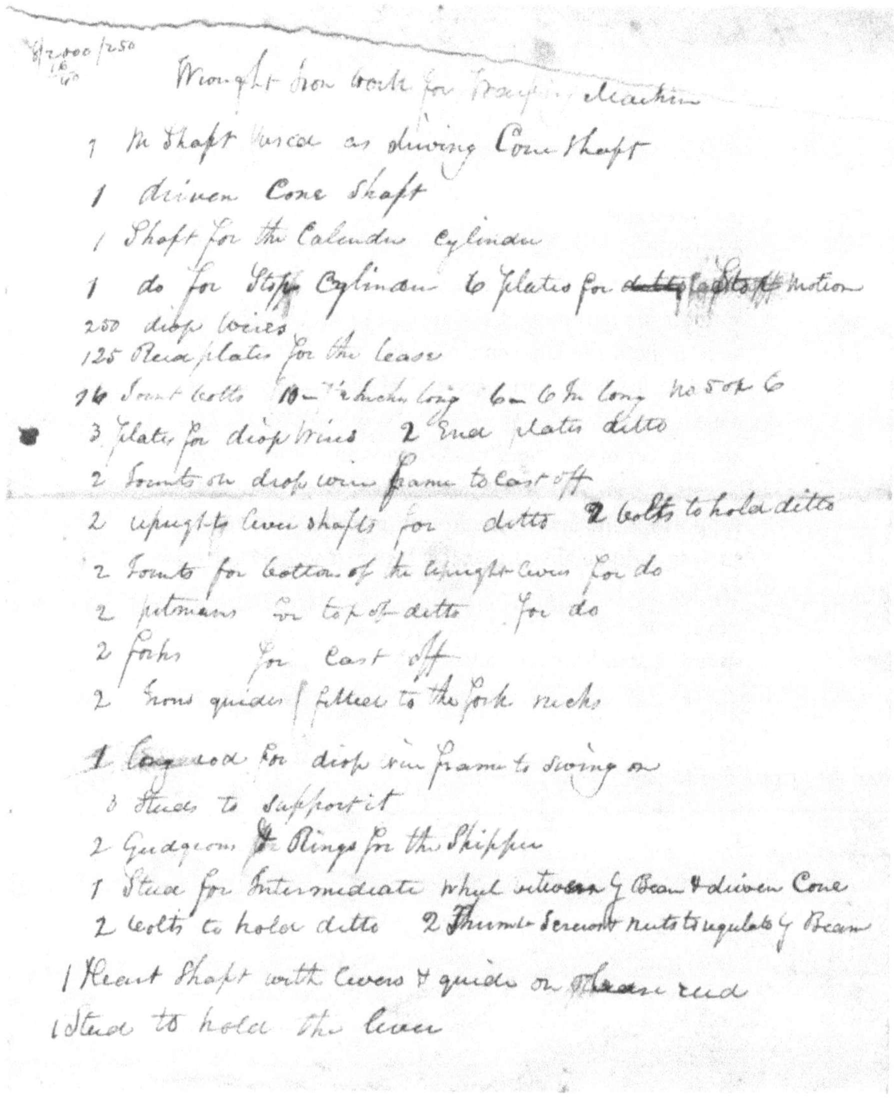

4/2000/250 16 40

Wrought Iron Work for Warping Machine

1 M Shaft [illegible] as driving Cone Shaft

1 driven Cone Shaft

1 Shaft for the Calender Cylinder

1 do for Stop Cylinder 6 plates for ~~ditto~~ Stop motion

250 drop wires

125 Reed plates for the lease

16 Joint bolts 10 — ½ inches long 6 — 6 In long no 5 of 6

3 plates for drop wires 2 End plates ditto

2 Joints on drop wire frame to cast off

2 upright lever shafts for ditto 2 bolts to hold ditto

2 Joints for bottom of the upright levers for do

2 pitmans for top of ditto for do

2 forks for cast off

2 Irons guides fitted to the fork necks

1 Long rod for drop wire frame to swing on

3 Studs to support it

2 Gudgeons & Rings for the Skipper

1 Stud for Intermediate wheel between Beam & driven Cone

2 bolts to hold ditto 2 Thumb Screws & nuts to regulate Beam

1 Heart Shaft with levers & guides on ~~these~~ reed

1 Stud to hold the lever

FIGURE 6.16 Warping machine.

Catalogue number: IEM 35.2a

Document Description

Measurement: Full sheet 40.5 x 24 cm, folded in half; includes Figures 6.16, 6.19, and 6.25

Medium: Ink

Condition: Torn across top of page; stains

Transcription

Wrought iron work for warping machine

1 M shaft used as driving cone shaft

1 driven cone shaft

1 shaft for the calendar cylinder

1 ditto for stop cylinder

6 plates for stop motion[15]

250 drop wires

125 reed plates for the lease

96 joint bolts

10 7½ inches long; 6 6 inches long, no 5 or 6

3 plates for drop wires

2 end plates ditto

2 joints on drop wire frame to cast off

2 upright lever shafts for ditto

2 bolts to hold ditto

2 joints for bottom of the upright levers for ditto

2 pitmans (?) for top of ditto for ditto

2 forks for cast off

2 iron guides fitted to the fork necks

1 long rod for drop wire frame to swing on

3 studs to support it

2 gudgeons & rings for the shipper

1 stud for intermediate wheel between Y [yarn] beam & driven cone

2 bolts to hold ditto

2 thumb screws & nuts to regulate Y beam

1 heart shaft with levers & guide on lease reed

1 stud to hold the lever

[15]The drop wires here suggest that the stop-motion was of the drop wire and rocker type. With 250 drop wires, it may have been similar in size to the warping frame built at Waltham by the Boston Manufacturing Co. in June 1819 when their warping frame had an average of about 220 drop wires per machine. Boston Manufacturing Co., Accounts Current, vol. 34 (1816–1822), p. 192, Baker Library; Jeremy, "Innovation in American Textile Technology," 63–72; Jeremy, *Transatlantic Industrial Revolution*, 62, 99, 183–195, 306 n40. See Appendix 3.

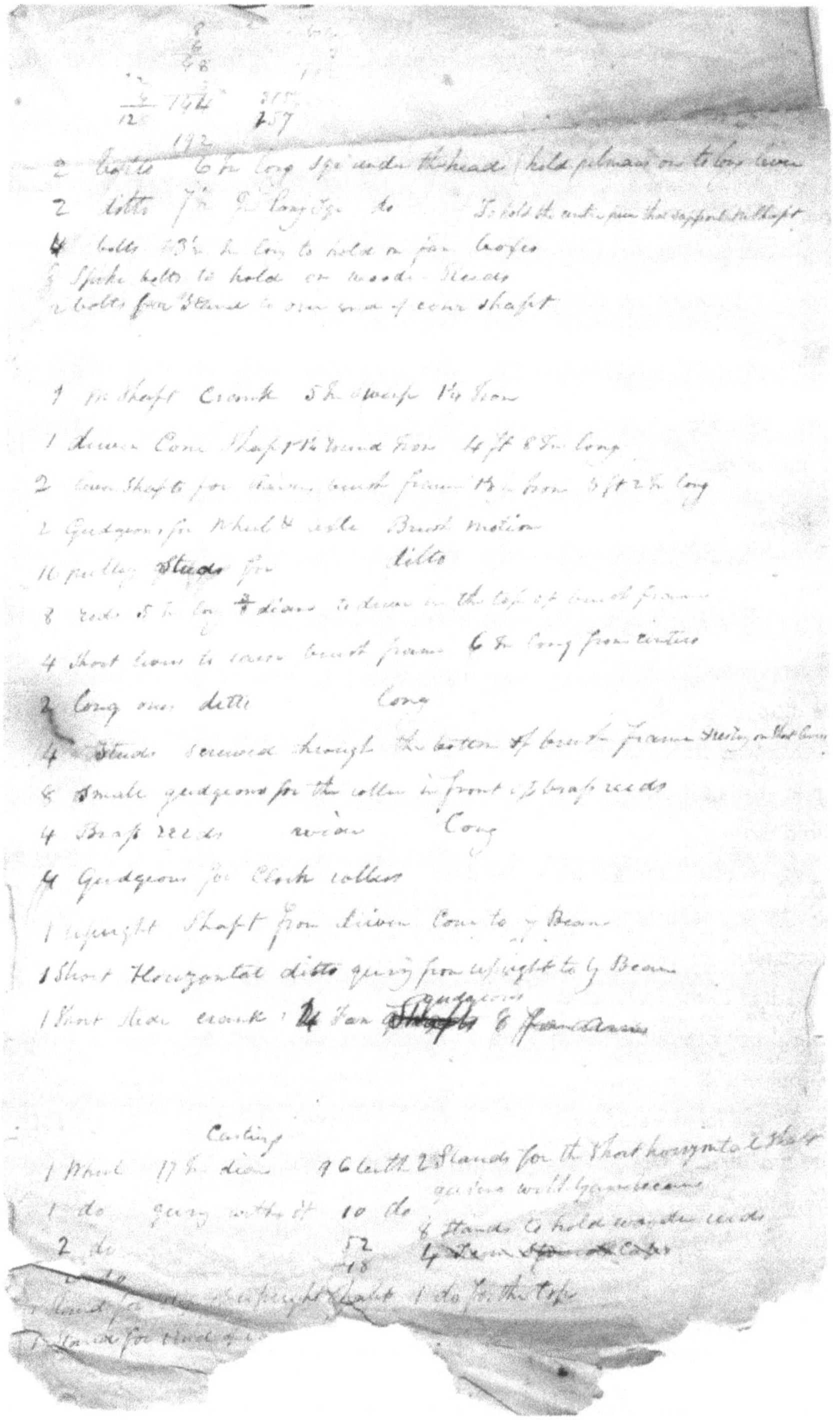

FIGURE 6.17A Dressing machine, metal work, and wooden parts.

Catalogue number: 1EM 35.3a

Document Description

Measurement: 32.7 x 20.2 cm

Medium: Ink

Condition: Pieces torn off bottom of page; reverse of Figure 6.17B

Transcription

2 bolts 6 In long sqr under the head (hold pdman [fitment?]) on to long lever

2 ditto $7^1/_2$ In long sqr do to hold the centre piece that supports the shaft

4 bolts $3^1/_2$ In long to hold on fan boxes

8 spike bolts to hold on the wooden Reeds

2 bolts for stand to one end of cone shaft

1 in Shaft crank 5 In Sweep $1^1/_4$ Iron

1 driver cone shaft $1^1/_4$ round Iron 4 ft 8 in long

2 lever shafts for raising brush frame $1^1/_8$ In Iron 5 Ft 2 in long

2 Gudgeons for wheel & axle brush motion

16 pulley studs for ditto

8 rods 5 in long $^3/_8$ diam to drive in the top of brush frame

4 short levers to raise brush frame six inches long from centres

2 long ones ditto long

4 studs screwed through the bottom of brush frame and resting on short levers

8 small gudgeons for the collar in front of brass reeds

4 Brass reeds wide long

4 gudgeons for clock collars

1 upright shaft from driven cone to Y Beam

1 short Horizontal ditto going from upright shaft to Y Beam

1 short slide crank 4 Fan gudgeons 8 Fan arms

Castings

1 wheel 17 in diam	96 teeth 2 stands for the short horizontal shafts	
1 ditto geering with it	10 do geering with yarn beam	
2 ditto	52	8 stands to hold wooden
	18	4 DELETION
	Shaft	1 do for the top

ILLEGIBLE

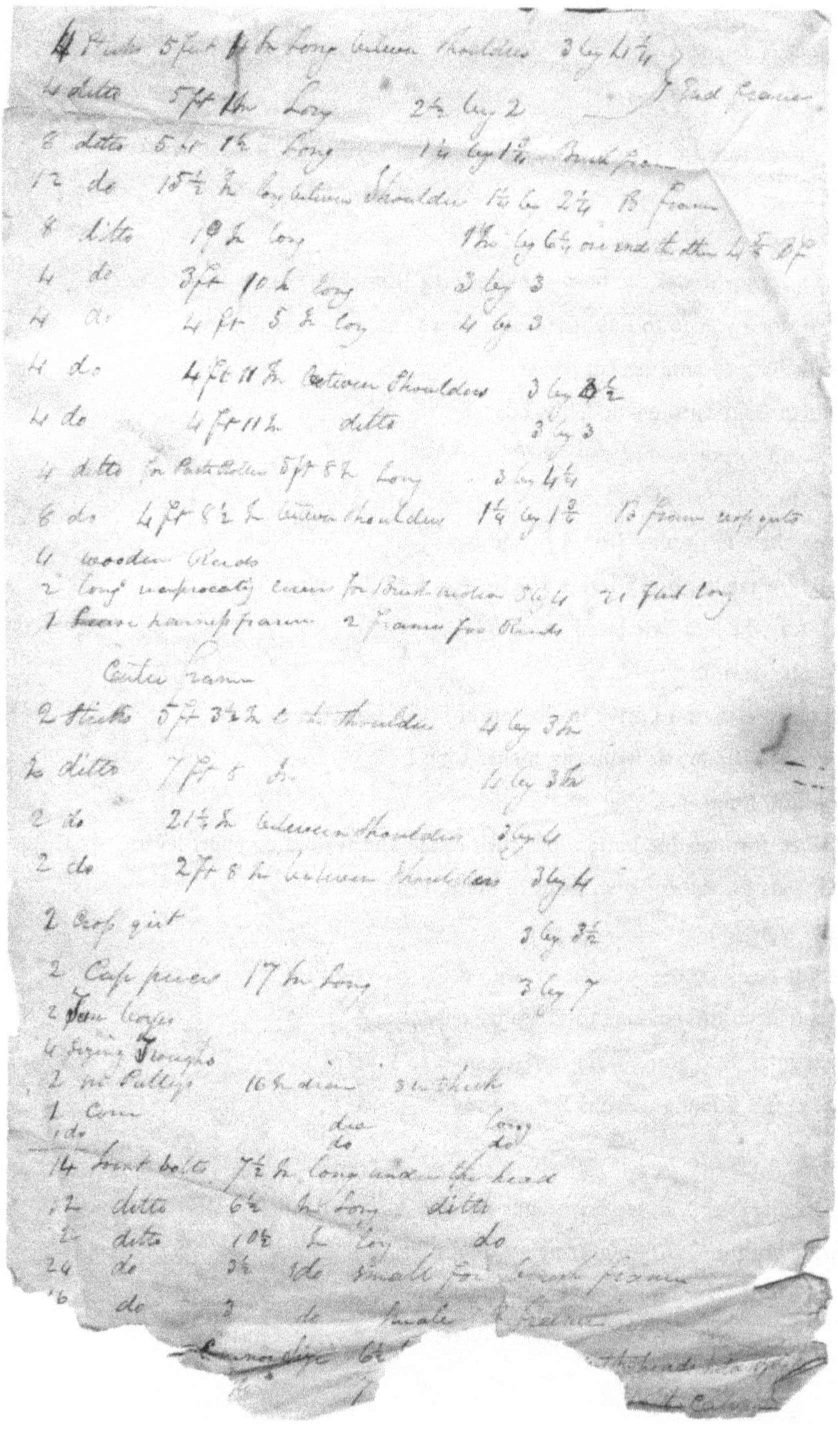
4 ditto 5 ft 1 in Long 2½ by 2
8 ditto 5 ft 1½ Long 1¼ by 1¾
12 do 15½ in long between Shoulders 1½ by 2¼ B frame
4 do 3 ft 10 in long 3 by 3
4 do 4 ft 5 in long 4 by 3
4 do 4 ft 11 in between Shoulders 3 by 3½
4 do 4 ft 11 in ditto 3 by 3
4 wooden Rods
Center frame
2 ditto 7 ft 5 in 4 by 3 in
2 do 21½ in between Shoulders 3 by 4
2 do 2 ft 8 in between Shoulders 3 by 4
2 Cap pieces 17 in long 3 by 7
4 Sizing Troughs
2 ditto 6½ in Long ditto
2 ditto 10½ in Long do

FIGURE 6.17B **Dressing machine, metal work, and wooden parts.**

Catalogue number: IEM 35.3b

Document Description

Measurement: 32.7 x 20.2 cm

Medium: Ink

Condition: Pieces torn off bottom of page; reverse of Figure 6.17A

VOOD]

4 sticks	5 ft 6(?) In long between shoulders	3 by 4 1/4	
4 ditto	5 ft 1 in Long	2 1/2 by 2	End frame
8 ditto	5 ft 1 1/2 Long	1 1/4 by 1 3/4	Brush frame
12 do	15 1/2 In long between shoulders	1 1/4 by 2 1/4	B frame
8 ditto	19 in long	1 In by 6 1/4 one end the other 4 5/8 B F	
4 do	3 ft 10 In long	3 by 3	
4 do	4 ft 5 In long	4 by 3	
4 do	4 ft 11 In between shoulders	3 by 4 1/2	
4 do	4 ft 11 do	3 by 3	
4 ditto	for paste rollers 5ft 8 In long	3 by 4 1/4	
8 do	4 ft 8 1/2 between shoulders	1 1/4 by 1 3/4 B frame cross guts (?)	
4	Wooden Reeds		
2	long reciprocating levers for Brush motion	3 by 4 21 feet long	
2	harness frames	2 frames for Reeds	
Center frame			
2 sticks	5 ft 3 1/2 In to the shoulder	4 by 3 In	
2 ditto	7 ft 8 In	4 by 3 In	
2 do	21 1/4 In between shoulders	3 by 4	
2 do	2 ft 8 In between shoulders	3 by 4	
2 cross gut (?)		3 by 3 1/2	
2 cap pieces	17 In long	3 by 7	
2 fan boxes			
4 sizing Troughs			
2 no (?) pulleys	16 In diam	3 in thick	
1	dia	long	
1 do	do	do	
14 Joint bolts	7 1/2 In long under the head		
32 ditto	6 1/2 In long	ditto	
2 ditto	10 1/2 In long	do	
24 do	3 1/2 do small for brush frame		
16 do	3 do small B (?) frame		

ILLEGIBLE and MISSING

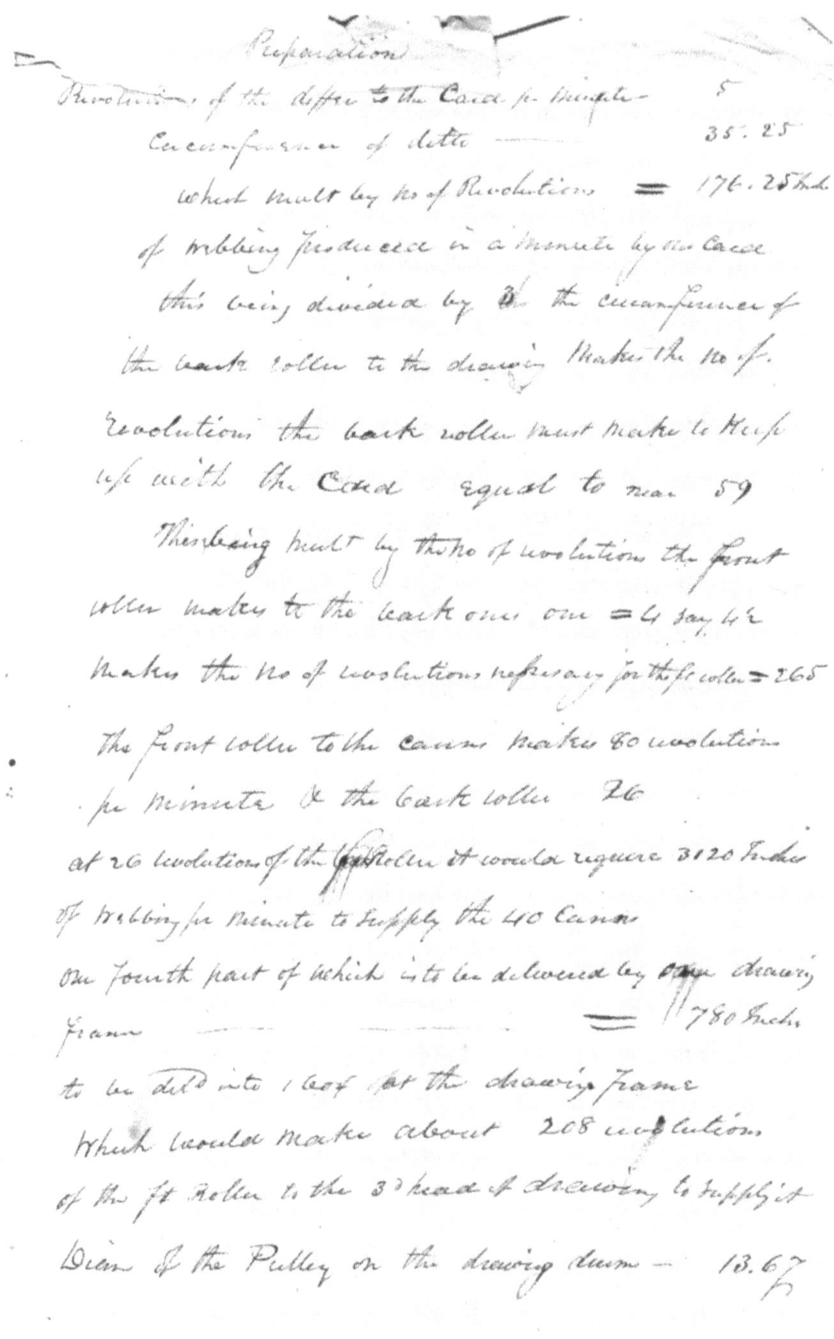

Preparation

Revolutions of the doffer to the Card per minute — 5

Circumference of ditto —— 35.25

which mult by No of Revolutions = 176.25 Inch

of webbing produced in a minute by the Card

this being divided by the the circumference of

the back roller to the drawing makes the No of

revolutions the back roller must make to keep

up with the Card equal to near 59

This being mult by the No of revolutions the front

roller makes to the back one one = 4 say 4½

makes the No of revolutions necessary for the ft roller = 265

The front roller to the cans makes 80 revolutions

per minute & the back roller 26

at 26 revolutions of the back Roller it would require 3120 Inches

of webbing per minute to supply the 40 Cans

One fourth part of which is to be delivered by one drawing

Frame —— = 780 Inches

to be del'd into 1 box at the drawing frame

Which would make about 208 revolutions

of the ft Roller to the 3d head of drawing to supply it

Diam of the Pulley on the drawing drum — 13.67

FIGURE 6.18 Carding machine.

Catalogue number: IEM 35.6

Document Description

Measurement: 33 x 20.2 cm

Medium: Ink

Condition: Some tears at top of page

Transcription

Preparation

Revolutions of the doffer to the card per minute 5

Circumference of ditto 35.25

Which multiplied by number of revolutions = 176.25 inches
of webbing produced in a minute by one card, this being divided by 3, the circumference of the back roller to the drawing makes the number of revolutions the back roller must make to keep up with the card equal to near 59

This being multiplied by the number of revolutions the front roller makes to the back one = 4 say $4^1/_2$

Makes the number of revolutions necessary for a front roller = 265

The front roller to the cans makes 80 revolutions per minute and the back roller 26

At 26 revolutions of the back roller it would require 3120 inches of webbing per minute to supply the 40 cans

One fourth part of which is to be delivered by one drawing frame = 780 inches

to be delivered into 1 box at the drawing frame

Which would make about 208 revolutions of the front roller to the 3rd head of drawing to supply it

Diameter of the Pulley on the drawing drum 13.67

FIGURE 6.19 Lap winder.

Catalogue number: IEM 35.2d

Document Description

Measurement: Full sheet 40.5 x 24 cm, folded in half; includes Figures 6.16, 6.19, and 6.25

Medium: Ink

Condition: Stains

Transcription

2 Sticks	4 ft 3 In long	4 by 2 3/4 In long
2 do	3 feet long	4 by 2 3/4
4 Gerts (?)	2 ft Long	
4 Lags	5 feet long for 6 In cylinder	
4 long Gerts (?)	5 ft 6 In long	3 by 4 Timber
[3 of them	3 by 4	one 2 by] DELETED

pulleys for cones

Lap Winder 9 ft 3 In long 26 In wide 2 ft 6 In high

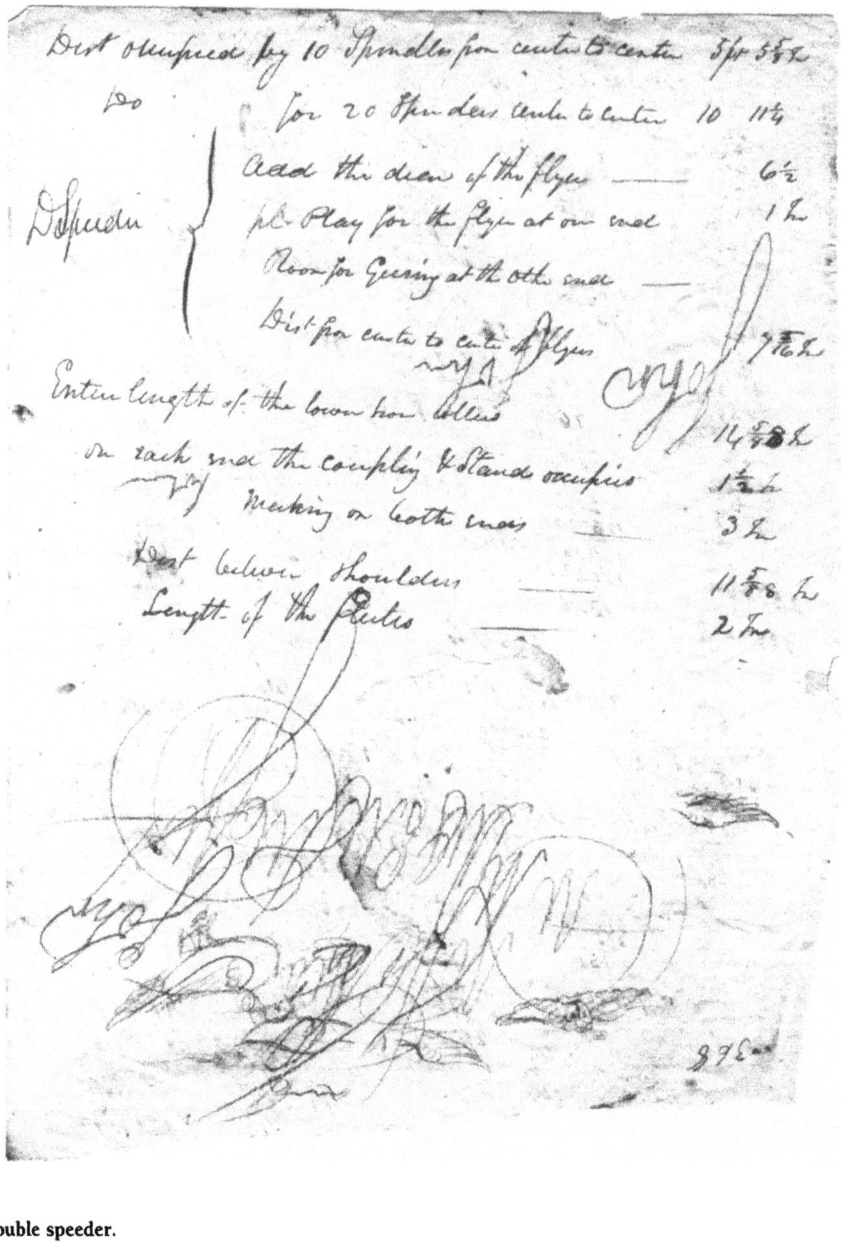

FIGURE 6.20A Double speeder.

Catalogue number: IEM 35.4a

Document Description

Measurement: 23 x 18 cm

Medium: Ink and pencil

Condition: Stains, torn edges; on reverse of Figure 6.20B

Transcription

Dist occupied by 10 Spindles from center to center		5 ft	$5\frac{5}{8}$ In
Do	for 20 Spindles center to center	10	$11\frac{1}{4}$
D[ouble]Speeder	add the diam of the flyer		$6\frac{1}{2}$
	Play for the flyer at one end	1 In	
	Room for Geering at the other end		
	Dist from center to center of flyers		$7\frac{5}{16}$
Entire length of the lower iron collars			$14\frac{5}{8}$ In
On each end the coupling & stand occupies			$1\frac{1}{2}$ In
Making on both ends			3 In
Dist between shoulders			$11\frac{5}{8}$ In
Length of the flutes (?)			2 In

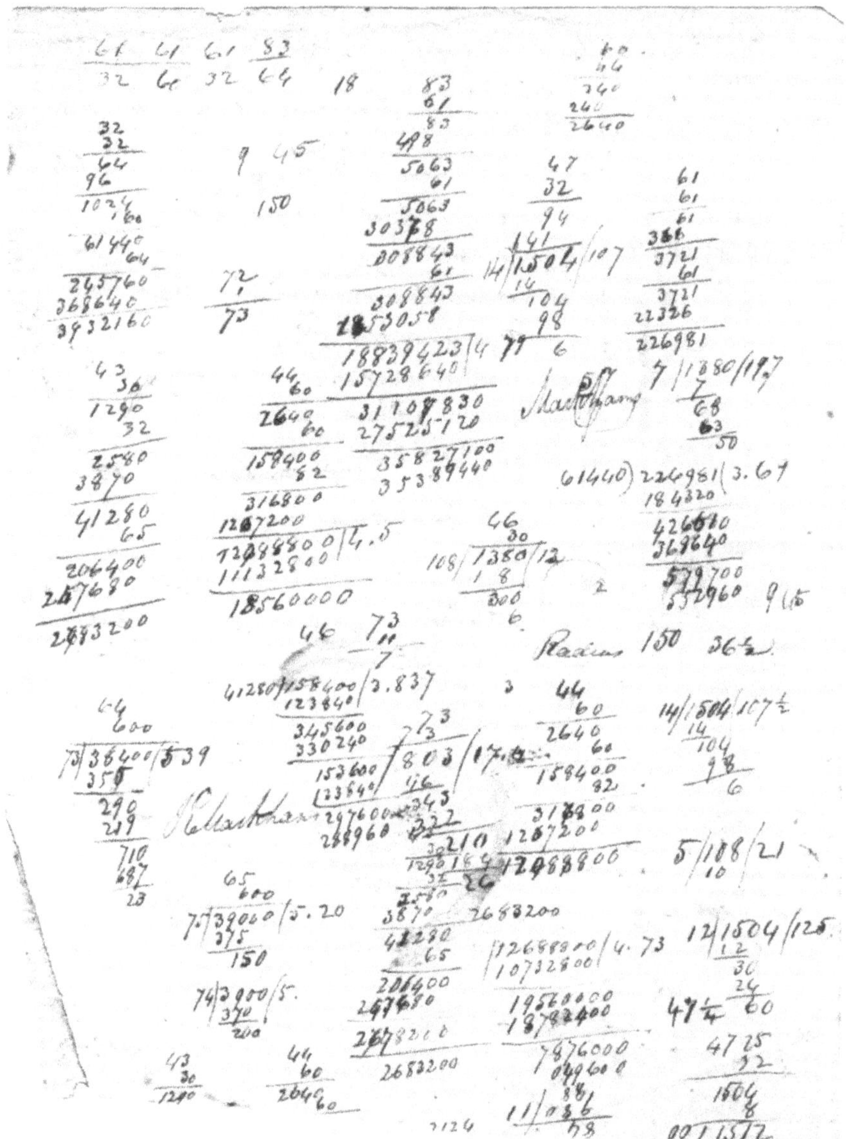

FIGURE 6.20B Rough calculations.

Catalogue number: IEM 35.4b

Document Description

Measurement: 23 x 18 cm

Medium: Ink and pencil

Condition: Stains, torn edges; on reverse of Figure 6.20A

Transcription

No transcription has been attempted.

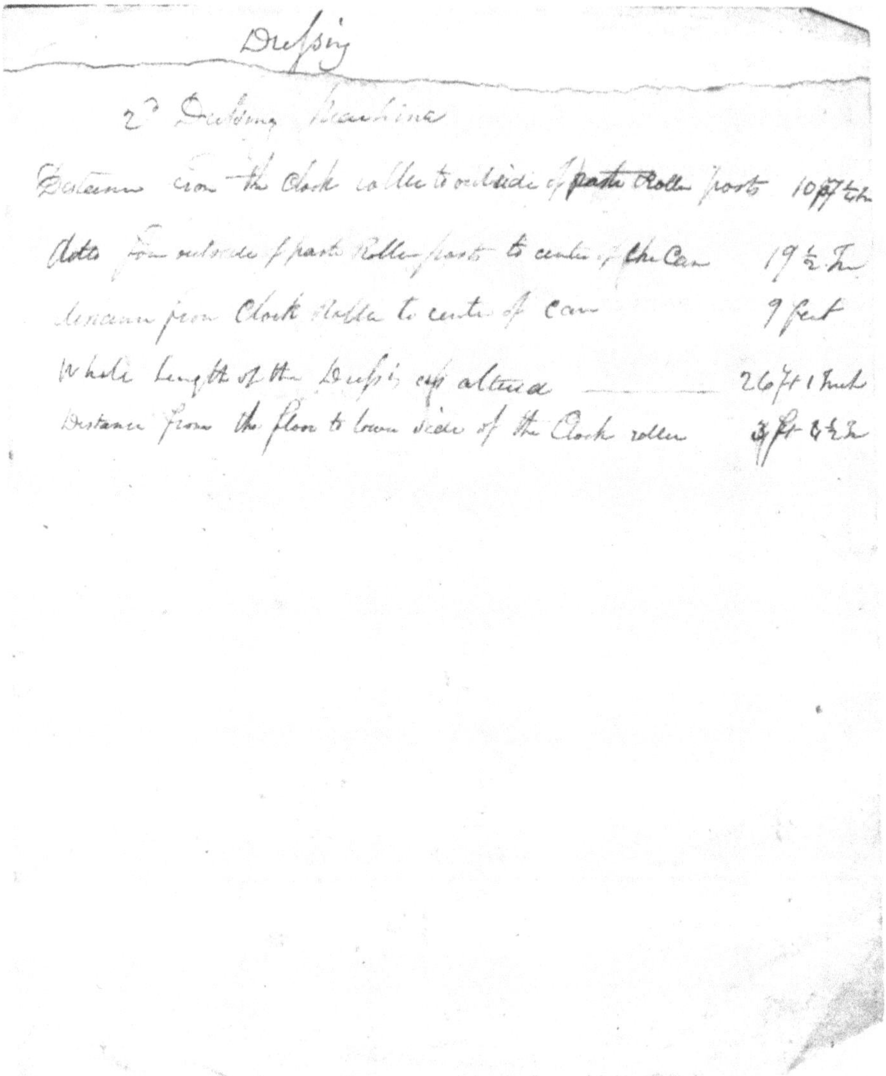

Dressing

2d Dressing Machine

Distance from the Clock roller to outside of ~~paste Roller~~ posts 10 ft 4 In

Ditto from outside of paste Roller posts to center of the Cam 19½ In

Distance from Clock Roller to center of Cam 9 feet

Whole length of the Dressing as altered 26 ft 1 Inch

Distance from the floor to lower side of the Clock roller 3 ft 8½ In

FIGURE 6.21 Dressing machine.

Catalogue number: IEM 35.5b

Document Description

Measurement: Full sheet 40.5 x 24 cm, folded in half; includes Figure 6.22

Medium: Ink

Condition: Stain, creases; torn across top of page

Transcription

2nd Dressing Machine

Distance from the clock roller to outside of paste roller posts	10 ft $7^1/_4$ in
Ditto from outside of paste roller posts to centre of the cam	$19^1/_2$ in
Distance from clock roller to centre of cam	9 ft
Whole length of the dressing frame as altered	26 ft 1 inch
Distance from the floor to lower side of the clock roller	3 ft $4^1/_2$ in

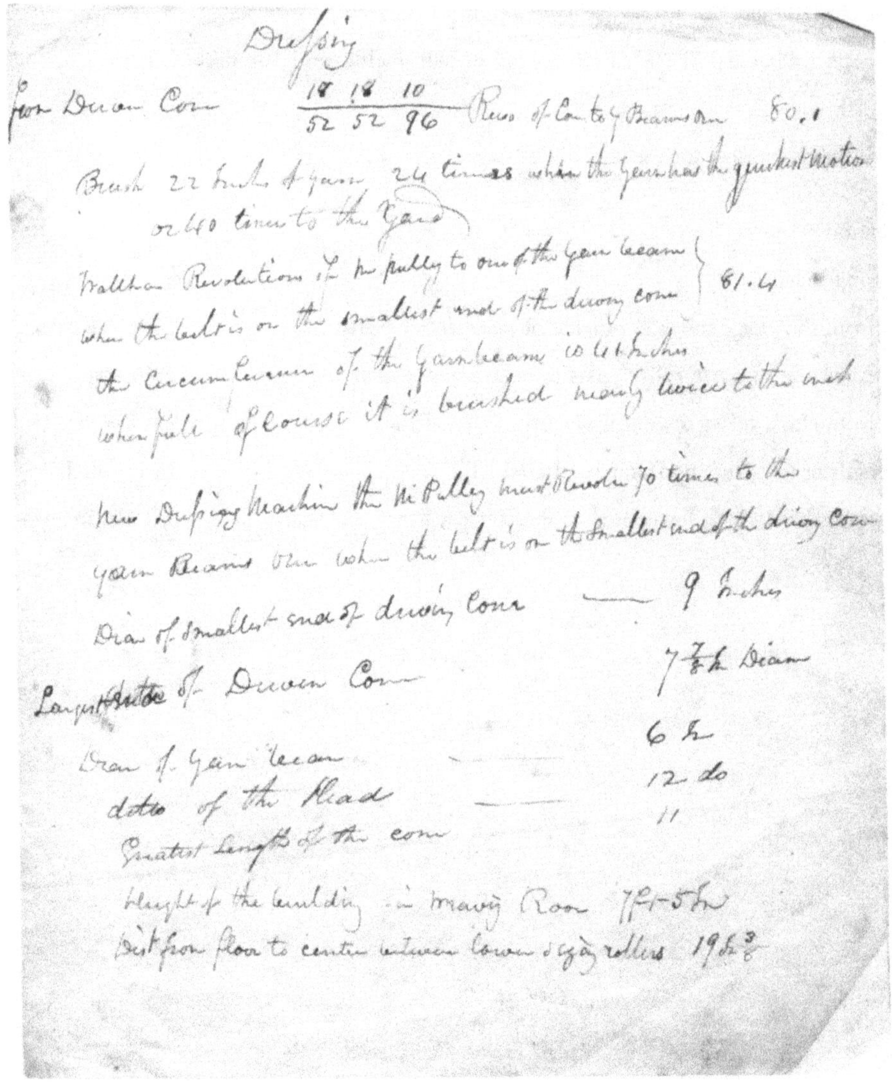
Dressing

from Drum Cone 18 18 10 / 52 52 96 Revs of Cone to Yarn Beam once 80.1

Brush 22 inches of yarn 24 times when the Yarn has the quickest motion or 40 times to the Yard

Waltham Revolution of the pulley to one of the yarn beam when the belt is on the smallest end of the drum cone } 81.4

the Circumference of the yarnbeam is 44 inches when full of course it is brushed nearly twice to the inch

New Dressing Machine the ... Pulley must Revolve 70 times to the yarn Beams once when the belt is on the smallest end of the drum Cone

Dia of smallest end of driving Cone —— 9 Inches

Largest ... of Driven Cone 7 7/8 in Diam

6 In

Dia of Yarn beam —— 12 do

ditto of the Head —— 11

Greatest Length of the cone

Height of the building in Weaving Room 7 ft 5 in

Dist from floor to center between lower dressing rollers 19 In 3/8

FIGURE 6.22 Dressing machine.

Catalogue number: IEM 35.5a

Document Description

Measurement: Full sheet 40.5 x 24 cm; folded in half; includes Figure 6.21

Medium: Ink

Condition: Stain, creases

Transcription

Dressing			
From Dover Co	18 18 10	[Revs?] of cone to Y Beams	80.1
		own	
	52 52 96		

Brush 22 inches of yarn 24 times when the yarn has the quickest motion or 40 times to the yard

Waltham revolutions of the pulley to one of the yarn beam
when the belt is on the smallest end of the driving cone
the circumference of the yarn beam is 41 inches
when full of course it is brushed nearly twice to the inch } 81.4

New dressing machine the [?] pulley must revolve 70 times to the yarn beam's one when the belt is on the smallest end of the driving cone

Diameter of the smallest end of driving cone	9 inches
Largest end [??] of driving cone	$7^7/_8$ in diam
Diamenter of yarn beam	6 in
Ditto of the head	12 ditto
Greatest length of the cone	11
Height of the building in weaving room	7 ft 5in
Distance from floor to centre between lower of sizing rollers	19in 3/8

Costs of Building Machines

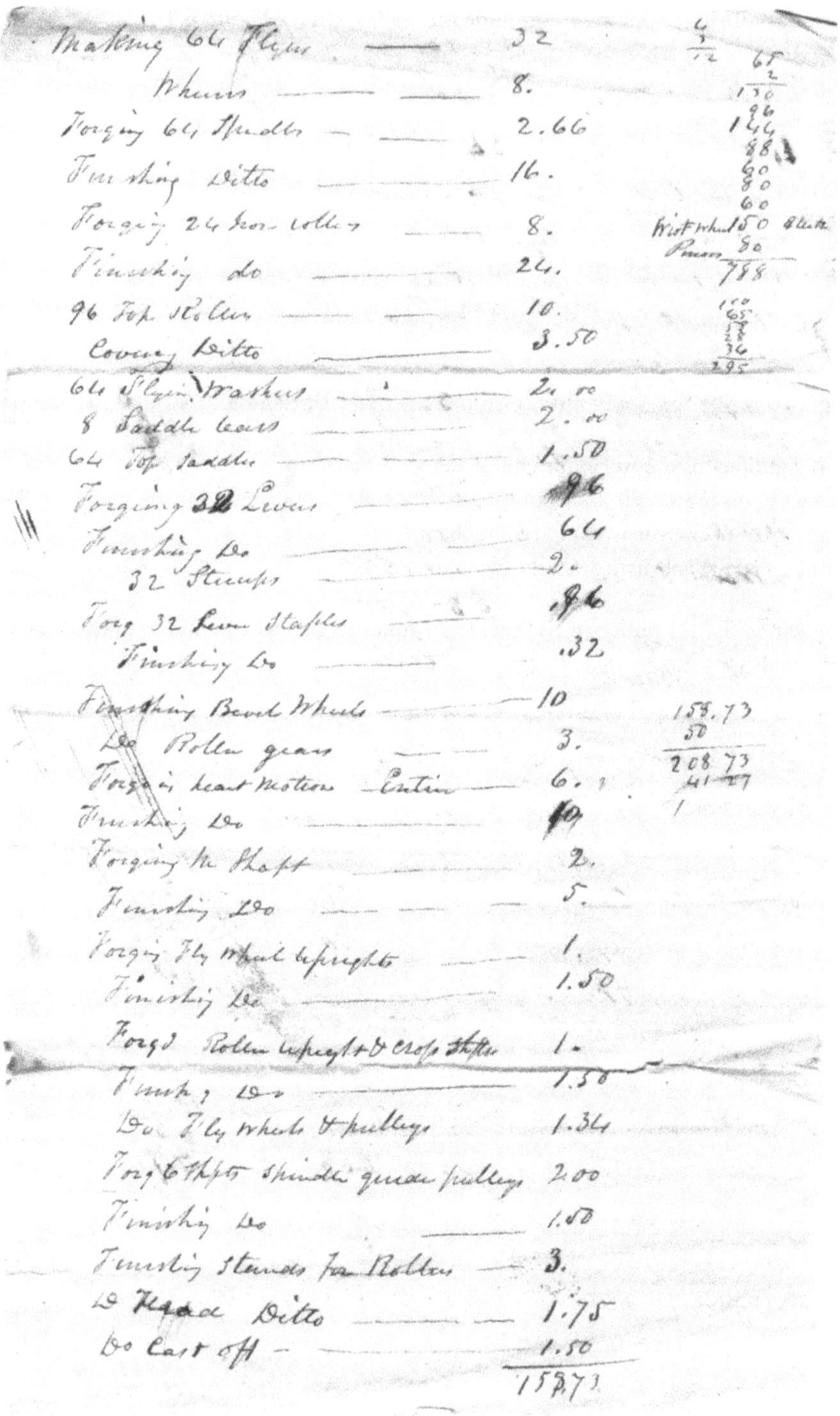

Making 64 Flyers	32
Wheels	8.
Forging 64 Spindles	2.66
Finishing Ditto	16.
Forging 24 Iron rollers	8.
Finishing do	24.
96 Top Rollers	10.
Covering Ditto	3.50
64 Flyer Washers	2.00
8 Saddle bars	2.00
64 Top saddles	1.50
Forging 32 Levers	.96
Finishing Do	64
32 Stirrups	2.
Forg 32 Lever Staples	.96
Finishing Do	.32
Finishing Bevel Wheels	10
Do Roller gears	3.
Forging heart Motion Entire	6.
Finishing Do	10
Forging M Shaft	2.
Finishing Do	5.
Forging Fly wheel upright	1.
Finishing Do	1.50
Forgd Roller upright & cross stiffs	1.
Finishing Do	1.50
Do Fly wheels & pulleys	1.34
Forg 6 Shafts spindle guide pulleys	2.00
Finishing Do	1.50
Finishing Stands for Rollers	3.
Do Head Ditto	1.75
Do Cast off	1.50
	15[illegible]73

158.73
50
208.73

FIGURE 6.23A Spinning frame.

Catalogue number: IEM 35.7a

Document Description

Measurement: 32.5 x 20 cm

Medium: Ink

Condition: Tears; on reverse of Figure 6.23B

Transcription

Making 64 Flyers	32
Whurrs (?)	8
Forging 64 spindles	2.66
Finishing Ditto	16
Forging 24 iron rollers	8
Finishing do	24
96 top rollers	10
covering Ditto	3.50
64 Flyer Washers	2.00
8 saddle bars	2.00
64 top saddles	1.50
Forging 32 Levers	9.6
Finishing Do	64
32 Stirups	2
Forging 32 Lever staples	0.96
Finishing Do	0.32
Finishing Bevel wheels	10
Do Roller gears	3
Forging heart Motion, entire	6
Finishing Ditto	10
Forging the Shaft	2
Finishing Do	5
Forging Fly wheel uprights	1
Finishing Do	1.50
Forging Roller upright & cross shaft	1
Finishing Do	1.50
Do Flywheels & pulleys	1.34
Forging 6 shfts spindle guide pulleys	2
Finishing Do	1.50
Finishing stands for Rollers	3
D Head Ditto	1.75
Do cast off	1.50
	158.73

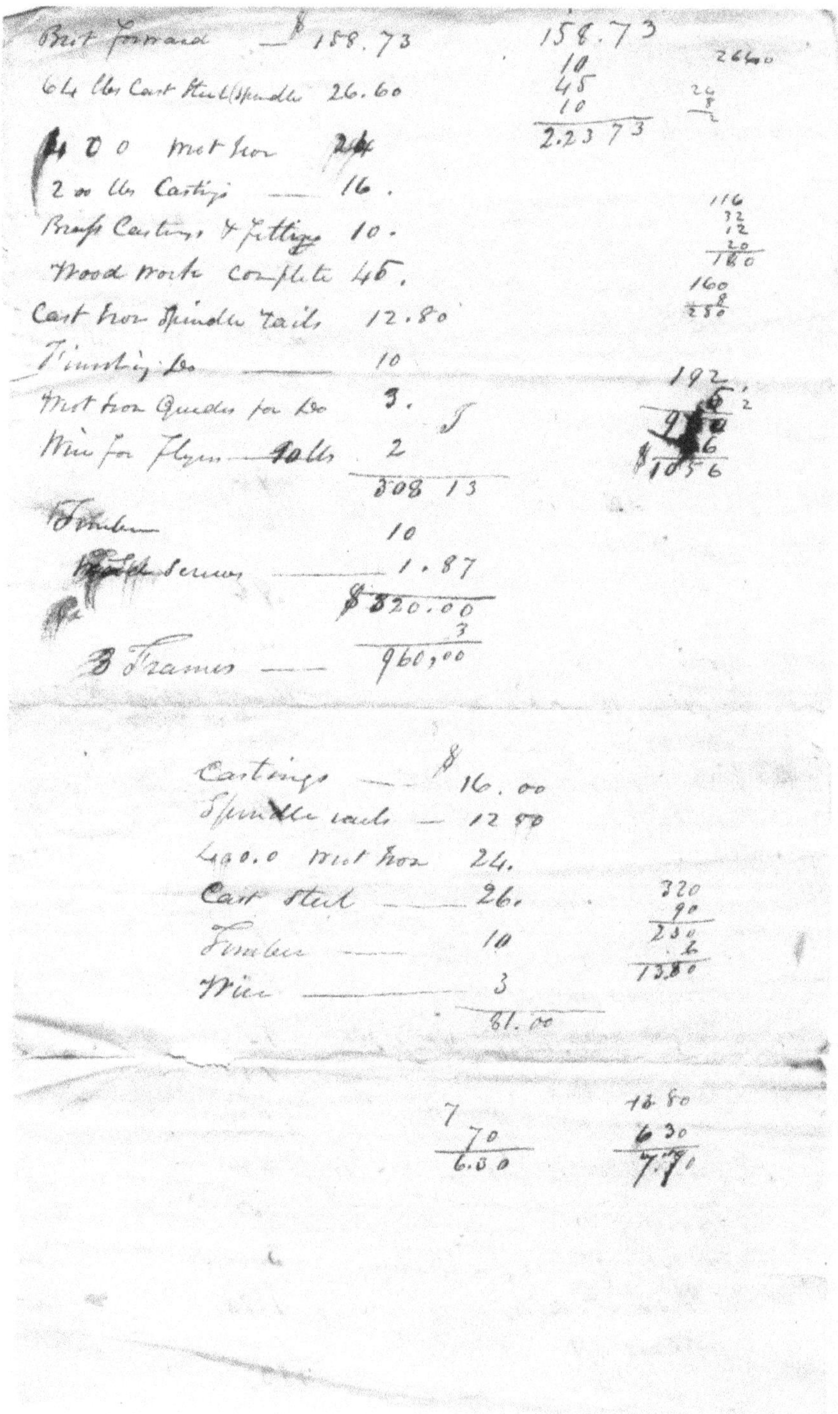

Brt forward — $ 158.73
64 lbs Cast Steel Spindles 26.60
400 Wrot Iron 24
200 lbs Castings — 16.
Brass Castings & fittings 10.
Wood work complete 46.
Cast Iron Spindle rails 12.80
Finishing Do — 10
Wrot Iron Guides for Do 3.
Wire for Flyers 40 lbs 2
308 13
Timber 10
Screws — 1.87
$320.00
3
3 Frames — 960,00

Castings — $ 16.00
Spindle rails — 12 80
400.0 Wrot Iron 24.
Cast steel — 26.
Timber — 10
Wire — 3
81.00

FIGURE 6.23B Spinning frame.

Catalogue number: IEM 35.7b

Document Description

Measurement: 32.5 x 20 cm

Medium: Ink

Condition: Tears; on reverse of Figure 6.23A

Transcription

Brot forward	158.73
64 lbs cast steel (spindles)	26.60
400 wrot iron	24
200 lbs castings	16
Brass castings & fitting	10
Wood work complete	46
Cast iron spindle rails	12.80
Finishing Do	10
Wrot iron guides for Do	3
Wire for Flyers 90 lbs	2
	308.13
Timber	10
Wood & screws	1.87
	320
	3
3 Frames	960.00
Castings	16.00
Spindle rails	12.00
400 wrot iron	24.00
Cast steel	26.00
Timber	10.00
Wire	3.00
	81.00

Output of Machines

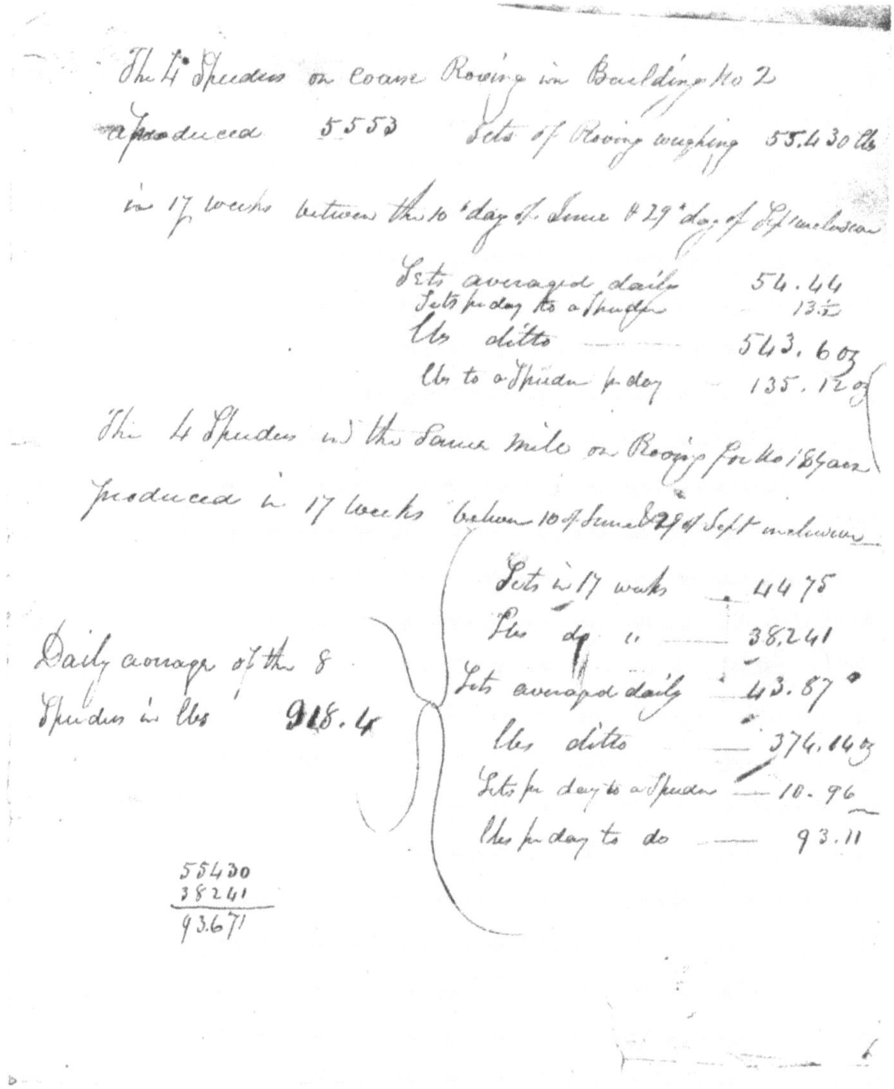

The 4 Speeders on Coarse Roving in Building No 2
produced 5553 Sets of Roving weighing 55.430 lbs
in 17 weeks between the 10 day of June & 29 day of Sept inclusive

Sets averaged daily	54.44
Sets per day to a Speeder	13½
lbs ditto	543.6 oz
lbs to a Speeder per day	135.12 oz

The 4 Speeders in the same mill on Roving for No 18 yarn
produced in 17 weeks between 10 of June & 29 of Sept inclusive

Sets in 17 weeks	4475
lbs do "	38.241
Sets averaged daily	43.87
lbs ditto	374.14 oz
Sets per day to a Speeder	10.96
lbs per day to do	93.11

Daily average of the 8 Speeders in lbs 918.4

55430
38241
93671

FIGURE 6.24A Speeder.

Catalogue number: IEM 35.8a

Document Description

Measurement: 24 x 20.4 cm

Medium: Ink

Watermark: E Burbank

Condition: Slight tears along edges, stains; on reverse of Figure 6.24B

Transcription

The 4 Speeders on a coarse Roving in Building No 2 produced 5,553 sets of Roving weighing 55,430 lbs in 17 weeks between the 10th day of June & 29th day of September inclusive.

Sets averaged daily	54.44
Sets per day to a speeder	13½
lbs ditto	543.6 oz
lbs to a speeder per day	135.12 oz

The 4 speeders in the same mill on Roving for No. 18 yarn produced in 17 weeks between 10 of June and 29 of Sept inclusive:

Sets in 17 weeks	4475
lbs ditto	38,241
Sets averaged daily	43.87
lbs ditto	374.14 oz
Sets per day to a speeder	10.96
lbs per day to do	93.11

Daily average of the 8 speeders in lbs: 918.4

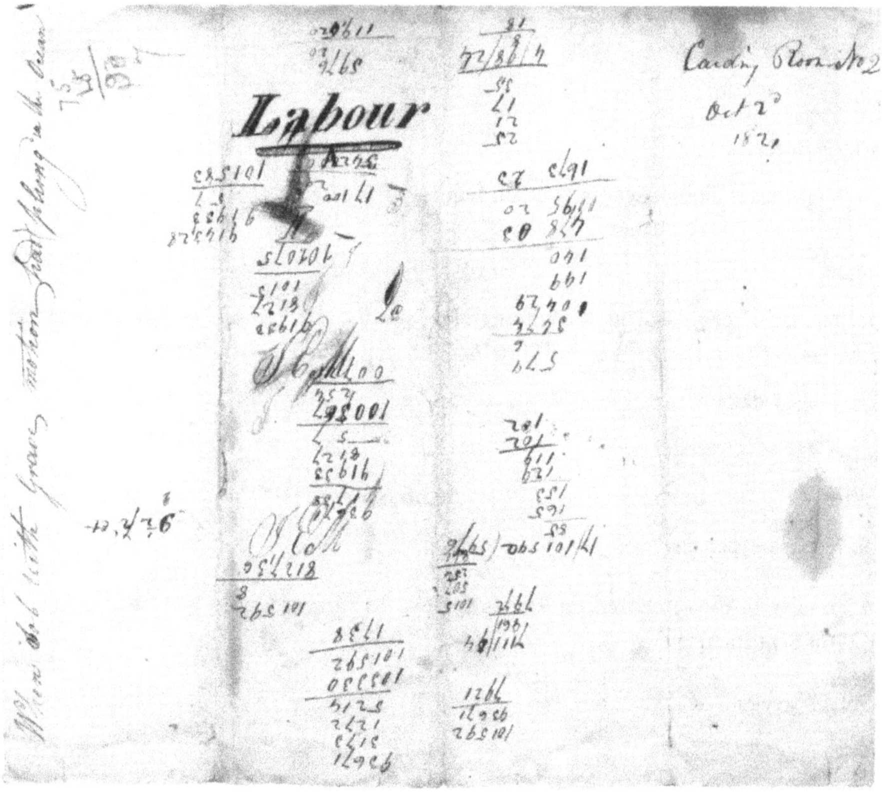

FIGURE 6.24B Loose sheet.

Catalogue number: IEM 35.8b

Document Description

Measurement: 24 x 20.4 cm

Medium: Ink

Watermark: E Burbank

Condition: Slight tears along edges, stains; reverse of Figure 6.24A

Transcription

Carding Room No 2
Oct 2^{d} 1821

Calculations: No transcription has been attempted.

When Job with Grave motion had flung in the Ocean

Commentary

Was this last bit of rhyme scribbled by Markham in a distracted moment or was it from a poem well known at the time?

FIGURE 6.25 Doodles.

Catalogue number: IEM 35.2b

Document Description

Measurement: Full sheet 40.5 x 24 cm, folded in half; includes Figures 6.16, 6.19, and 6.25

Medium: Pencil, ink

Condition: Stains

2. SKETCHBOOK

The evidence that Markham owned and compiled the Sketchbook is that it was found with the other Markham drawings, that the watermarks on the paper were in use during Markham's working life, that the same drawing techniques were used, and that the Sketchbook appears to have been created by the same hand that executed the workshop drawings.

It is a remarkable record of engineering iconography. Comprising over 100 miniature sketches of elevations, an occasional plan, outlines, and three-dimensional views of textile machines and machine parts, power systems and components, the sketches are reminiscent of the pages of machines and machine parts to be found in the encyclopedias and mechanics' handbooks of the late eighteenth and early nineteenth centuries. Yet these are different. They are free-hand sketches. No scale, no dimensions are given. Items are out of proportion with each other and they are unidentified. On the other hand, each page treats a separate processing stage following for the most part the order of factory production (carding engine, roving frame, dresser, power loom) and provides sketches of key components. So the sketches were not randomly compiled or assembled haphazardly. The ordered sequence implies that the volume had some use as a reference work to which the author could return, perhaps as evidence of training or skill, or to show a potential customer. Many subsidiary components are clearly made of metal, which Markham became skilled in working

Were these obligatory exercises made by a trainee mechanic? Were they openly sketched or secretly prepared in the draftsman's spare time? Are they careful observations of operating machines? Or are some drawings imaginative and fanciful, such as the bottom sketch on Figure 6.28 (see p. 190) (with two sets of twin cones operating in conjunction with heart-wheel cams), or the top sketch on Figure 6.31 (p. 196) (which seems to have only one cone pulley and puzzling diagonal arms behind the framing)? Do they depict machines at Waltham or Dover or Middlebury?

Whatever their origin, whatever their purpose, these sketches hold a delight of their own. Choice miniatures, freshly and sharply sketched, they are informed by the eye of an artist, in their execution and their balanced composition on each page. The question remains: Do these sketches record the early-nineteenth-century New England mechanic as observer or as dreamer? In view of the manner in which early-nineteenth-century American engineers like Latrobe and Markham organized their design and production operations (see Appendix 5), it is likely that this was the first type of design document, a record of imaginative ideas, though it may have been begun while Markham was still training.

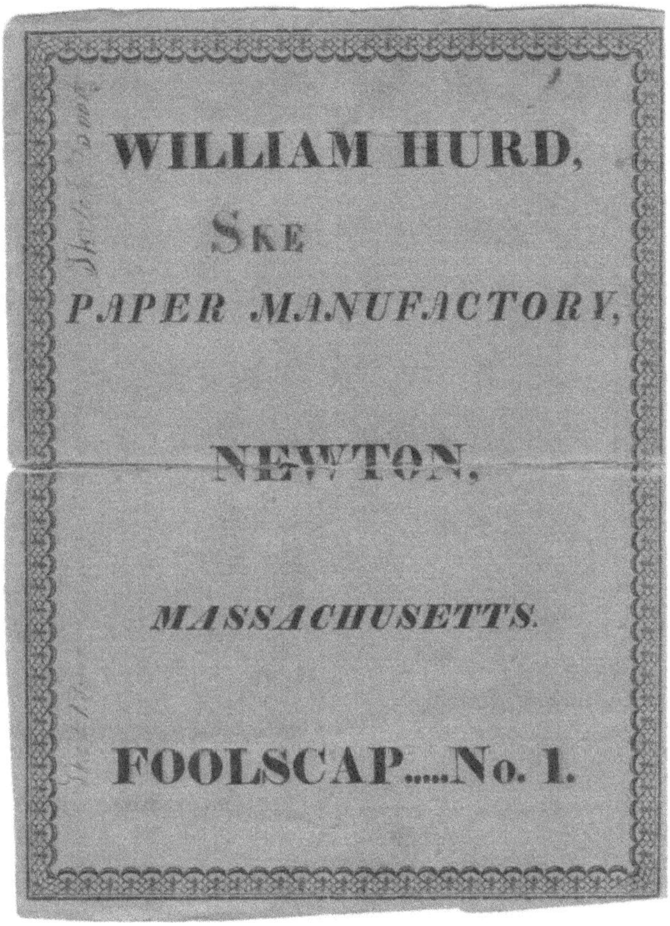

FIGURE 6.26 Sketchbook, cover.
Catalogue number: IEM 28

Document Description

Measurement: 25 x 20 cm

Medium: Ink

Condition: Folded in half and sewn through center, shown opened up.

Commentary

If Hurd was the paper manufacturer who sold sketchbooks such as this, did he usually print his name to fill the cover?

The owner of the book clearly began to write "sketchbook" in capital letters and then decided to place the words in the top left corner. The latter endorsement looks like Markham's hand.

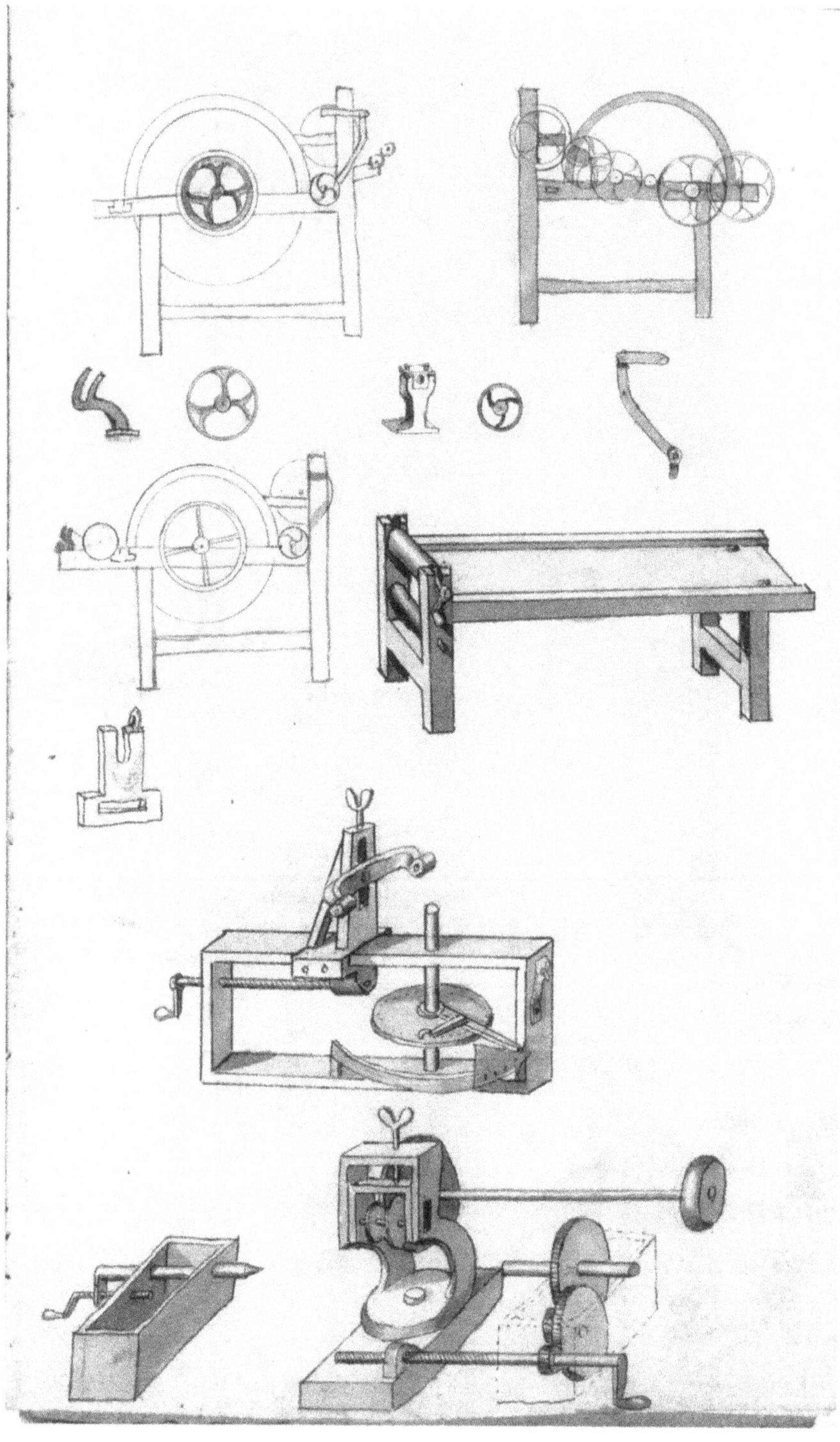

FIGURE 6.27 Sketchbook, page 1. Carding machine sketches.

Catalogue number: IEM 28.1

Document Description

Measurement: 20 x 12.4 cm

Medium: Ink and wash, some pencil

Watermark: Dove

Commentary

Sketches of 13 machine components and mechanisms. The upper part of the page relates to carding machines (*top, middle*).

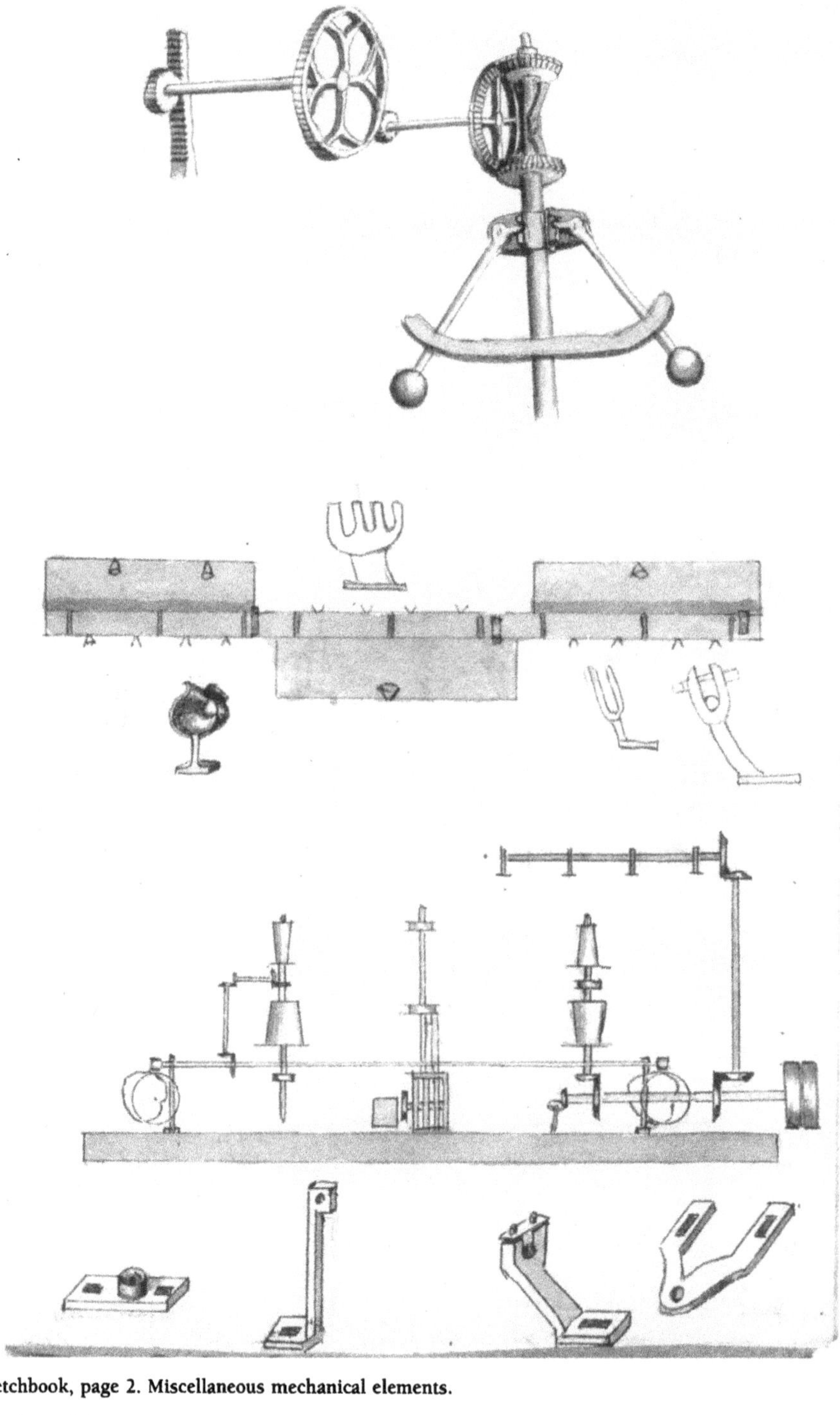

FIGURE 6.28 Sketchbook, page 2. Miscellaneous mechanical elements.

Catalogue number: IEM 28.2

Document Description

Measurement: 20 x 12.4 cm

Medium: Ink and wash, erased pencil marks

Watermark: Dove

Commentary

Sketches of 11 machine components and mechanisms for a factory power system, including a governor (*top*) and stands for drafting rollers (*middle*).

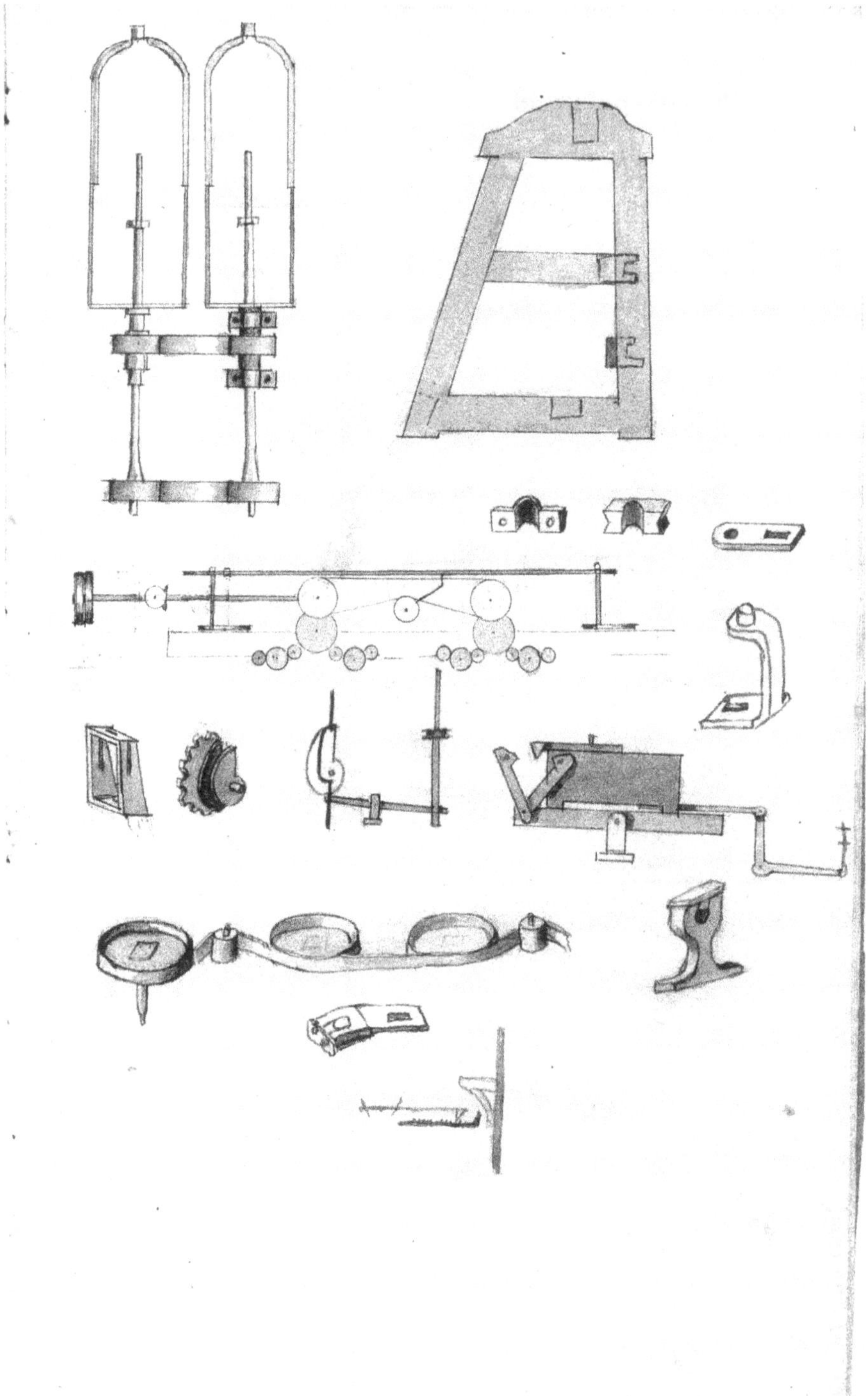

FIGURE 6.29 Sketchbook, page 3. Roving-frame sketches.

Catalogue number: IEM 28.3

Document Description

Measurement: 20 x 12.4 cm

Medium: Ink and wash

Watermark: Dove

Commentary

Sketches of 15 machine components and mechanisms, relating to the roving frame, including two flyer and spindle arrangements (*top*).

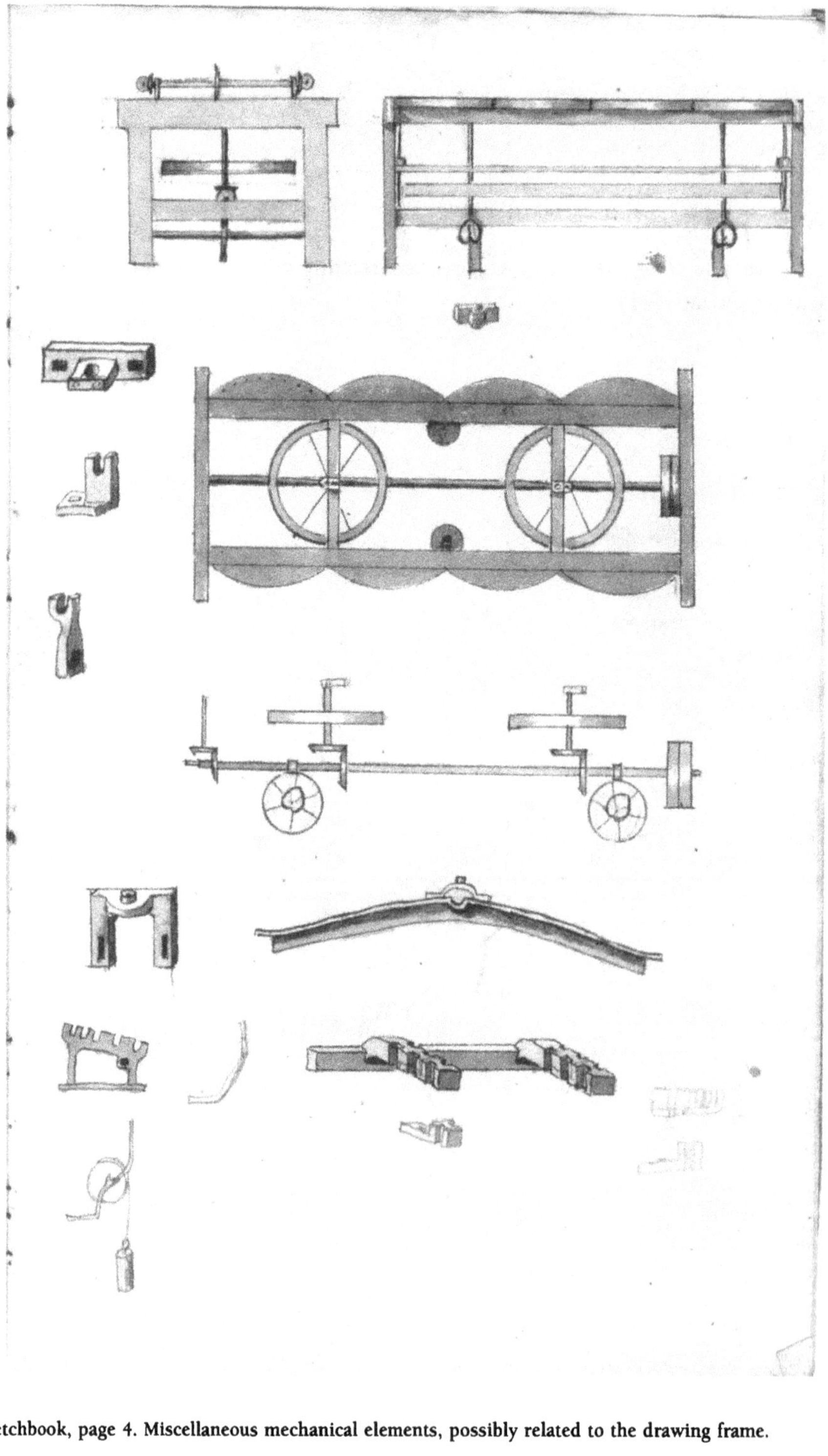

FIGURE 6.30 Sketchbook, page 4. Miscellaneous mechanical elements, possibly related to the drawing frame.

Catalogue number: IEM 28.4

Document Description

Measurement: 20 x 12.4 cm

Medium: Ink, wash and pencil

Watermark: "C Baker & Co"

Commentary

Sketches of 15 machine components and mechanisms, possibly for a drawing frame (*top, middle*).

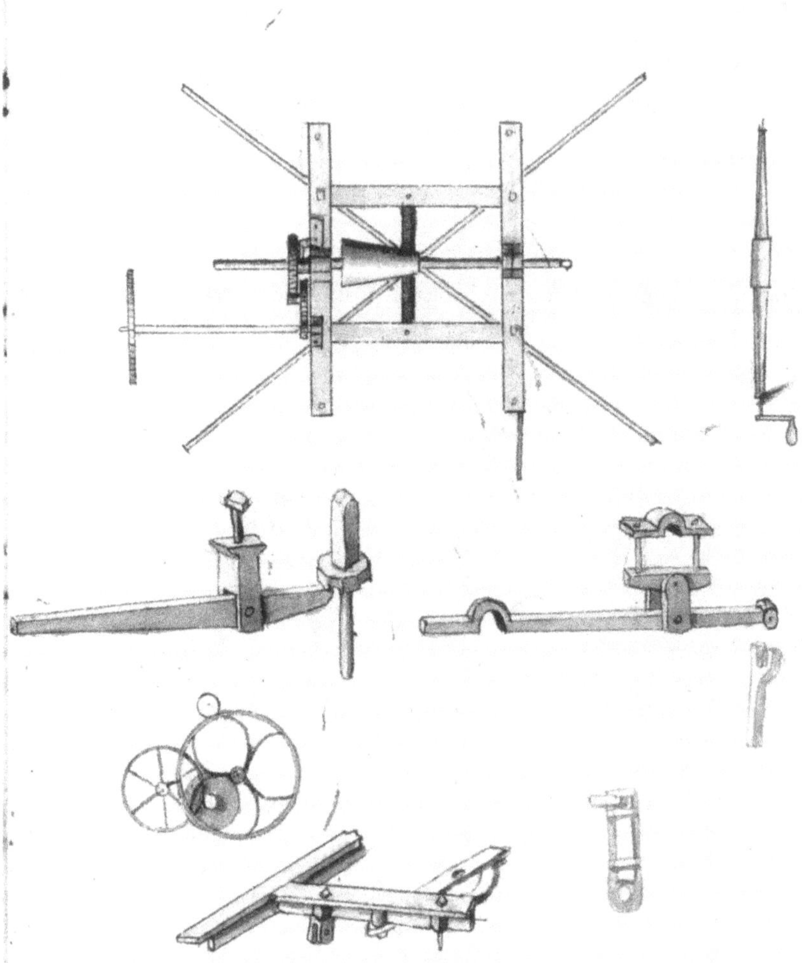

FIGURE 6.31 Sketchbook, page 5. Miscellaneous mechanical elements.

Catalogue number: IEM 28.5

Document Description

Measurement: 20 x 12.4 cm

Medium: Ink and wash, some pencil

Watermark: "C Baker & Co"

Condition: Smears

Commentary

Sketches of eight machine components and mechanisms, including a single-cone pulley (*top*).

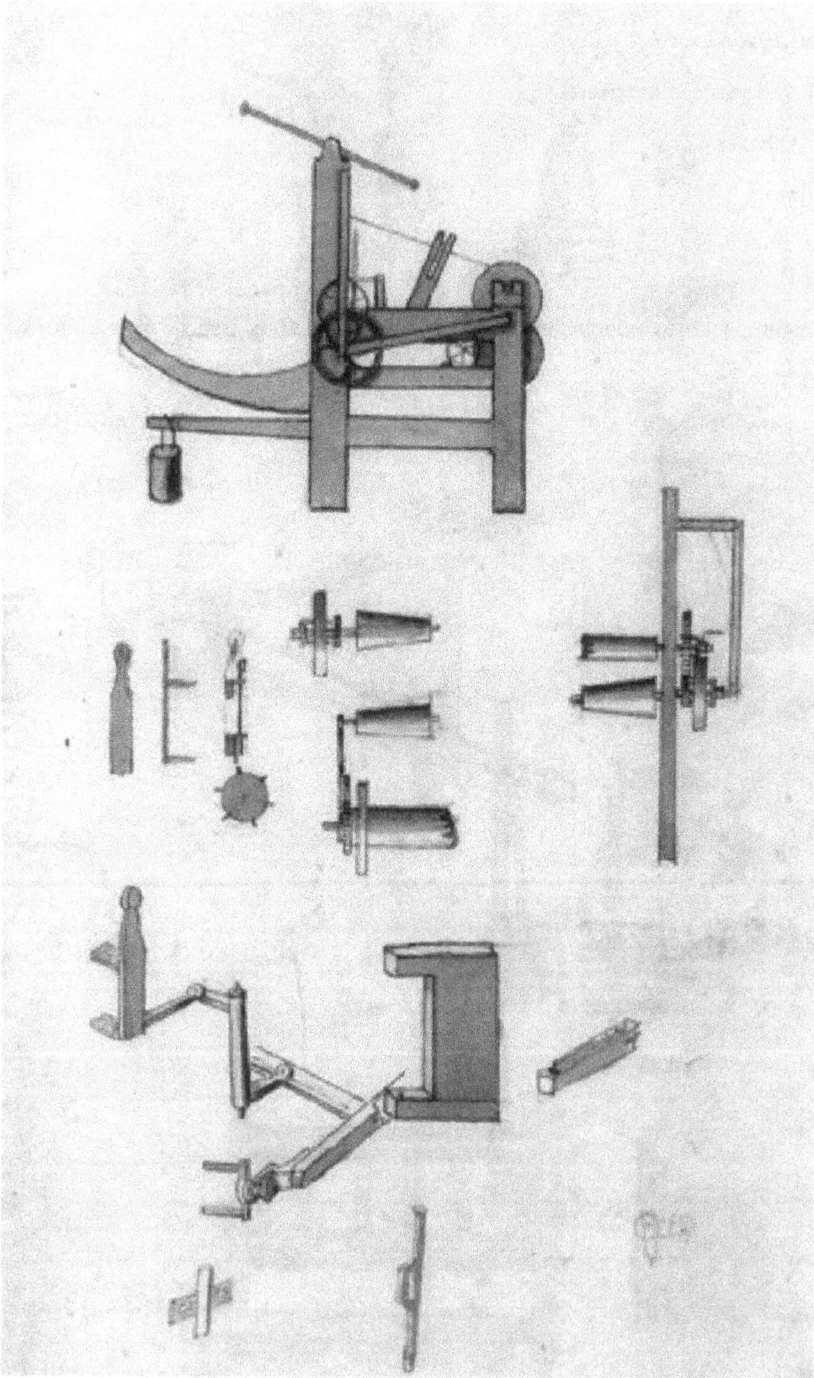

FIGURE 6.32 Sketchbook, page 6. Power-loom mechanisms.

Catalogue number: IEM 28.6

Document Description

Measurement: 20 x 12.4 cm

Medium: Ink, wash and pencil

Watermark: "C Baker & Co"

Commentary

Sketches of 12 machine components and mechanisms, relating to a power loom, including the side elevation of a power loom (*top*) with a curious shedding motion and an odd curved arm (*left*). The lever arrangement may be related to the beating-up motion seen in the power loom at Figure 6.34 (p. 202).

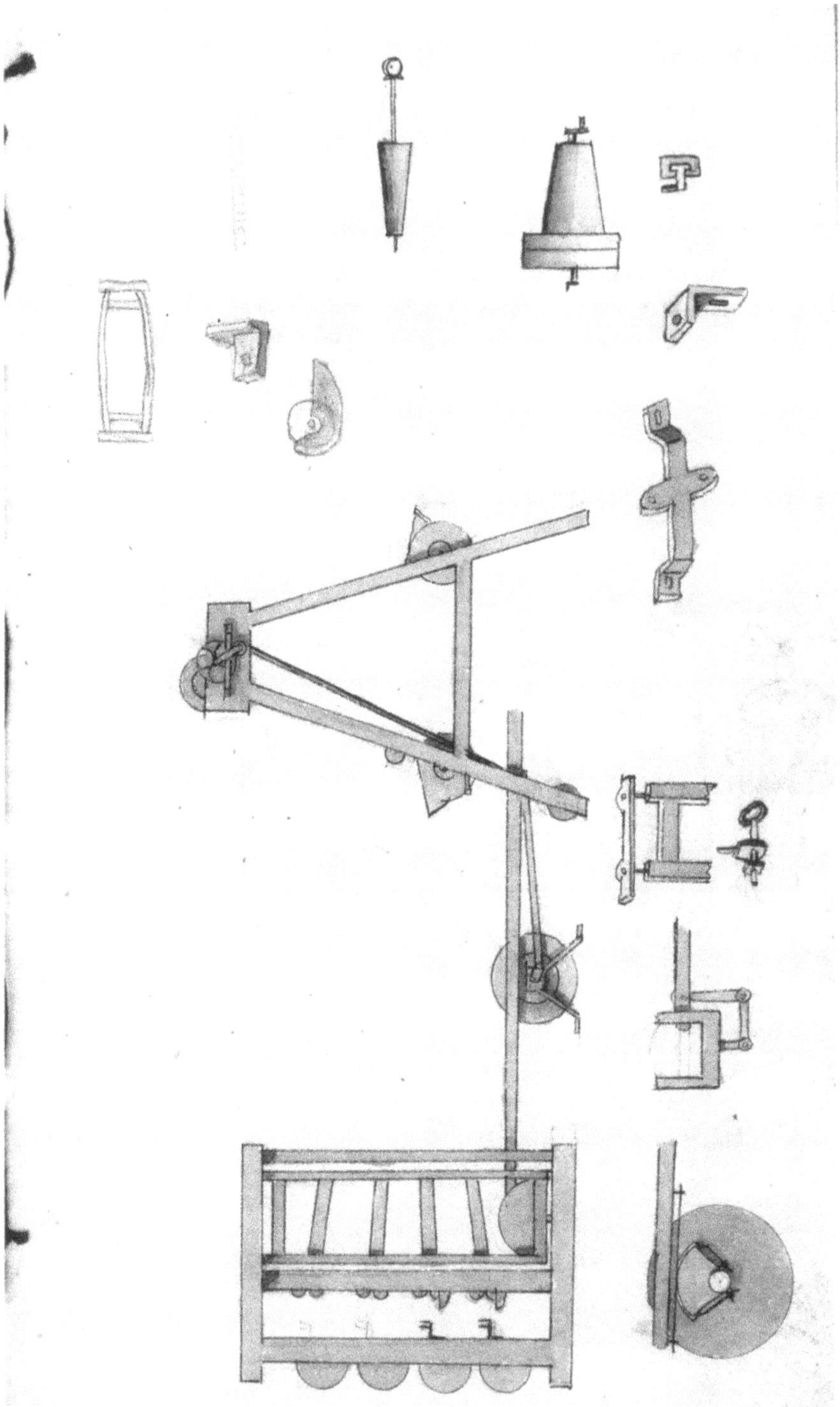

FIGURE 6.33 Sketchbook, page 7. Dresser mechanisms.

Catalogue number: IEM 28.7

Document Description

Measurement: 20 x 12.4 cm

Medium: Ink, wash and pencil

Watermark: "C Baker & Co"

Commentary

Sketches of 13 machine components and mechanisms related to the dresser (*center*). The sketch below the frame holding the four section beams shows the cam arrangement by which the long sweep moved the frame of brushes up and down. Compare this to the scale drawing in Figure 6.75 (p. 294).

The double-cone pulley at the top of the page implies that it was related to the dresser, probably as explained in Appendix 3.

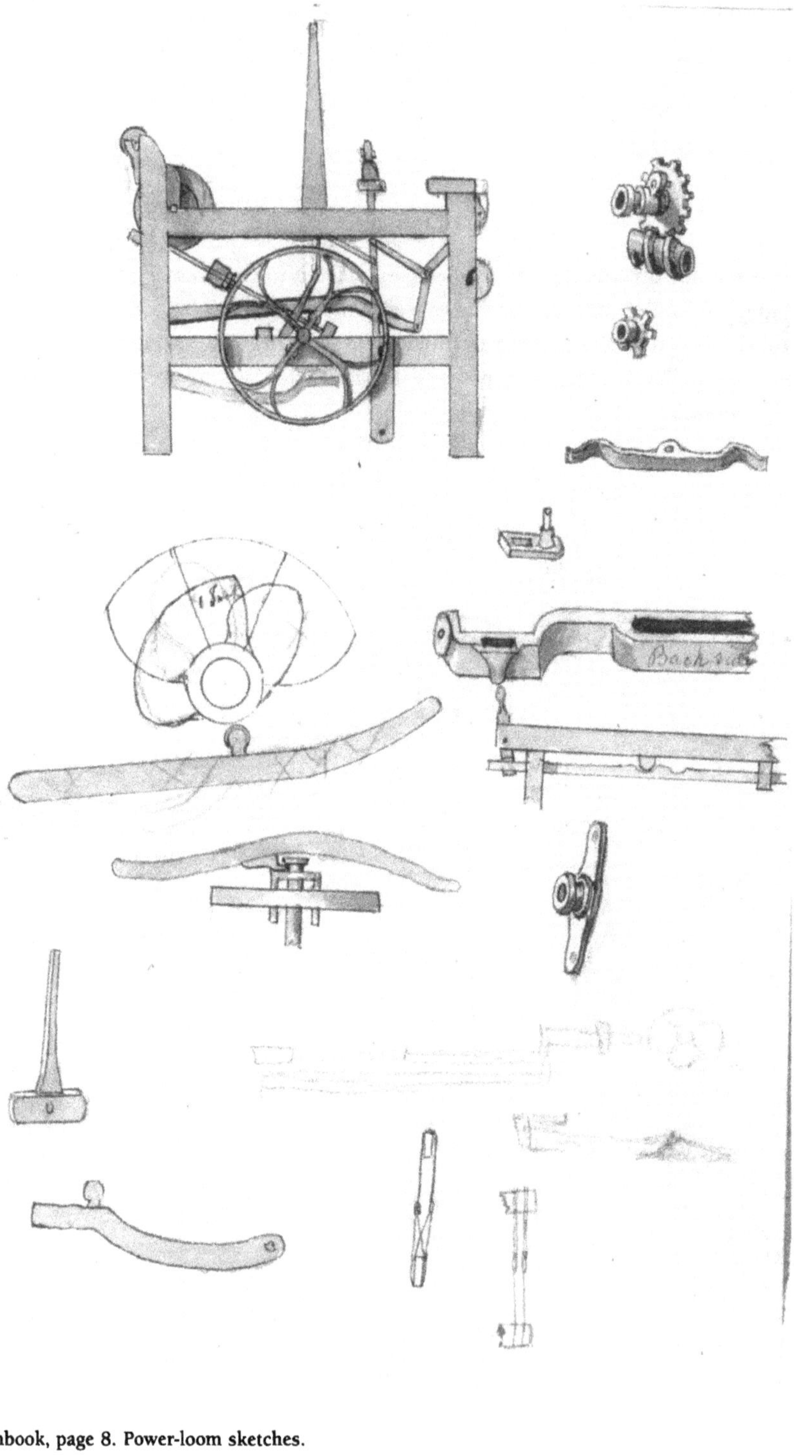

FIGURE 6.34 Sketchbook, page 8. Power-loom sketches.

Catalogue number: IEM 28.8

Document Description

Measurement: 20 x 12.4 cm

Medium: Ink, wash and pencil

Watermark: "C Baker & Co"

Commentary

Sketches of 13 machine components and mechanisms, evidently related to a power loom, including the side elevation of a power loom (*top*) and shedding cams for a power loom (*middle*). The top sketch is very similar to the scale drawing, Figure 6.79 (p. 306). In one respect it is very different. Here there seems to be an attempt to record a compound lever for the beating-up motion. If so, it may well have been inspired by the compound lever in the power loom with a variable batten speed motion patented in England by William Horrocks in 1813 (see Fig. A3.18, figures 4 and 6). The nice perspective sketch of a worm gear (*top right*) and the gear/cam below it appear again in the perspective of a power loom, Figure 6.82, p. 312, possibly part of a let-off motion. One of the components is marked "back side" and could be part of the harness (shedding) mechanism. The cam and treadle (middle left) are interesting because they show a roller on the treadle that would absorb the wear and tear otherwise imposed by the cam on the treadle. For cam patterns see Figures 6.83A, 6.83B, and 6.89 (pages 314, 315, and 325, respectively).

FIGURE 6.35 Sketchbook page 9, "Harness Slides." Power-loom mechanisms.

Catalogue number: IEM 28.9

Document Description

Measurement: 20 x 12.4 cm

Medium: Ink and wash, erased pencil

Watermark: Dove

Commentary

Sketches of two mechanisms: "Harness Slides(?)" and cam.

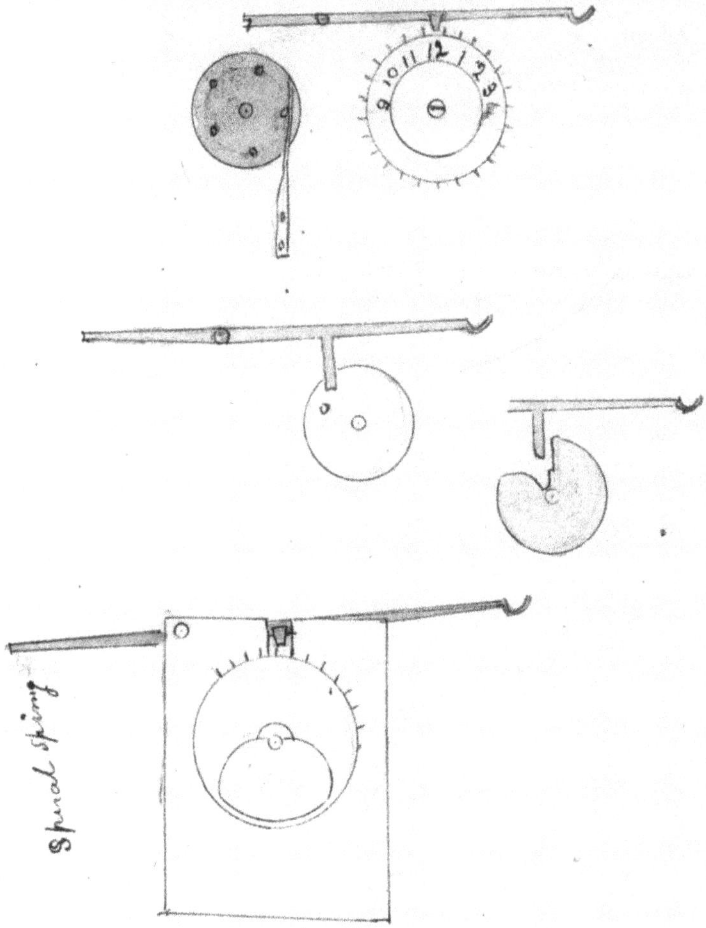

FIGURE 6.36 Sketchbook, page 10, "Spiral Spinning." Spinning-mule elements.

Catalogue number: IEM 28.10

Document Description

Measurement: 20 x 12.4 cm

Medium: Ink and wash

Watermark: Dove

Commentary

Given the similarity of these five views to some of the components in John Wood's mule patent of 14 June 1803 (GP patent 2711), then this page illustrates part of the winding-up and cop-building mechanism (involving the mendoza wheel) of the spinning mule (Wood's patent, Figure 3; the drawings of which are reproduced here at Fig. 3.19, p. 67).

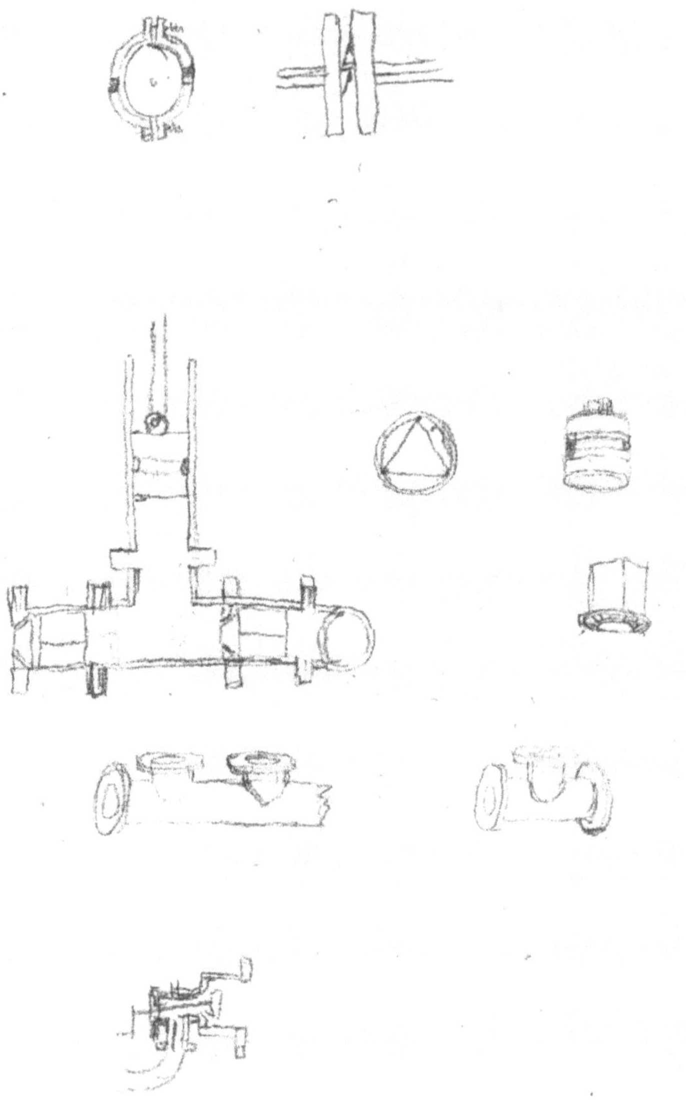

FIGURE 6.37 Sketchbook, page 11. Hydraulic sketches.

Catalogue number: IEM 28.11

Document Description

Measurement: 20 x 12.4 cm

Medium: Pencil

Watermark: Dove

Commentary

These sketches are suggestive of hydraulic experiments.

3. DRAWINGS: CARDING

Carding is the operation in which textile fibers are straightened and massed either in a sheet or in a thick, continuous tube. The alignment of the textile fibers is performed much like a brush-and-comb action, by passing them between two surfaces covered with hooked wires (the wires are set in leather, known as card clothing). One set of card clothing is fixed to the surface of a very large roller, the main cylinder. The other set of card clothing is fixed either on stationary flat bars mounted over the top of the main roller or else fixed to the surfaces of a series of small rotating rollers, also positioned over the top of the main cylinder.

All cards had three mechanical components: a feeding device, a carding mechanism, and a delivery system. Likewise, all early cards, as part of the delivery system, had a slow-moving, large roller (the doffer) that removed the fibers from the main roller. These arrangements are the key to understanding what is being depicted in these drawings of carding machines.

In English carding machines derived from Arkwright's patent of 1775, loose cotton, previously roughly torn apart by the picking machine, was placed on a moving feed cloth that took it onto the carding machine's main roller. On the other side of the main roller the cotton was taken off by the doffer cylinder from which it was in turn removed, now as a diaphanous sheet of fibers, by a reciprocating serrated-blade knife, the doffer plate. Then it was taken through a funnel and then a pair of weighted rollers to emerge as a thick tube of cotton fibers. In short, by the 1790s, the cotton factories normally had one type of carding machine.

This is illustrated in Rees's *Cyclopaedia* (see Fig. A.3.1, p. 349). Published in England in 1808, it may be surmised that this was the design known to American manufacturers by the time of the War of 1812, if not earlier. It used flats (or tops) rather than rollers over the top of the main cylinder to perform the carding operation. Rollers were preferred in woolen cards (see Fig. A3.2, p. 350). Speaking from his perspective as a Glasgow cotton manufacturer, Montgomery reported in 1832 that stationary flats were generally preferred to rollers (known as urchins in Britain) to work against the main cylinder.[16]

Thorough carding was fundamental in the formation of cotton yarn. As Montgomery instructed in 1832:

> If any machine in the whole process of cotton spinning be of more use and importance than another, it is the carding engine. ...the process of cotton spinning (properly speaking) begins only at the carding; for all the previous departments of the process are merely preparatory to this, and consist chiefly, in mixing, cleaning, and opening the cotton, so as that their cards may take the best effect upon it.
>
> Previous to the cotton being put through the cards, the fibres may be lying in every direction into which they may accidentally be thrown; but the use of the carding engine is to draw out the fibres of the cotton, to straight, and lay them side by side, and form them into a thread commonly called an end; and this is the first formation of the thread of yarn. ...
>
> When the fibres are properly straighted, and the end equally formed at the cards, there is good reason to expect a superior quality of yarn; but failing this, an inferior quality is unavoidable; for no skill or attention applied to any subsequent department of the process, can altogether remedy the injuries the cotton may have sustained in this; hence, it is an object of the highest importance in cotton spinning, to have the cards always properly set, and adjusted to suit the particular kind of cotton used, and the quality of the yarn required.[17]

---

[16]Montgomery, *Carding and Spinning Master's Assistant*, 62.
[17]Montgomery, *Carding and Spinning Master's Assistant*, 63–64.

To improve the carding function it was possible to have an arrangement whereby the fibers were carded twice over. In the woolen industry two main cylinders were incorporated in a single machine. In the cotton industry two different carding machines were used to produce a better quality of product: the breaker card formed a sheet or lap; the finisher card, a tube of fibers or end. A contemporary diagram of this arrangement was published in the United States.[18] Markham seems to have preferred two, rather than one, carding machines in his processing line, judging by these drawings.

---

[18]Nicholson, *Operative Mechanic* I, 388–389.

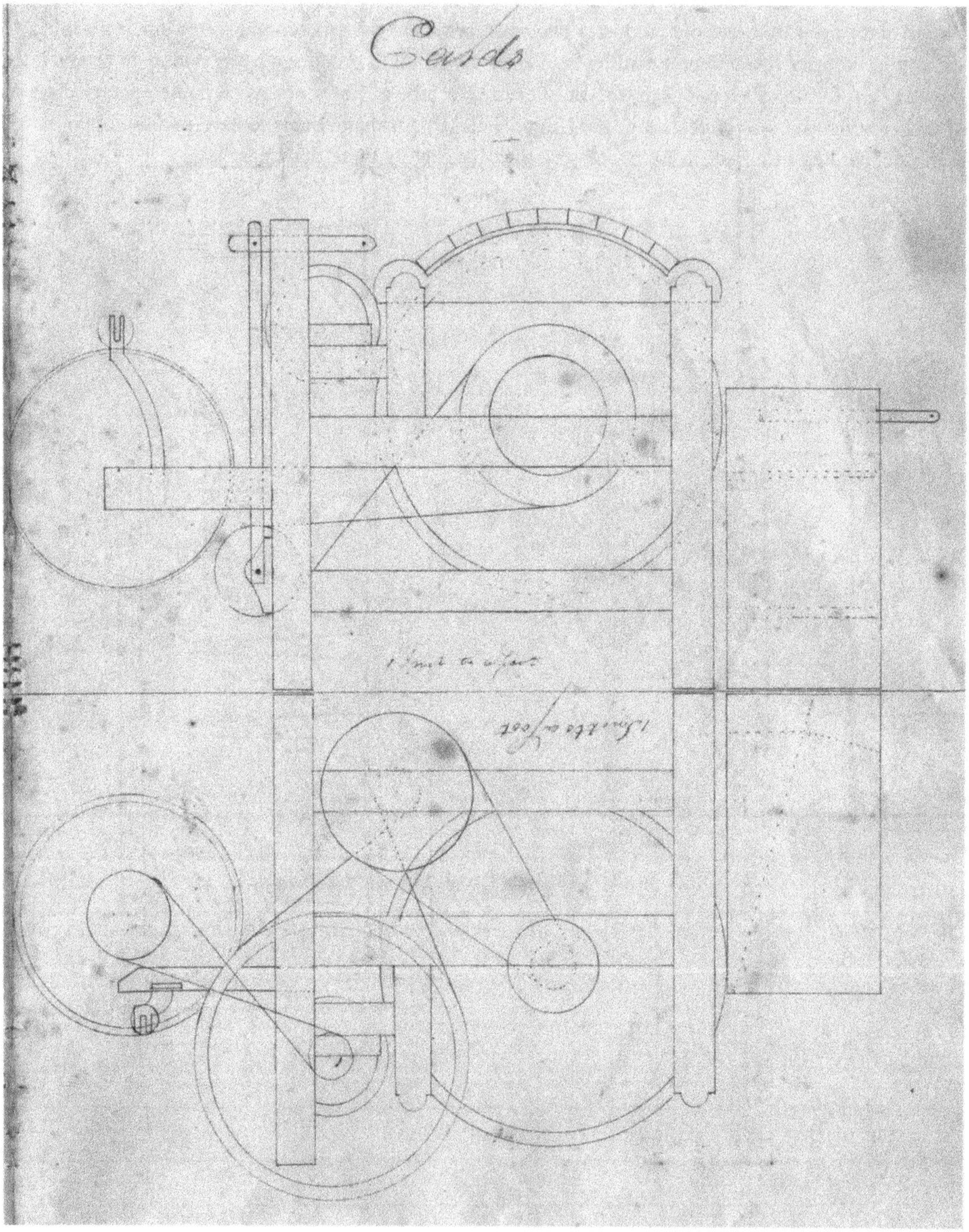

FIGURE 6.38 "Cards." Scale drawing, side elevations. Cotton carding engine, breaker, side elevations.

Catalogue number: IEM 2.3

Document Description

Measurement: 26.9 x 22.3 cm (one page of a sheet 22.3 x 44.5 cm folded in half and sewn together with another sheet, making eight pages in all; see Figs. 6.38, 6.46–6.48, 6.52, 6.54, 6.55, 6.57)

Medium: Ink

Condition: Pinholes in top corners, foxing, stain in upper right corner, light bleed-through, erasure, sewn

Commentary

The first carding-machine drawing, Figure 6.38, is that of a breaker cotton card; the second, Figure 6.39, seems to be that of a finisher card. The identification of the breaker card is suggested by a feeding receptacle (holding picked cotton) on the right-hand side of the main cylinder and a delivery arrangement producing a lap that was wound onto the second largest cylinder in the machine (the lapper cylinder) on the far left.

In Figure 6.38 the two side elevations share the same baseline. The upper elevation shows the fiber-processing elements of the machine, the lower elevation the power arrangement of pulleys and belts. The upper elevation indicates a frame for holding a doffer plate, to the left of the doffer cylinder. Left again, the cotton lap is held in place on the powered lapper roller by a friction-turned wooden roller. As the lap is wound onto the lapper roller, the circumference of the lapper roller increases. To compensate for this, the axle of the wooden roller is supported by two forked arms, which evidently allows the small wooden roller to slide upward. When about twenty rotations of the lap have been wound onto the lapper roller, they are cut free to form a thick sheet ready for the finisher card. "In this operation we are presented with the first act of plying or doubling, which is introduced in the process of spinning in order to obtain equality in the strength and thickness of the yarn."[19]

Comparators

Drawings: published: J. Thomson, "Manufacture of Cotton" (1808); Nicholson, *Operative Mechanic*. See Figure A3.1 (p. 349).

Artifacts:

British:

Arkwright-type machines in Science Museum, London; North West Museum of Science and Industry, Manchester; Strutt's North Mill, Belper, Derbyshire.

American:

National Museum of American History, Smithsonian Institution, Washington, DC; Old Slater Mill Museum, Pawtucket, RI.

---

[19]Nicholson, *Operative Mechanic*, 388–389.

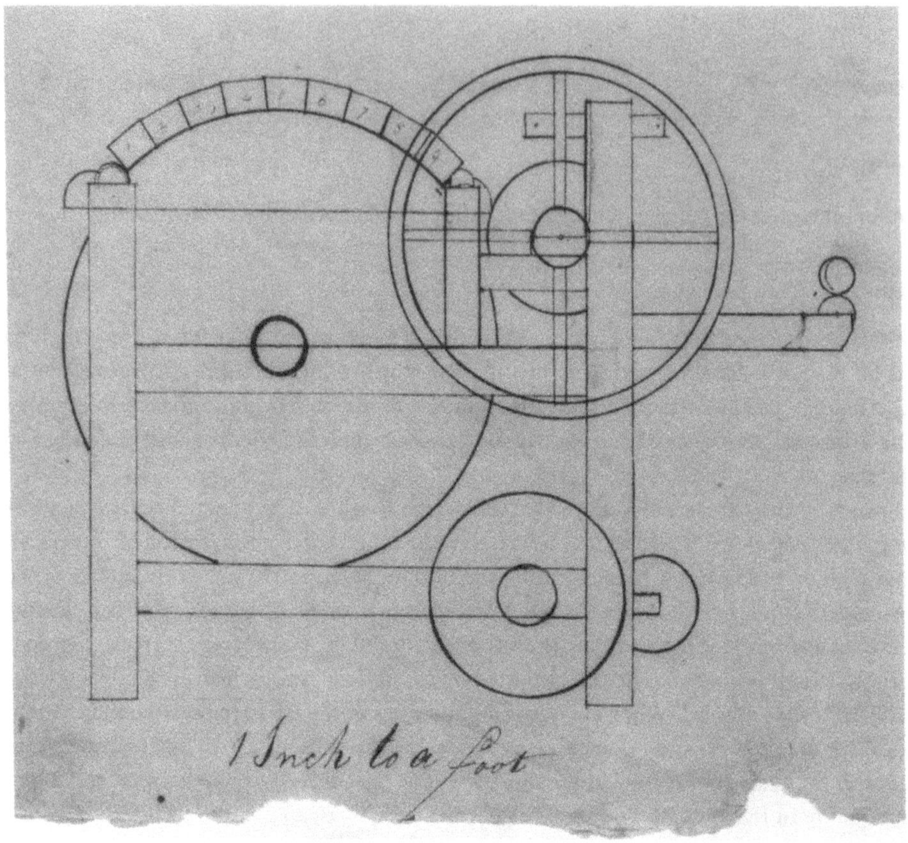

FIGURE 6.39 Flat-top cotton-carding machine. Scale drawing, side elevation. Cotton-carding engine, finisher, side elevation.
Catalogue number: IEM 6

Document Description

Measurement: 23 x 28 cm

Medium: Ink

Condition: Top 3.2 cm folded over, pinhole in crease, ragged bottom edge, stains on back corners

Commentary

This is probably a finisher carding machine. The feeding end may be assumed on the left because the flats are numbered from the left. If the delivery end is on the right, the drawings must be incomplete because the lap would have been formed into a thick tube by a funnel positioned (but not shown) between the doffer roller and the two rollers on the far right.

Comparators

Same as Figure 6.38.

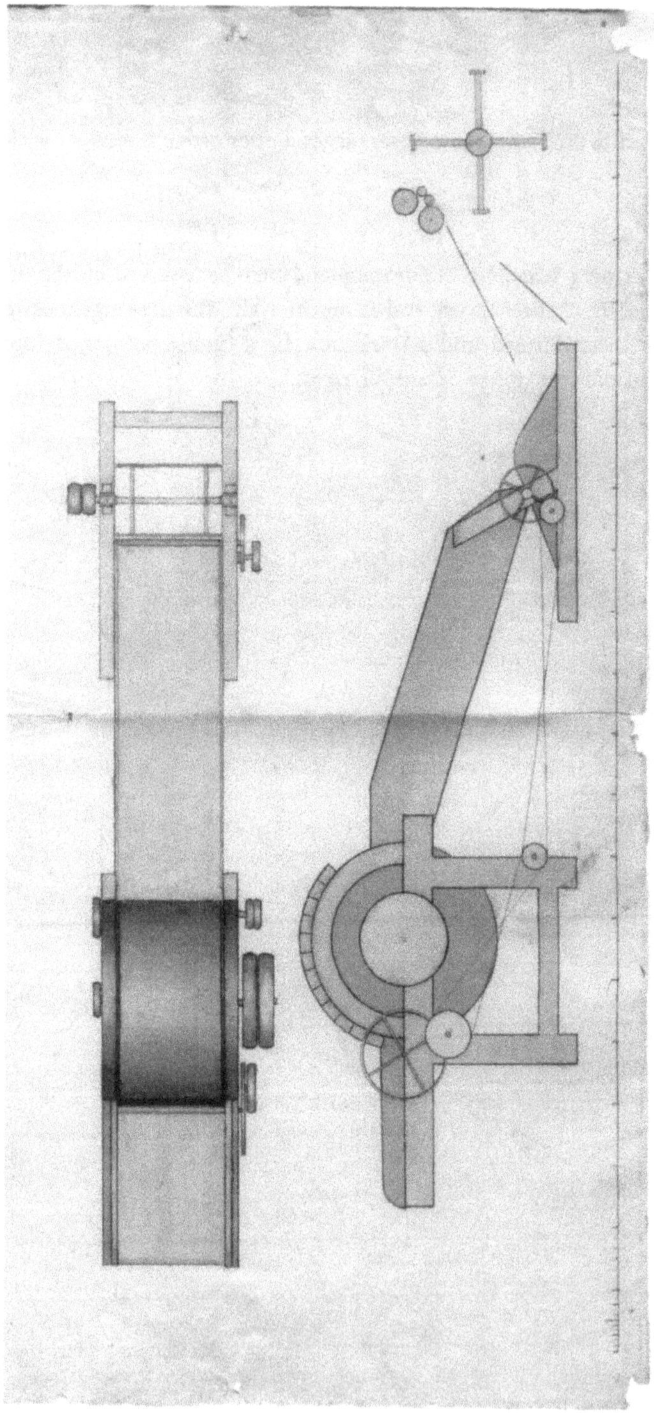

FIGURE. 6.40 Carding machine. Scale drawing, plan and side elevation. Scale partly indicated. Carding engine, elevator feeding system, plan and side elevation.

Catalogue number: IEM 7

Document Description

Measurement: 27 x 42.7 cm

Medium: Pencil, ink and wash

Watermark: T G & Co.

Condition: Sheet folded in half, pinholes in corners and centers of edges, corner losses, minor stains

Commentary

Here Markham appears to have recorded an experimental machine. The plan (*above*) and the section (*below*) show a carding machine with main cylinder and flats. The novel arrangement is on the right of the drawing. What is not clear is whether this is a feeding arrangement or a delivery one.

If it is a feeding arrangement, it would have shown a pile of loose cotton from the picking machine about to be driven by the motion of rotating paddle bars onto a long feeding cloth. Presumably the length of the feeding cloth was necessary to accommodate the angle of its incline. The design would avoid the need, in the customary carding machine, to lift the cotton up onto the horizontal feeding cloth. Additionally, the rotating bars would further open the cotton stock, separating fibers a little more than the picking machine would have done. One puzzling element shown in the section drawing is the juxtaposition of two lines and four rollers near the paddle-wheel section.

If the machine was working in the opposite direction, the feeding cloth is on the left and the long trunk on the right is delivering a carded lap of cotton. However, the lap would need to be rolled up and not collected in a pile. Nor does this explain where the four rollers belong. They could be drafting or carding rollers.

The balance of visual evidence points toward this design as being one for a novel feeding arrangement for a breaker carding machine.

Comparators

See Figure A3.1 (p. 349).

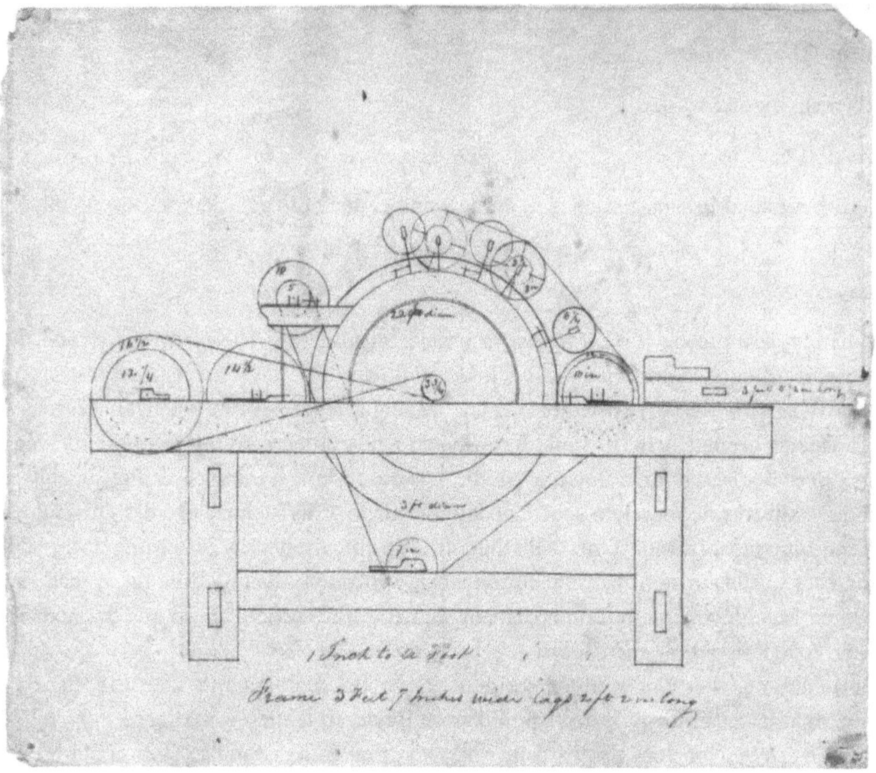

FIGURE 6.41A Wool-carding engine. Scale drawing, side elevation.

Catalogue number: IEM 8.1

Document Description

Measurement: 22 x 26 cm (one page of folded sheet 26 x 44 cm that includes Figs. 6.41A, 6.41B, 6.43, 6.61)

Medium: Ink

Condition: Pinholes and tears in corners, red substance (wax?) in corners, minor bleed-through, foxing

Commentary

Figures 6.41A and 6.41B are side elevations, arranged to mirror each other, of a carding machine using rollers rather than flats. Figure 6.42, which is very similar, suggests that this is a set of diagrams for a woolen carding engine.

The feeding cloth, acting like a moving belt, would be on the right of Figure 6.41A and to the left of Figure 6.41B.

The purpose of the diagram is clear from the dimensions given for the rollers and pulleys in the machine: to record the data essential for establishing a correct set of relationships between the various rollers performing the carding action. Those relationships can be briefly summarized.

Central to the operation of the carding machine were the two pairs of small rollers positioned over the top of the main roller. Between the main cylinder and the two (or more) pairs of rollers occurred the two prime movements of the carding machine, namely, carding and stripping.

In each pair, one roller is larger than the other. This is the worker roller. The wire teeth of the card clothing with which the worker is covered are bent in a direction opposite to the teeth of the card clothing on the main cylinder. This achieves the carding action, involving the tearing apart and opening tufts of fiber.

The small roller (known as the stripper, clearer, or cleaner) in each pair had its card clothing teeth bent in the same direction as the teeth of the card clothing on the worker. However, the stripper rotated at a much higher speed than the worker, enabling it to lift the textile fibers out of the teeth of the worker and return them to the main cylinder, which carried them on to the next pair of workers-and-strippers.

Of the other rollers surmounting the main cylinder, that on the right, between the main cylinder and the revolving feeding apron, is the lickerin, which takes the fibers into the teeth of the main cylinder. The lickerin also has a stripper above it, to clear it.

On the opposite, delivery, side of the main cylinder is a roller supported by a bracket from the arch frame. This appears to be what was later known as the fancy. The function of the fancy was neither carding nor stripping. Instead, it was to raise the fluffy and voluminous stock of fibers collected in the card wires of the main cylinder. The card clothing wires of the fancy were long and reached into the wires on the main cylinder, allowing them to catch the fibers and to bring them up to the points of the main cylinder wires in readiness for removal by the doffer cylinder, the one below it. This was accomplished by having the card wire teeth of the fancy bent in the same direction as the teeth of the main cylinder but driving the fancy at a much higher speed than the main cylinder.

As explained in Figure 6.38, the last two rollers on the left are the doffer (next to the main cylinder) and the lapper. The former removes the sheet of fibers from the main cylinder, whereas the latter rolls it up ready for the next processing stage.

Belt drives allowed for some slippage, as when tufts of fiber clogged between two moving surfaces in the carding machine. The distance, or tolerance, between the main cylinder's wires and those on any of the small rollers was adjusted by "screw props," as Ure called them, supporting the bearings in which the necks of the small cylinders rested.[20]

Neither J. Thomson, "Manufacture of Cotton" (1812), nor Nicholson, *Operative Mechanic*, showed a machine of this design in their articles on cotton manufacturing. However, Montgomery, in his first treatise on cotton manufacturing, observed:

> Carding engines have sometimes been made to consist of one large cylinder, and a number of smaller ones, called urchins, disposed of at proper distances over above the main cylinder, and revolving in opposite directions to it, but nearly in contact; by which means the cotton was delivered from cylinder to cylinder, until it came to the finishing cylinder called the doffer, from which it was taken off by the comb. At present, carding engines are generally made to consist of only two cylinders, sometimes three; one at the feeding rollers, called the lickerin. But the main cylinder is covered with a kind of arch, composed of several pieces of wood called tops, which have no motion, having sheet cards fixed on them, and nearly in contact with the main cylinder.[21]

Comparators

Drawings:

Markham's design is reminiscent of woolen-carding engines commonly used in Britain: Samuel Godwin, GB Patent 2766 (30 May 1804), drawings, sheet 5. Bakewell, "Woollen Manufacture"; see Figure A3.2 (p. 368).

Artifacts:

Similar woolen-carding engines were first built by the Scholfield brothers in the United States in the 1790s. Examples in the Smithsonian Institution, Washington, DC; Pennsylvania State Museum, Harrisburg, PA; American Museum of Textile History at Lowell, MA.

---

[20] Ure, *Cotton Manufacture* II, 32.
[21] Montgomery, *Carding and Spinning Master's Assistant*, 62.

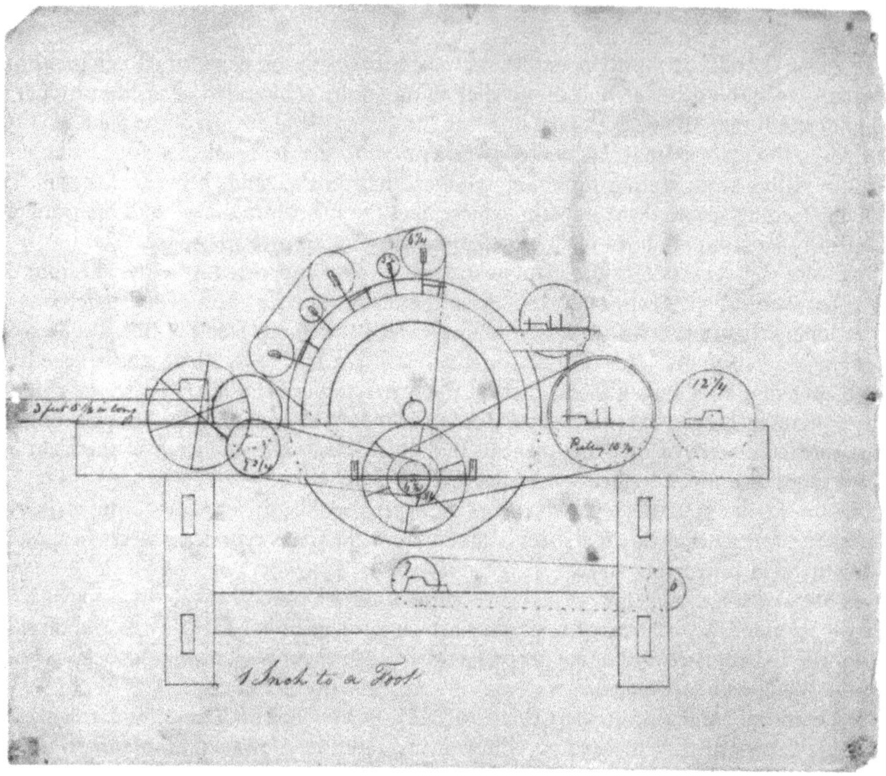

FIGURE 6.41B Wool-carding engine. Scale drawing and exactly the same scale as in Fig. 6.41A of which it is the mirror and opposite side elevation.

Catalogue number: IEM 8.4

Document Description

Measurement: 22 x 26 cm (one page of folded sheet 26 x 44 cm that includes Figs. 6.41A, 6.41B, 6.43, 6.61)

Medium: Ink

Condition: Corner tears, stains, bleed-through, pinholes

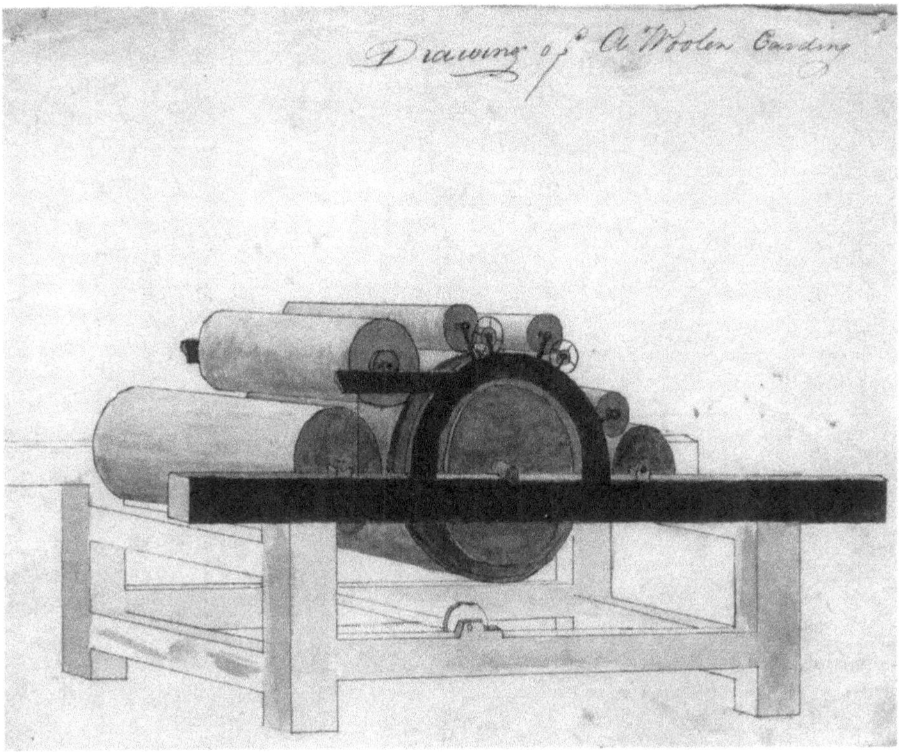

IGURE 6.42 "Drawing of a Woolen Carding [Engine]." Perspective sketch.

:atalogue number: IEM 10.4

Document Description

Measurement: 22 x 26.8 cm (half of folded sheet 26.8 x 44 cm that includes Figs. 6.42, 6.45A, 6.45B, 6.65)

Medium: Ink and wash

Condition: Pinholes in corners, foxing

Commentary

Roller type of carding engine, similar to that shown in drawing 6.41A and 6.41B, tending to confirm the identification of Figure 6.41 as a woolen-carding machine.

Comparator

Bakewell, "Woollen Manufacture"; see Figure A3.2 (p. 350).

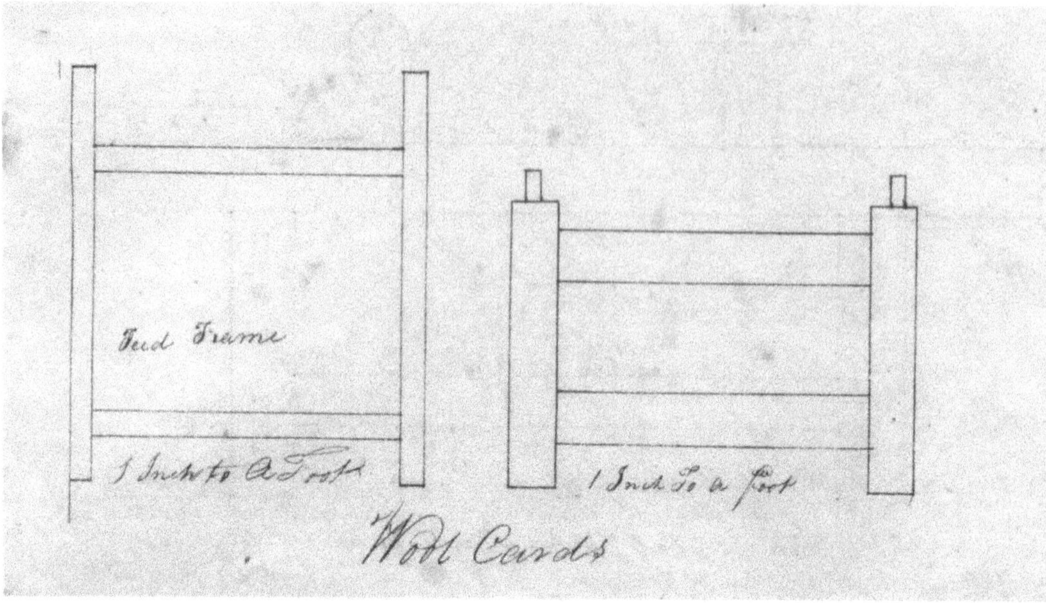

FIGURE 6.43 "Wool cards"; frame members. Scale drawing, elevations.

Catalogue number: IEM 8.2

Document Description

Measurement: 22 x 26 cm (half of folded sheet 26 x 44 cm that includes Figs. 6.41A, 6.41B, 6.43, 6.61)

Medium: Ink

Condition: Pinholes and tears in corners, foxing, tiny losses on edges, bleed-through

Commentary

Shows elevations of the frame members of a woolen-carding engine, at its feed end (diagram to left) and for the rest of the machine.

Comparators

Drawings:
Markham's timber-framing design is reminiscent of woolen-carding engines commonly used in Britain: Samuel Godwin, GB Patent 2766 (30 May 1804), drawings, sheet 5.

Artifacts:

In the Smithsonian Institution, Washington, DC; Pennsylvania State Museum, Harrisburg, PA; American Museum of Textile History at Lowell, MA.

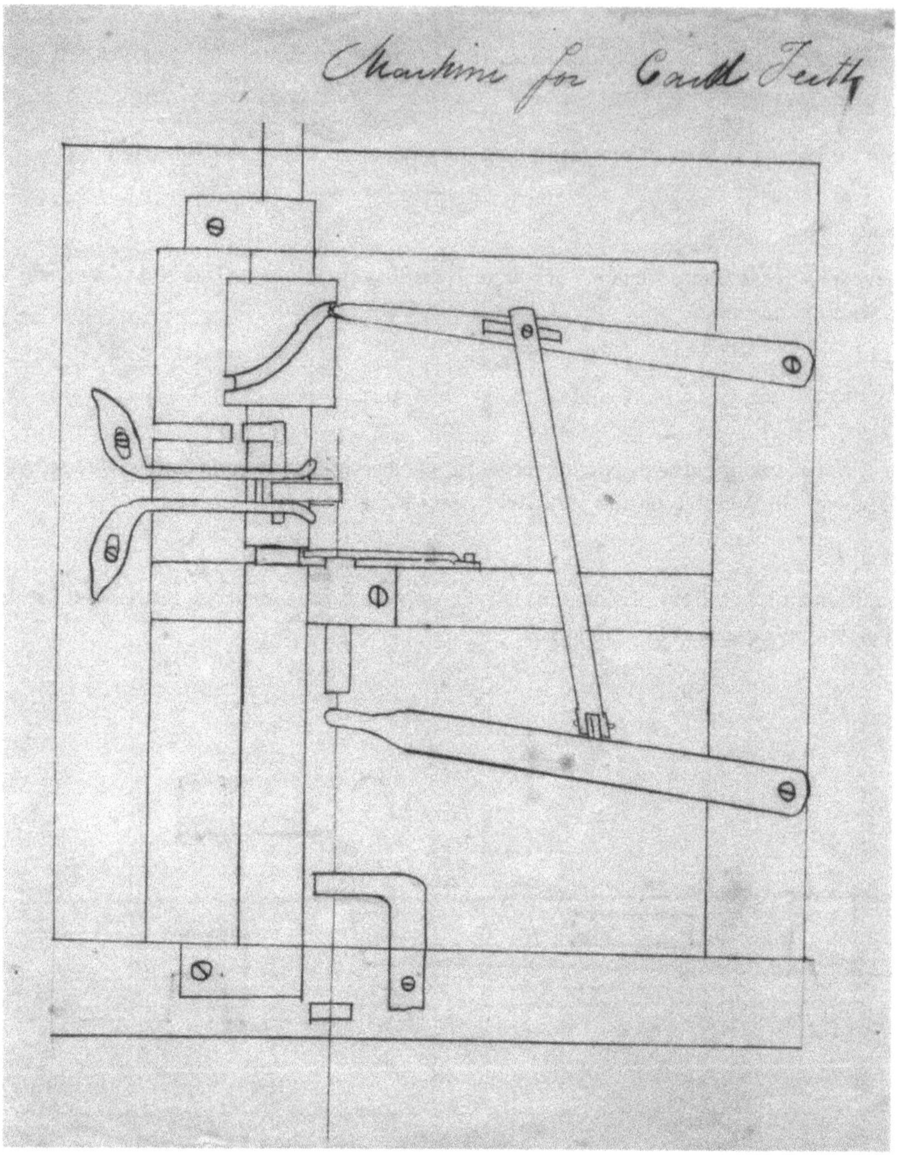

FIGURE 6.44 "Machine for Card Teeth." Drawing, no scale.

Catalogue number: 9.1

Document Description

Measurement: 27.5 x 27.3 cm

Medium: Ink

Condition: Pinholes in corners, foxing, bleed-through; reverse side of Figure 6.62

Commentary

The title of the drawing indicates that this machine was related to the manufacture of card clothing. Card clothing was made in two forms: sheet, which could be nailed onto a main cylinder or a doffer roller; and fillet, which was wound in spiral fashion around one of the secondary rollers (3). In both kinds the structure of the card clothing was the same, consisting of a layer of leather pierced by holes through which wire staples were driven and then bent at an angle.

Mechanizing the manufacture of card clothing required the complex combination of mechanical movements. In England, John Kay, the inventor of the fly shuttle, had two machines in operation by 1754 and his sons united them into one machine by 1794. Other machines were invented in Cumbria in the 1770s. Of these early machines, one pricked the leather, the other cut and crooked the wire. Both machines were introduced to the United States by 1785. However, the setting of the staple-shaped wire teeth in the leather was done by hand until, in 1799, Amos Whittemore, of Cambridge, Massachusetts, pirating the ideas of other New England mechanics, patented an automatic machine that made sheet-card clothing. A complementary machine for making fillet-card clothing was patented by Pliny Earle of Leicester, Massachusetts, in 1803.[22]

What then does Markham's drawing depict? Is it an attempt to improve the early single-functioning machine that cut and bent the wire teeth? Or, is it part of a bigger, integrated machine designed to rival Whittemore's or Earle's? If the latter, why was Markham searching for an alternative to the patented machines? To be sure, the patented machines caused protracted wrangles about claims to invention in the Boston courts between 1810 and 1815—which was hardly surprising because card clothing was critical in textile manufacturing and monopoly rights for its manufacture would have been very profitable.[23]

If the drawing is obscure, one element seems clear. The two linked and parallel levers serve to tug the card wire into a cutting and bending zone. The reciprocal motion, giving the to-and-fro movement needed to pull forward a new length of wire, is provided by the spiral cam guiding the tip of one of the levers.

Comparators

Artifacts:

British

Bendigos, or wire-bending implements, are exhibited in the Bankfield Museum, Halifax, Yorkshire. A range of card-making machines is to be found in the Science Museum, West Kensington, London.

---

[22]Mann, "Cardmaking Machine"; Jeremy, *Transatlantic Industrial Revolution*, 13–14, 176–178.

[23]Jeremy, *Transatlantic Industrial Revolution*, 13–14, 176–178.

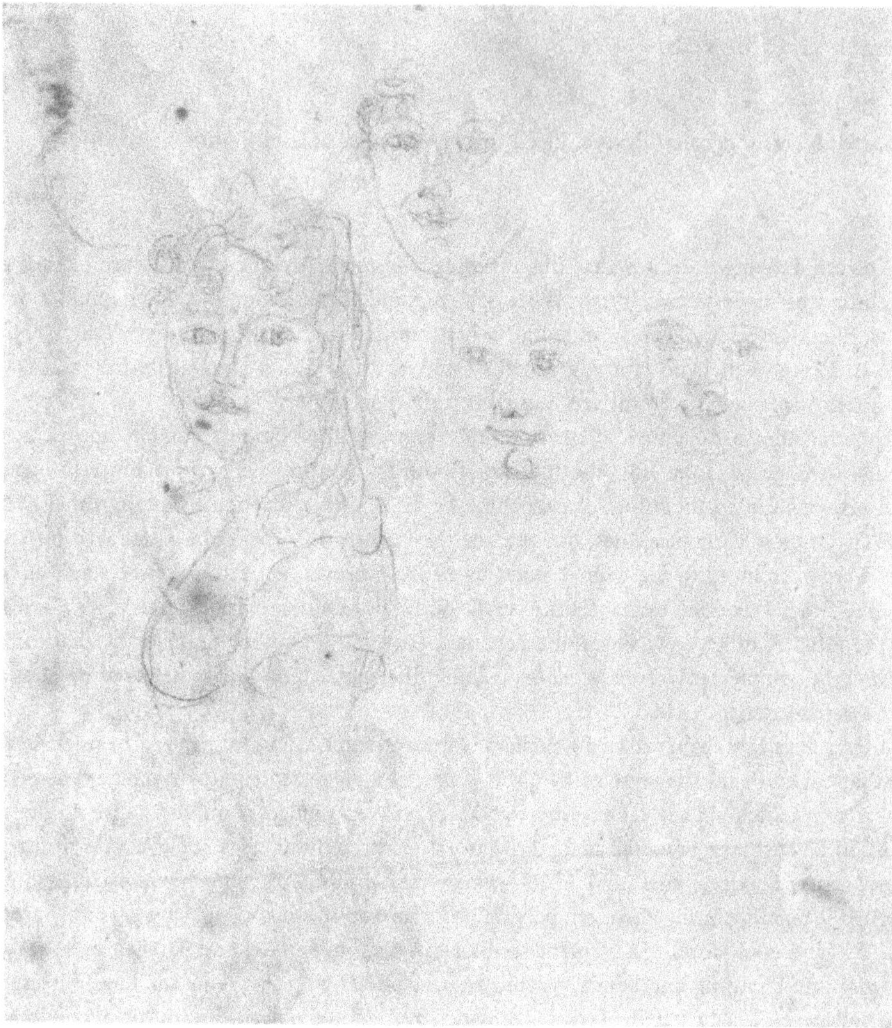

FIGURE 6.45A **Doodles. Located in same original file as the carding-machine drawings.**
Catalogue number: IEM 10.3

Document Description

Measurement: 26.8 x 22 cm (half of folded sheet 26.8 x 44 cm that includes Figs. 6.42, 6.45A, 6.45B, 6.65)

Medium: Pencil

Condition: Pinholes in corners, foxing, bleed-through

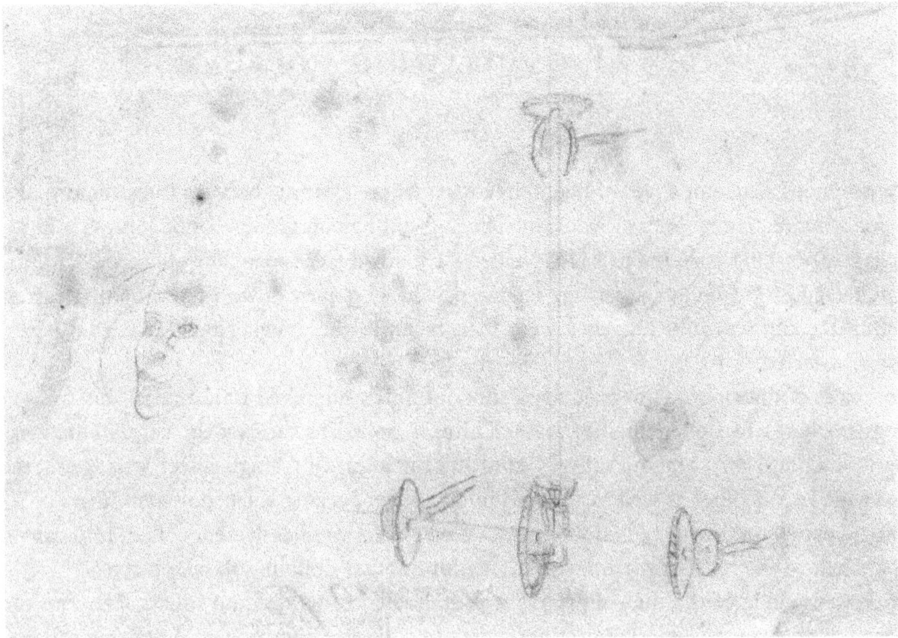

IGURE 6.45B Doodles. Located in same original file as the carding-machine drawings.

atalogue number: IEM 10.2

Document Description

Measurement: 22 x 26.8 cm (half of folded sheet 26.8 x 44 cm that includes Figs. 6.42, 6.45A, 6.45B, 6.65)

Medium: Pencil

Condition: Stains

4. DRAWINGS: DRAWING AND ROVING

Drawing

Carding produced a tube or sliver of fibers. In cotton manufacturing, because the structure of the cotton fiber made it more spinnable than wool, the sliver could be continuous. Until John Goulding patented his (American) piecing machine in 1826, woolen slivers had to be manually spliced. By contrast, cotton slivers delivered by the common carding engine passed by gravity down into a rotating can where they coiled until the can was full. The cans were then taken to the drawing frame where the next stage of processing occurred.

The object of drawing was to produce a sliver of much improved uniformity. This was achieved by doubling (merging) two or more slivers (each already coiled in cans by the carding engine) and then drafting (stretching) the combined slivers until the thickness of a single sliver was again reached. The process yielded a smoother as well as a more uniform sliver because the component fibers of cotton were increasingly parallelized with each doubling. Drafted slivers were delivered, coiled, into cans that could easily be taken to the back of the drawing frame for another doubling to take place.

The key principle of the drawing frame was embodied in its drafting rollers, between each pair of which the cotton sliver passed. Because there were two, three, or four pairs of drafting rollers, each pair rotating at a successively higher speed, the sliver was stretched in proportion to the difference in speed between each pair of rollers.[24]

Drafting rollers were first patented, though not necessarily built, by Lewis Paul in England in 1738.[25] A successful design for drafting rollers, followed by widespread construction, was incorporated in the waterpowered spinning frame patented in England in 1769 by Richard Arkwright (see Fig. 3.8, p. 57).

In the early industrial decades the number of doublings was reported as four or five. By the early 1830s, however, a No 36 yarn required at least 64 doublings, producing a very even yarn.[26]

Montgomery's comments on the significance of drawing are apposite.

> But after passing the cards, perhaps there is no machine in the whole process so well adapted for straighting the fibres as the drawing frame, and the means here employed for that purpose, is the doubling of two or more ends together, and drawing them out again to their former grist. And when the whole doubling and drawing throughout the process are judiciously arranged to suit the particular kind of cotton that may be used, a superior quality of yarn may be reasonably expected. ...Too many doublings may be as injurious to the yarn as too few, because it is not so much from the mere multiplicity of doublings that it is to be improved, as from a suitable arrangement of them in the proper places, and their adaptation to the quality of the cotton and the yarn required. Cotton that is long and strong in the staple or fibre, requires to be oftener doubled than that which is short and weak; and for spinning very low numbers from Waste, Surat, or Madras cottons, the fewer doublings the better.[27]

Roving

Roving is the process by which the sliver or "rope" of fibers produced by the drawing frame is reduced in diameter to a size suitable for spinning. Not only is the sliver narrowed in diameter by passing it

[24] J. Thomson, "Manufacture of Cotton."

[25] British Patent 562; Hills, *Power,* 32–38; Wadsworth and Mann, *Cotton Trade,* 419.

[26] J. Thomson, "Cotton Manufacture"; Montgomery, *Carding and Spinning Master's Assistant,* 88.

[27] Montgomery, *Carding and Spinning Master's Assistant,* 87–88.

through drafting rollers but also it then receives some degree of twist as it is wound onto the bobbin by the spindle-and-flyer.

Thomson noted:

> The preparation of such rovings as shall be perfectly regular in size, and have an equal quantity of twist in every part, and which shall be exceedingly soft, is a most essential point in cotton spinning. As it is impossible to correct these imperfections in the spinning, they will be given to the thread."[28] Montgomery echoed this judgment: "And here it may be remarked, that, to make good rovings, is a most important object in cotton spinning; they ought to be uniformly level and evenly in their grist, having an equal degree of twist in all their parts, and no more than is necessary to give the requisite strength.[29]

The central problem in mechanizing the roving operation arose from the fragility of the sliver (now called a roving, from "roping") emerging from the drafting rollers. After drafting, it had to be simultaneously twisted and wound into a package ready for spinning. However, unless it was laid onto the bobbin at the same rate that it came from the drafting rollers, it would be uneven or perhaps break. With each layer of roving the diameter of the bobbin package increased in size. Consequently, the surface velocity of the bobbins would have to be decelerated each time a layer of roving (a traverse) was completed. Two broad solutions were tried. One, using rotating cans instead of the bobbin with the spindle-and-flyer familiar in the water frame, was a crude approach and the first solution. The second, using the bobbin with the spindle-and-flyer, required a more sophisticated mechanical technology.

By 1812 several designs of roving frame were available in Britain. Oldest was the roving-can frame, first introduced by Arkwright (see Fig. A3.4, p. 352). Its essence was twist insertion by the rotation of the cans into which the drafted slivers fell.[30] This was still favored by mule spinners in the 1820s.[31] Its main drawback was that it did not communicate twist equally and uniformly to the sliver.[32]

Fragile rovings taken from the can frame risked injury when wound onto bobbins by a winding block. The skeleton frame was devised to reduce this problem. It was similar to the can frame except that metal cages (skeletons) rotated the cans, which were now removable, minimizing disturbance to the roving until it was positioned in the next machine, the stretching frame (a version of the mule).[33]

Yet another roving machine was the jack frame. However, it was too complex a mechanism and consequently was subject to a greater degree of failure than its alternatives. As Montgomery observed, the jack frame was "a very ingenious but extremely complicated machine."[34]

Before the early 1820s the most satisfactory roving frame was variously known as the double-speeder,[35] the bobbin-and-flyer roving frame,[36] or the fly frame[37] (see Figs. A3.5 [p. 352] and A3.6 [p. 353]). Thomson claimed that it was perfect: "A great number of different constructions of roving-frame have been in repute, at different periods, among cotton spinners; but it is only lately that by a machine, called the double speeder, it has been brought to perfection."[38]

---

[28] J. Thomson, "Manufacture of Cotton."
[29] Montgomery, *Carding and Spinning Master's Assistant,* 104–105.
[30] J. Thomson, "Manufacture of Cotton."
[31] Nicholson, *Operative Mechanic* I, 390.
[32] J. Thomson, "Manufacture of Cotton"; Montgomery, *Carding and Spinning Master's Assistant,* 106.
[33] J. Thomson, "Manufacture of Cotton"; Nicholson, *Operative Mechanic* I, 391; Montgomery, *Carding and Spinning Master's Assistant,* 107.
[34] Montgomery, *Carding and Spinning Master's Assistant,* 108.
[35] J. Thomson, "Manufacture of Cotton."
[36] Nicholson, *Operative Mechanic,* I, 391.
[37] Montgomery, *Carding and Spinning Master's Assistant,* 107.
[38] J. Thomson, "Manufacture of Cotton."

In winding the roving directly from the drafting rollers onto the bobbins, two variable movements needed to be mechanized. One motion powered the up-and-down action of the lifter rail, which raised and lowered (traversed) the bobbins inside the flyers. A second motion changed the rotational speed of the bobbins, relative to the rotational speed of the flyers, so that roving was wound on the bobbin at the same speed as that of the delivery pair of drafting rollers. The U-shaped flyers, fixed to the top of the spindles, inserted twist. One arm of each flyer was a hollow tube through which the roving passed. As the flyer revolved round the bobbin, the roving was simultaneously twisted and laid onto the bobbin. Flyers always rotated at a constant speed to impart a uniform number of twists per inch of roving.

Because the roving had to be wound onto the bobbins, there must be relative rotation between the flyers and bobbins and this led to a phenomenon that puzzled contemporaries. The relative rotation could be achieved either by rotating the bobbins faster than the flyers or vice versa. In the case of "bobbin lead," with the bobbin traveling faster than the flyer, roving was wound onto the bobbin in a direction of rotation opposite to that of both bobbin and flyer. Consequently, roving was laid onto the bobbin behind the flyer. With each traverse, the diameter of the bobbin increased, requiring the absolute rotational speed of the bobbin to be decreased so as to maintain a constant roving take-up speed by the bobbins.[39] In mathematical terms, bobbin lead appears thus,

$$(n-N)\Pi d=V$$

Where

n = bobbin speed (reducing with each traverse)

N = flyer speed (constant, and slower than bobbin speed)

d = bobbin diameter (increasing with each traverse)

V = surface velocity of bobbin (constant, but same as delivery drafting rollers)

In the case of "flyer lead," with the flyers driven faster than the bobbins, roving was wound onto the bobbin in the same direction of rotation as that of the bobbin and flyer. Consequently (and contrary to expectations) bobbin speed was accelerated with each traverse—otherwise the roving take-up speed by the bobbins would progressively exceed the rate at which the roving emerged from the flyer.

In mathematical terms, flyer lead appears thus:

$$(N-n)\Pi d=V$$

Where

n = bobbin speed (increasing with each traverse)

N = flyer speed (constant, and faster than bobbin speed)

d = bobbin diameter (increasing with each traverse)

V = speed of the roving being wound on the bobbin (constant, but same as delivery drafting rollers)

Descriptions in Rees and by Montgomery suggest that flyer lead was preferred in the early nineteenth century.[40]

Mechanizing these two movements (of lifter rail and variable bobbin speed, which may be why the double speeder was given its name), was not achieved overnight. In 1772, a bobbin-lifter rail was patented

[39]I am grateful to Dr. R. Hugh Gong, Reader in Textile Technology, School of Materials, University of Manchester, for explaining to me the puzzle of flyer lead.

[40]J. Thomson, "Manufacture of Cotton"; Montgomery, *Carding and Spinning Master's Assistant*, 110–111; *Cyclopaedia of Textile Work* I, 191–194.

by Coniah Wood of Nottingham, England. Heart-shaped cams, used to impart a traverse motion to the lifter rail, were trialed by Jedediah Strutt of Belper, Derbyshire, in the 1780s in early spinning machines.[41]

Variable bobbin speed with the spindle-and-flyer came later. The first solution using the bobbin was patented in England by Aaron Garlick of Dukinfield, near Manchester, in 1797 (see Fig. A3.12, p. 359). He deployed two cones, one reversed to the other and linked by a belt, which, controlled by a traversing "screw," moved across the two cones. One cone, turning at a constant speed, took power to the fliers and also drove the up-and-down motion of the fliers (rather than of the bobbins in this case). The second cone imparted a variable motion, which (because of flier lead) progressively increased the speed of the bobbins. This arrangement sufficed for more than two decades, but it would have caused uneven winding and defective roving due to the slippage of the cone belt.[42]

When Joseph Raynor of Sheffield patented his complex device for controlling the cone belt in 1813 (see Fig. A3.13, p. 360), he observed of his invention:

> The regulation of the roving bobbin, so as to produce upon it a speed varying in the proportion required to meet the encrease [*sic*] of the coiling surface as the bobbins fill, has therefore occupied a large share of the attention of spinners and mechanics connected with them, from the infancy of machines spinning. ...Many experiments have been made. ...Among the minor, but not unimportant, advantages of this frame may be mentioned the facility with which it works; the little labour it requires in tending; and the almost entire absence of waste.[43]

Paul Moody, engineer to the Boston Manufacturing Co. of Waltham, Massachusetts, the prototype of the Waltham system of cotton manufacturing, patented an English-type speeder in 1821. His variation consisted in using two pairs of belt-driven cones, one pair to power the relative speed of the bobbin-and-flyer, the other to drive the traversing lifter motion. Interestingly, Moody explicitly said in his patent that he ran the flyers faster than the bobbins, that is, he preferred flyer lead.[44]

While Markham was busy in Vermont, the most elegant solution to the roving-frame problem was solved by the reinvention of the differential gear. This divided the power between the flyers and the bobbins by an ingenious arrangement of epicyclic gears in a manner that was both efficient and simple. Almost simultaneously, the differential was described in a book of machines by James White of Manchester in 1822 and, quite independently, was patented in the United States by Aza Arnold of Rhode Island in January 1823.[45] With the differential gear installed during the mid- and late-1820s, roving frames came in several forms, which Montgomery described in 1832.[46]

[41]Coniah Wood, GB Patent of 22 May 1772 (1018); White, *Samuel Slater*, 34.
[42]Aaron Garlick, GB Patent of 7 February 1797 (2166); Jeremy, *Transatlantic Industrial Revolution*, 59; English, *Textile Industry*, 170.
[43]Joseph Raynor, GB Patent of 1 January 1813 (3633); *RAMA* 2nd ser 23 (Nov 1813), 329–330. Intriguingly, though advertised, the drawing for Raynor's roving frame was omitted from the *RAMA*, at least from the edition in the Manchester Central Reference Library that I was able to use.
[44]Paul Moody, US Patent 19 January, 1821, in the First Circuit Court Case Files and Final Records, 1790–1846.
[45]White, *New Century of Inventions*; Jeremy, "Technological Diffusion."
[46]Montgomery, *Carding and Spinning Master's Assistant*, 109–126.

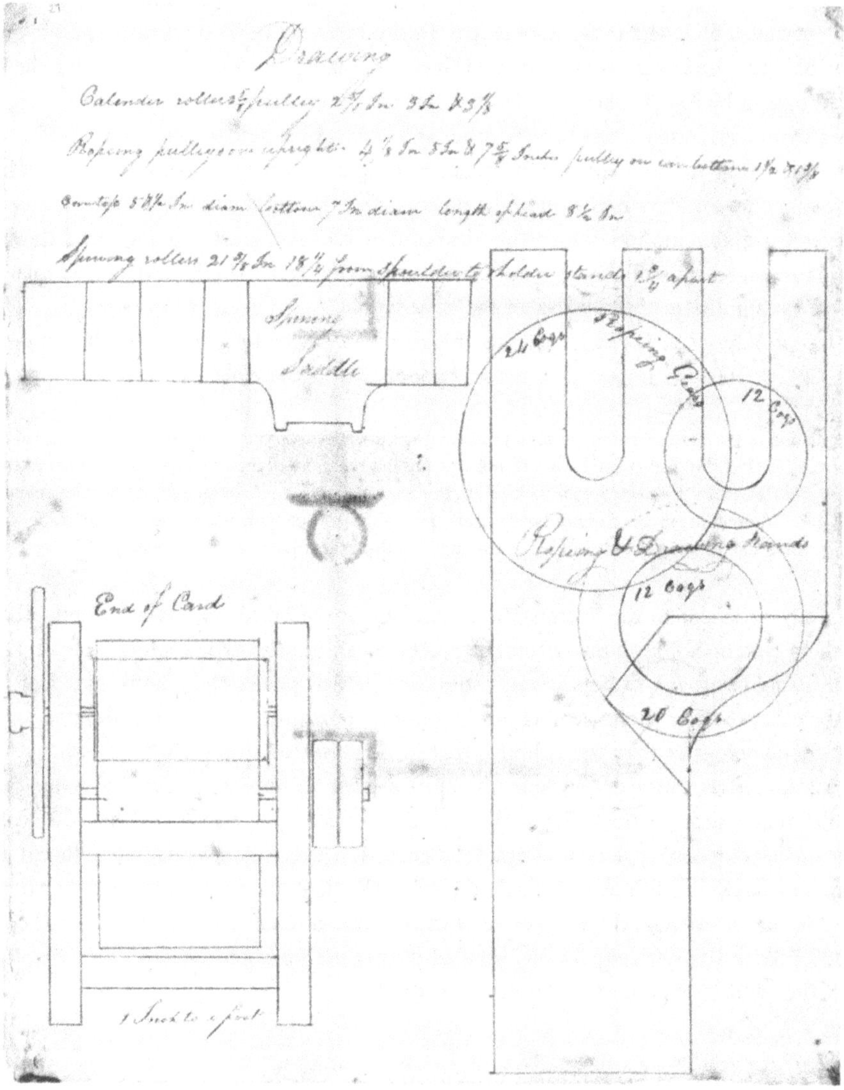

FIGURE 6.46 **Drawing frame and carding-machine components. Scale drawings, elevations.**

Catalogue number: IEM 2.6

Document Description

Measurement: 27.1 x 22.3 cm (one page of a sheet 22.3 x 44.5 cm folded in half and sewn together with another sheet, making eight pages in all; see Figs. 6.38, 6.46–6.48 6.52, 6.54, 6.55, 6.57)

Medium: Ink

Condition: Some bleed-through, foxing, staining in fold, pinholes in corners, sewn

Commentary

At Figure 6.46 are the front elevation of a spinning saddle, the side elevation of a drafting roller stand, and the elevation of the delivery end of a carding machine. The last is dealt with in previous notes.

The spinning saddle was part of the weighting system in a head of drafting rollers by which the upper rollers were held tightly against the lower rollers, thereby gripping the sliver or end of yarn as it passed between the rollers. The saddle was designed to distribute weight evenly on the necks of the drafting rollers. Pressure on the upper rollers was exerted by heavy weights hung up by stirrups from the saddle, as shown in Figure 6.47.

In recording the section of the drafting roller stand, Markham was clearly interested in establishing the dimensions of the gears driving the rollers.

Comparators

Drawings: See Figures A3.3 (p. 351) and A3.8 (p. 355).

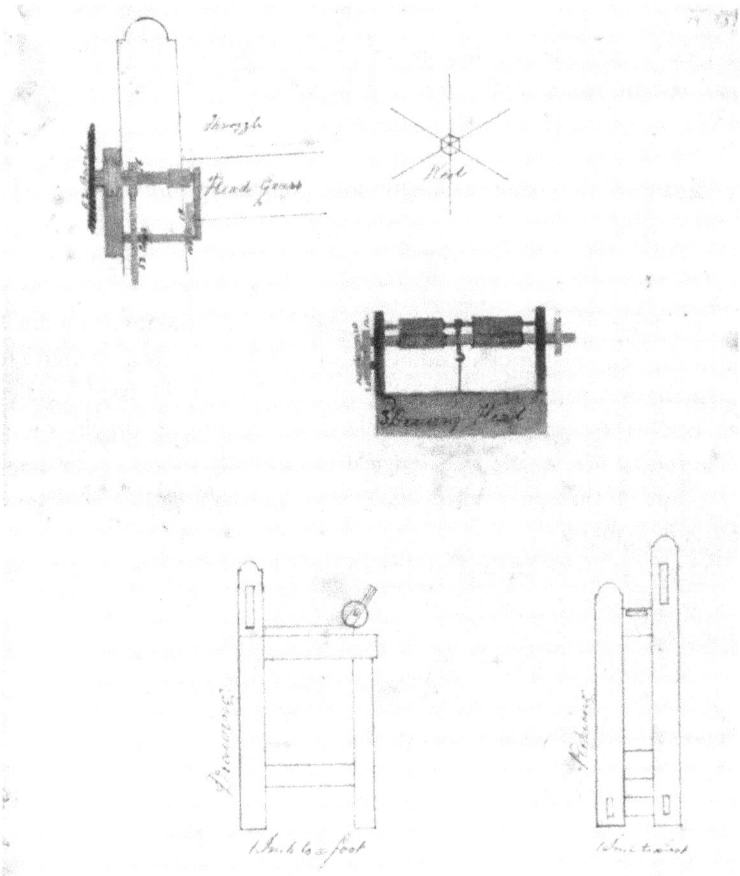

FIGURE 6.47 Drawing head (front and side elevations) for drawing and roving machines and for throstle spinning; framing for drawing and roving machines. Scale drawings, elevations.

Catalogue number: IEM 2.7

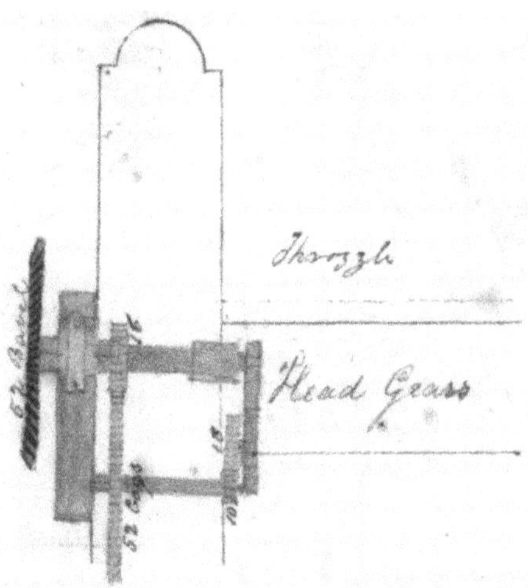

Detail of drawing head for throstle spinning.

Document Description

Measurement: 27.1 x 22.2 cm (one page of a sheet 22.3 x 44.5 cm folded in half and sewn together with another sheet, making eight pages in all; see Figs. 6.38, 6.46–6.48, 6.52, 6.54, 6.55, 6.57)

Medium: Ink wash, red and yellow watercolor

Condition: Pinholes in corners, foxing, stains in fold, torn edges, some transfer and bleed-through, sewn

Commentary

The upper-left colored drawing shows the front elevation of the "head" gear arrangement for driving a series of drafting heads, seen in the upper-right colored drawing. In each, numbers of gear teeth, and therefore relative speeds, are specified. The drawing on the upper right clearly shows the saddle on top of the rollers and the stirrup connecting to a weight unseen below.

Two of the other three drawings on the sheet present side elevations of machine frames (evidently made of wood) one for a drawing frame, the other for a roving frame.

The last item on the page is a sketch of the structure of a reel (*top, right*). This was used when the spun yarn had to be packed for market. The yarn was taken from bobbins and wound onto the reel, a frame consisting of six long horizontal bars positioned in a cylinder formation. It took yarn from a number of bobbins simultaneously. The circumference of the cotton reel was 1.5 yards. When the length of yarn supplied by each of the bobbins had been wound on 80 times, a bell rang. Seven more of these 80 revolutions completed the formation of a hank of yarn 840 yards long. The size or thickness of the yarn was then expressed as the number of hanks per pound avoirdupois, the higher the number, the finer the yarn.[47]

Comparators

Drawings: See Figures A3.3 (p. 351) and A3.8 (p. 355).

[47]Nicholson, *Operative Mechanic* I, 395–396; Jeremy, "British and American Yarn Count Systems."

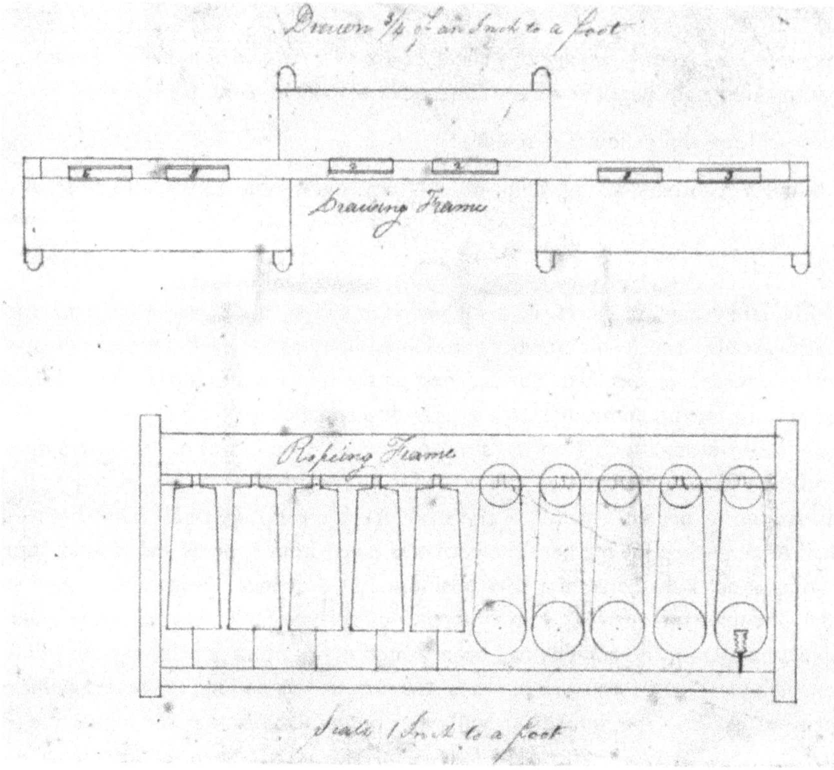

FIGURE 6.48 "Drawing Frame" and "Ropeing Frame." Scale drawings, plan and elevation.

Catalogue number: IEM 2.4

Document Description

Measurement: 22.3 x 26.9 cm (one page of a sheet 22.3 x 44.5 cm folded in half and sewn together with another sheet, making eight pages in all; see Figs. 6.38, 6.46–6.48, 6.52, 6.54, 6.55, 6.57)

Medium: Ink

Condition: Pinhole in one corner, minor bleed-through, foxing, some transfer from opposite page

Commentary

The upper drawing is the plan for a three-section drawing frame. The numbers indicate the location of the drafting heads. The lower drawing is a roving-can frame. The drawing appears to be split between a front elevation of the cans holding the slivers (taken from the drawing frame); and a plan of the pulley arrangement for driving the cans.

Comparators

Drawings: See Figures A3.3 (p. 351) and A3.4 (p. 352).

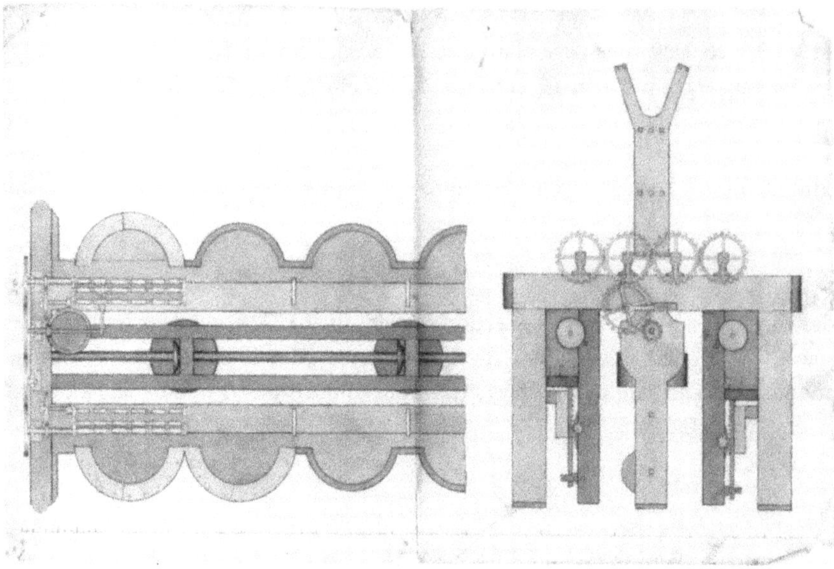

FIGURE 6.49 Drawing frame. Scale drawings, plan and elevation.

Catalogue number: IEM 3.1-2

Document Description

Measurement: 27 x 42.5 cm

Medium: Ink and wash

Watermark: "T G & Co"

Condition: Holes in corners and centers of edges, torn edges, minimal staining

Commentary

Part plan on the left, side elevation on right. Apparently, the main function of this machine was in passing carded slivers through heads of drafting rollers so as to stretch them and to impart greater elongated uniformity to the cotton fibers. Evidently, the drafted slivers were delivered into cans raised and lowered by a rack-and-pinion motion. The vertical frame member shown in this section would hold a rack of carded slivers.

Comparators

Drawings: See Figure A3.3 (p. 351).

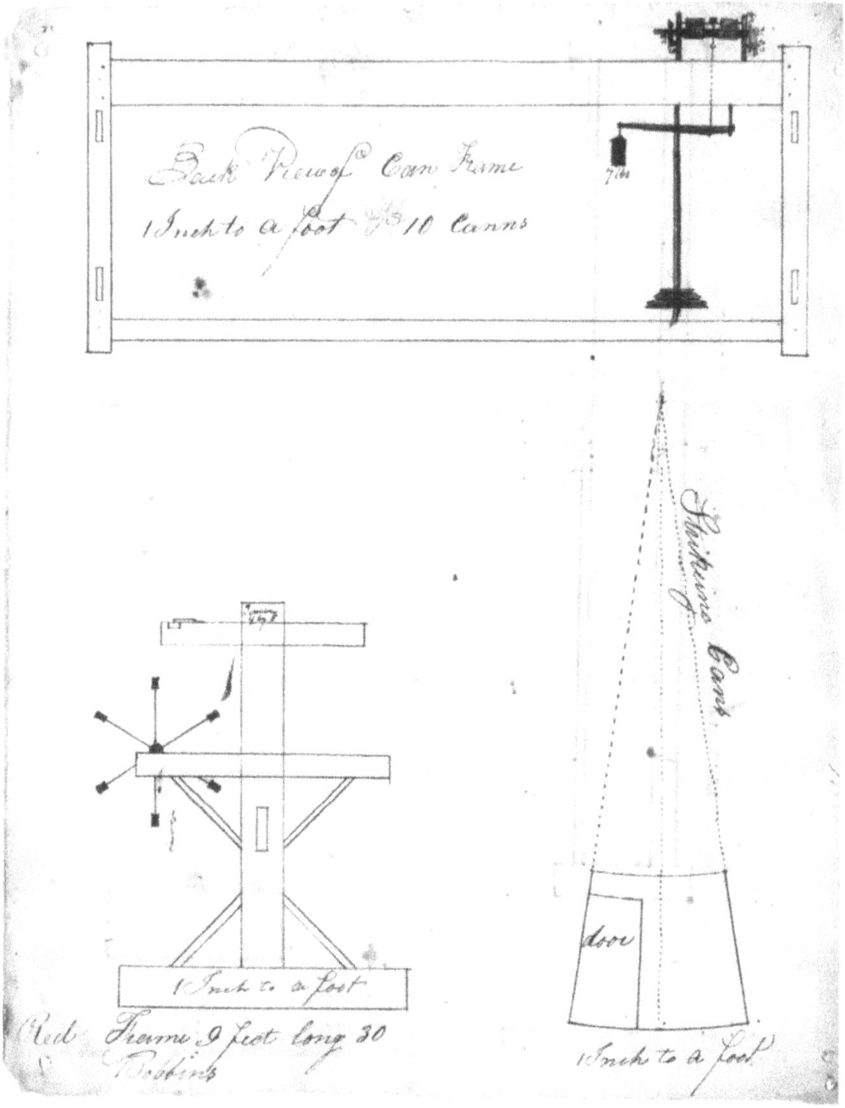

FIGURE 6.50 Roving-can frame, components, and reeling frame. Scale drawings, elevations.

Catalogue number: IEM 4.2

Document Description

Measurement: 27.7 x 22.8 cm (one page of sheet 27.7 x 45.5 cm folded in half to make four pages in all; see Figs. 6.50, 6.67, 6.69, 6.70)

Medium: Ink, wash, red and yellow watercolor

Condition: Pinholes in corners, small stains, ink smear, foxing, bleed-through

Commentary

At top, a rear elevation is clearly marked "Can Frame...for 10 Canns" and gives the weight to be applied to the saddle on the upper drafting rollers.

At lower left, "Reel frame 9 ft long 30 Bobbins," would have been used to take the yarn from 30 bobbins and wind it into 30 skeins formed on the spoked reel at the upper left of the reel drawing. See Figure 6.47 (p. 234).

The lower-right drawing, marked "Strikeing Cans," shows the cans, shaped like the frustum of a cone, for receiving the roving. The meaning of striking is unclear.

Comparators

Drawings: J. Thomson, "Manufacture of Cotton"; Nicholson, *Operative Mechanic* I, 390; see Fig. A3.4 (p. 352) for the roving-can frame; see Fig. A3.11 for the reeling frame (p. 358).

FIGURE 6.51A "Sketch of the Old English Speeder Oct 2d, 1821 Presented by Asahel Adams."
Catalogue number: IEM 5b

Document Description

Measurement: 21.5 x 30.4 cm

Medium: Ink, laid paper

Condition: Creases, stains, foxing, bleed-through; reverse side of Figure 6.51B

Commentary

Asahel Adams: possibly Markham's collaborator of that surname who built two dressing frames for the Dover Cotton Factory at Dover, New Hampshire. An 1820 inventory of the Dover Cotton Factory lists "2 warping machines with stop motions with 2 dressing of Adams and Markham's plan," all valued at $1,000.[48] Presumably the calculations were related to gear teeth.

[48]Agreement between Williams & Wendell and A. & J. Wendell, 29 June 1820, Dover Cotton factory file, Williams & Wendell, 1815–1831, Case 11, Wendell Collection, Baker Library.

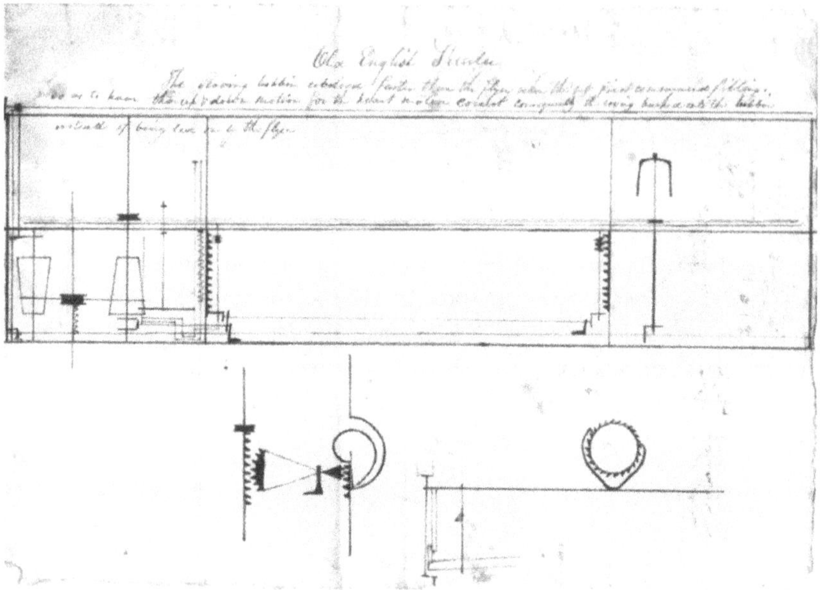

FIGURE 6.51B "Old English Speeder" (Double speeder). Drawing, not to scale, showing elevation and components.
Catalogue number: IEM 5a

Document Description

Measurement: 21.5 x 30.4 cm

Medium: Ink

Condition: Creases, stains, foxing, bleed-through; reverse side of Figure 6.51A

Transcription

"Old English Speeder

The Roving bobbin revolved faster than the flyer when the set first commenced filling so as to have the up and down motion for the heart motion correct consequently the roving backed onto the bobbin instead of being led on by the flyer."

Commentary

Adams used bobbin lead (as explained in the introduction to this section) so that he could use the constant speed of the cone driving the flyers to power the traverse of the bobbin-lifter rail, a regular up-and-down movement.

Whereas Joseph Raynor, in GB Patent of 1 January 1813 (3633) installed a cylinder of differentiated concentric gears, in conjunction with a sliding pinion, to slow the bobbins (see Figure A3.13, p. 360), here a rack and pawl are used.

The two lower drawings are less clear. That on the left shows a rack-and-pawl arrangement. That on the right apparently shows a rocking escapement in association with an inverted flyer and a bobbin-lifter rail.

See chapter 6, section 4 for a technical explanation of bobbin lead and flyer lead in roving.

Comparators

Drawings: Figures A3.5 (p. 352), A3.6 (p. 353), A3.12 (p. 359), A3.13 (p. 360).

5. DRAWINGS: SPINNING—CONTINUOUS AND INTERMITTENT

Spinning

To convert roving into yarn, all spinning implements and machines incorporate three distinct actions: drafting, twisting, and winding. In the ancient distaff all three were separate movements. In the medieval wool-spinning wheel, drafting and twisting operated concurrently, winding was a separate operation. In the later so-called flax wheel, which used the bobbin-and-flyer, drafting, twisting, and winding were concurrent.

The eighteenth-century textile revolution saw the development of waterpowered machines driving dozens and later hundreds of spindles simultaneously. Although there were a number of variant early spinning machines, the principles on which they ran developed into two rival systems.

Continuous-spinning machines allowed drafting, twisting, and winding to occur simultaneously. This technological breakthrough was achieved by Richard Arkwright.[49] His water frame, patented in 1769, successfully combined a head of drafting rollers with the flyer and bobbin (see Fig. A3.8, p. 355). Because drafting and twisting were physically separated in the machine, it produced low-count, hard yarn suitable for warps. To drive, it needed more power than a single human operative could supply and therefore originally relied on waterpower, hence the term "water twist" for its yarn.

A variant of the water frame appeared by the time Thomson wrote in 1812.[50] In the water frame, spindles were arranged in small groups (of four in the machines seen by Thomson in Strutts' mill at Belper in Derbyshire) each supplied by a head of drafting rollers and driven by pulleys on a vertical axle. By replacing the pulleys with a long, hollow, tin roller from which power to the spindles was taken by cord drives, many more spindles could be accommodated in the spinning machine. It was known as a throstle because, apparently, the whirling noise of the tin roller echoed the song of a thrush (see Figs. 3.8 and 3.9 (pp. 57, 58).

Rivaling the water frame was Samuel Crompton's mule, invented in the late-1770s. This machine combined part of Arkwright's water frame with part of James Hargreaves's spinning jenny of 1770 (see Fig. 3.3, p. 51). That is, Arkwright's drafting rollers (see Fig. 3.9, p. 58 and Fig. A3.8, p. 355) were used to insert some initial twist while Hargreaves's moving clasp or carriage provided much more twist. Working on the intermittent-spinning principle, in which drafting and twisting were simultaneous but winding was a separate movement, technically it was the most versatile of all spinning machines, capable of making all counts and twists of yarn and with maximum uniformity (see Fig. A3.10, p. 357).

Improvements to the mule before 1815 applied power to its drafting rollers, to the spindles on the outward run of the carriage (when the yarn was drafted and twisted simultaneously), and to the outward movement of the carriage. To the observer, as Thomson himself noted, "this machine is extremely complicated in its movements."[51] However, the inward run of the carriage and the winding action of the spindles defied mechanization and needed to be controlled by a skilled operative, the mule spinner.

During the 1820s, the militancy of skilled mule spinners in Lancashire drove mill owners to promote the quest for a self-acting mule in which the winding action was performed by mechanical movements.

---

[49]For whom see Fitton, *Arkwrights*.
[50]J. Thomson, "Manufacture of Cotton."
[51]J. Thomson, "Manufacture of Cotton."

In New England a parallel interest in improving the mule and other spinning machines was impelled by the shortage, and consequent relatively high wages, of mule spinners. Demand-side problems in Britain proved more effective than supply-side problems in the United States and the pioneering automatic mule was designed by Richard Roberts of Manchester and covered by his British patents of 1824 and 1830.[52]

[52]Hills, *Life and Inventions of Richard Roberts*; Hills, "Richard Roberts" in Cantrell and Cookson (eds.).

Spinning: Continuous Spinning

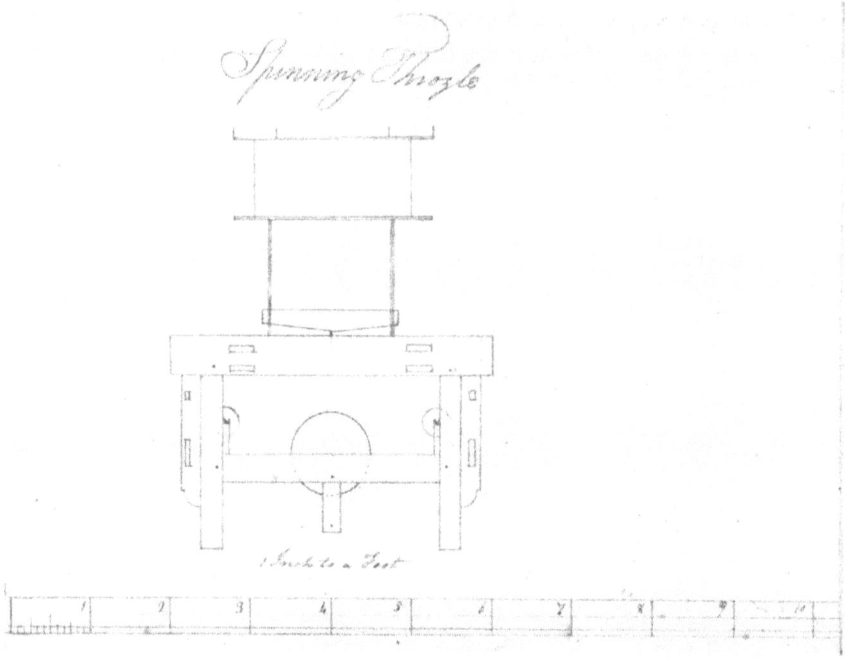

FIGURE 6.52 "Spinning Throzle." Scale drawing, side elevation.

Catalogue number: IEM 2.1

Document Description

Measurement: 22.2 x 26.9 cm (one page of a sheet 22.3 x 44.5 cm folded in half and sewn together with another sheet, making eight pages in all; see Figs. 6.38, 6.46–6.48, 6.52, 6.54, 6.55, 6.57).

Medium: Ink

Condition: Foxing, sewn, stitches on bottom are loose, pinholes in two corners

Commentary

The elements above the main frame represent a rack for holding roving bobbins. Within the lower part of the frame are two throstle rollers, one for each side of spindles. The circle in the center represents a drum that is positively driven by gearing (see Fig. 6.54) and which in turn drives the throstle rollers by means of cords or bands. This arrangement, as Thomson noted, facilitated piecing:

> The bands [driving the spindles] are very loose, and,...are inclined, so that their weight tends to draw them tight, and turn the spindles about with the proper velocity; but still the child attending the machine can, by pressing his knee against the whirl, as the pulley is called, stop the motion of any one spindle for a moment whilst a broken thread is repaired, the band slipping round it all the time.[53]

Comparators

Drawings: Figure A3.9 (p. 357).

Artifacts:

American:

Somewhat later versions, with iron frames, are in the National Museum of American History, Smithsonian Institution, Washington, DC, and the Old Slater Mill Museum, Pawtucket, Rhode Island.

[53] J. Thomson, "Manufacture of Cotton."

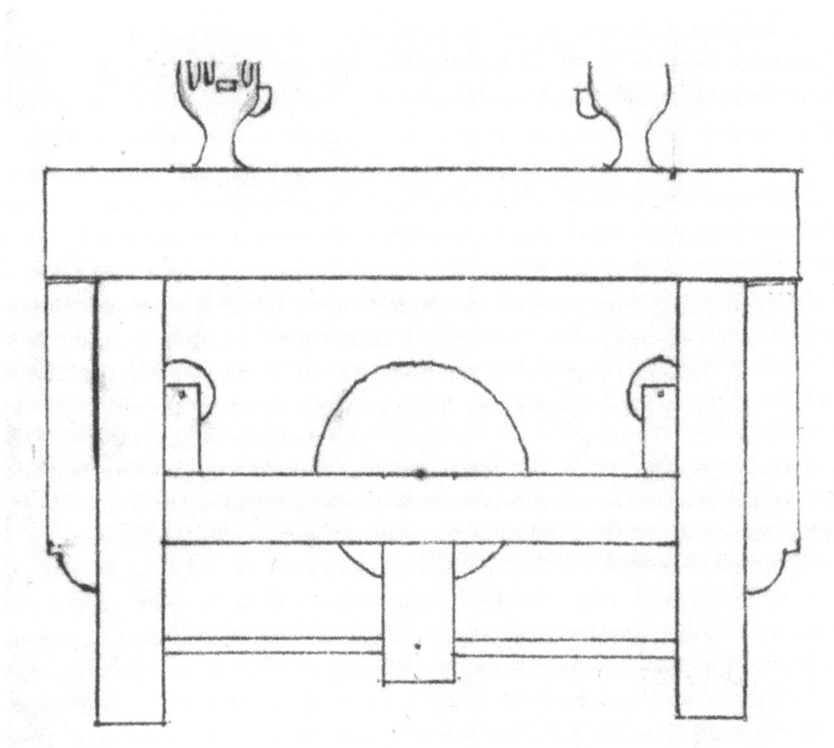

FIGURE 6.53 Throstle. Drawing, side elevation.

Catalogue number: IEM 21

Document Description

Measurement: 22.7 x 27.6 cm

Medium: Ink

Condition: Minor discoloring, page torn, some losses to paper, torn pinhole

Commentary

This shows the position of drafting roller stands, seated on the top horizontal frame member, and, below, the central pulley drive out to the throstle rollers, on the right and the left.

Comparators

Drawings: Figure A3.9 (p. 356).

Artifacts:

American:

Somewhat later versions, with iron frames, are in the National Museum of American History, Smithsonian Institution, Washington, DC, and the Old Slater Mill Museum, Pawtucket, Rhode Island.

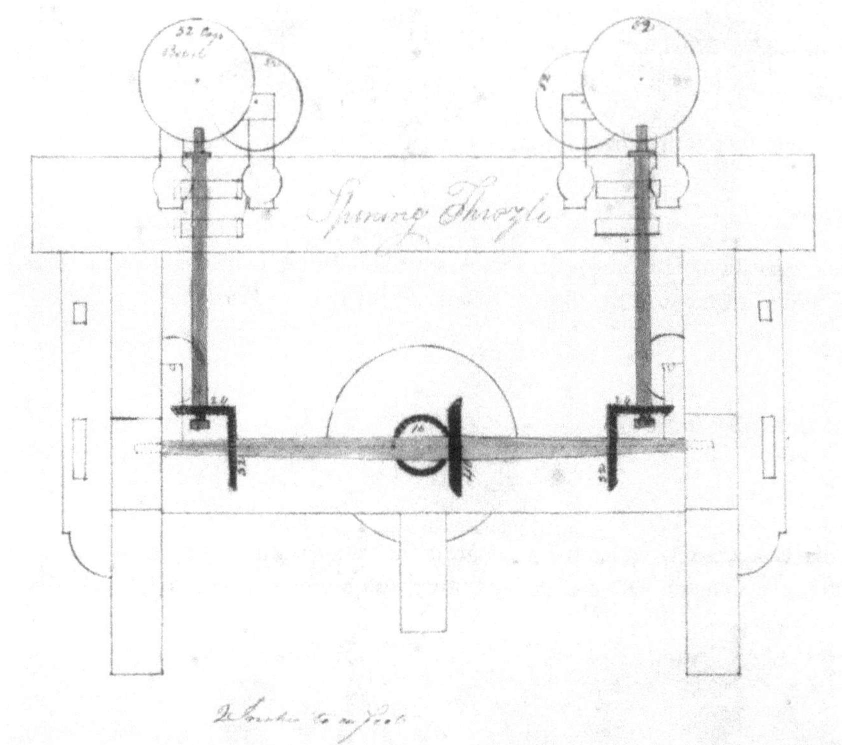

FIGURE 6.54 "Spinning Throzle." Scale drawing, side elevation.

Catalogue number: IEM 2.5

Document Description

Measurement: 22.3 x 26.9 cm (one page of a sheet 22.3 x 44.5 cm folded in half and sewn together with another sheet, making eight pages in all; see Figs. 6.38, 6.46–6.48, 6.52, 6.54, 6.55, 6.57)

Medium: Ink and wash

Condition: Foxing, pinholes in corners, bleed-through, sewn

Commentary

The drawing shows gearing (and therefore positive drives) to the drafting rollers (gear teeth numbered) and to the main driving roller powering the two throstle rollers (*extreme left and right*). Unseen are the cords taking power from the throstle rollers to the spindles. The cross-section of the (presumably wooden) shaft carrying the three driving gears is presumably asymmetrical so that the gears can be hammered or wedged securely into place.

Comparators

Drawings: Figure A3.9 (p. 356).

Artifacts:

American:

Somewhat later versions, with iron frames, are in the National Museum of American History, Smithsonian Institution, Washington, DC, and the Old Slater Mill Museum, Pawtucket, Rhode Island.

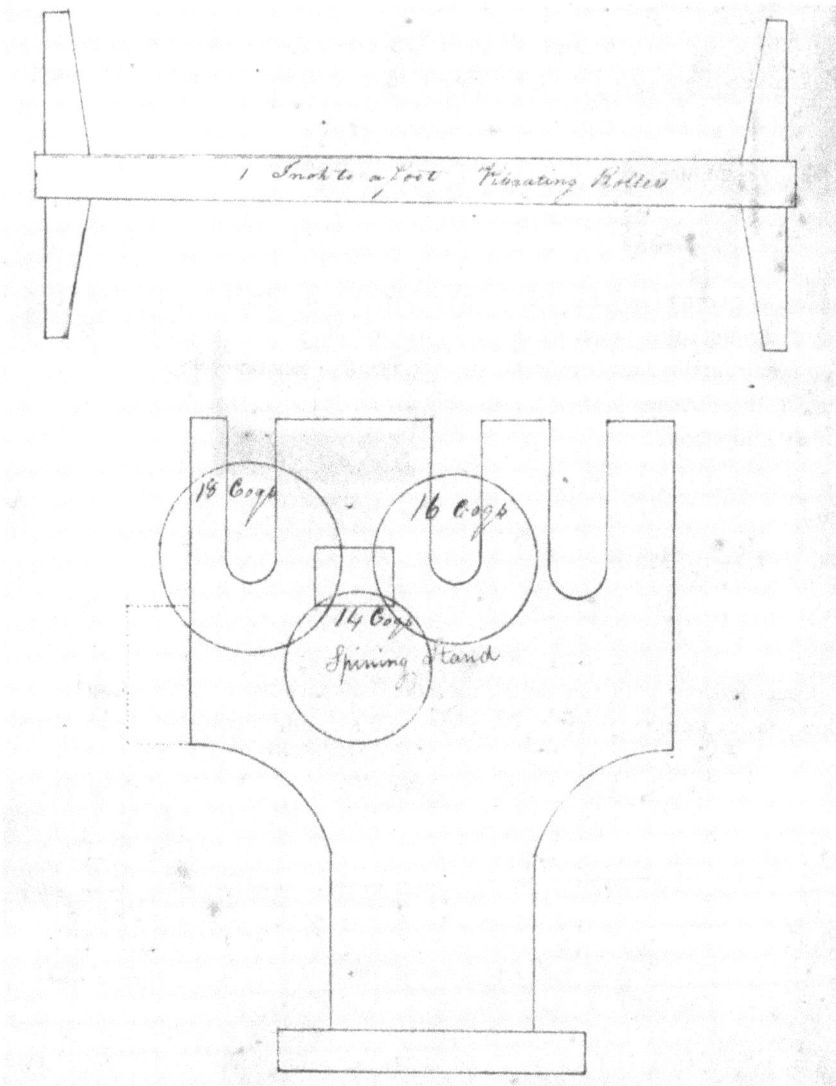

FIGURE 6.55 Throstle components: throstle roller (plan or front elevation) and drafting head (side elevation). Scale drawings.

Catalogue number: IEM 2.8

Document Description

Measurement: 27.1 x 22.2 cm (one page of a sheet 22.3 x 44.5 cm folded in half and sewn together with another sheet, making eight pages in all; see Figs. 6.38, 6.46–6.48, 6.52, 6.54, 6.55, 6.57)

Medium: Ink

Condition: Red substance (wax?) in corners, pinholes in corners, foxing, stained edges, bleed-through from obverse, tear on edge, sewn

Commentary

The upper drawing is of a plan of a vibrating or throstle roller (for which see introduction to this section); the lower, of a side elevation of a drafting roller (or spinning) stand, showing numbers of gear teeth to the lower front and back rollers. The upper front and back rollers would have been loosely slotted into the stand and pressed down on the lower ones by a weighted saddle.

Comparators

Drawings: Figure A3.9 (p. 356).

Artifacts:

American:

Somewhat later versions, with iron frames, are in the National Museum of American History, Smithsonian Institution, Washington, DC, and the Old Slater Mill Museum, Pawtucket, Rhode Island.

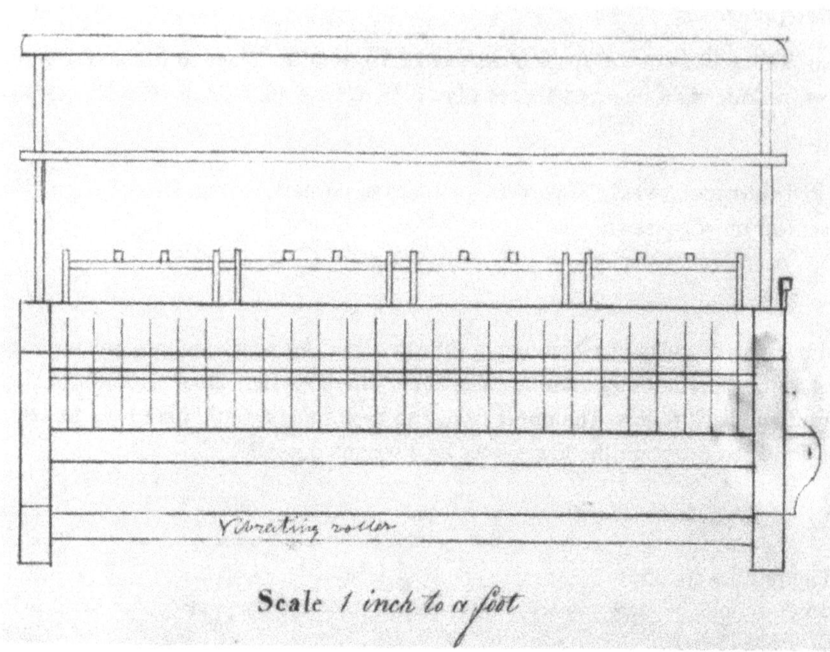

FIGURE 6.56 Throstle. Scale drawing, front elevation.

Catalogue number: IEM 20

Document Description

Measurement: 22.7 x 27.6 cm (one page of a sheet 22.3 x 44.5 cm folded in half and sewn together with another sheet, making eight pages in all; see Figs. 6.38, 6.46–6.48, 6.52, 6.54, 6.55)

Medium: Ink

Condition: Red substance (wax?) in corner, slight foxing, stain, torn edges

Commentary

This shows the vibrating or throstle roller, above it the position of the spindles (24 gaps suggesting that number of spindles), the drafting roller heads, and above them the frame for holding bobbins loaded with roving.

Comparators

Drawings: Figure A3.9 (p. 356).

Artifacts:

American:

Somewhat later versions, with iron frames, are in the National Museum of American History, Smithsonian Institution, Washington, DC, and the Old Slater Mill Museum, Pawtucket, Rhode Island.

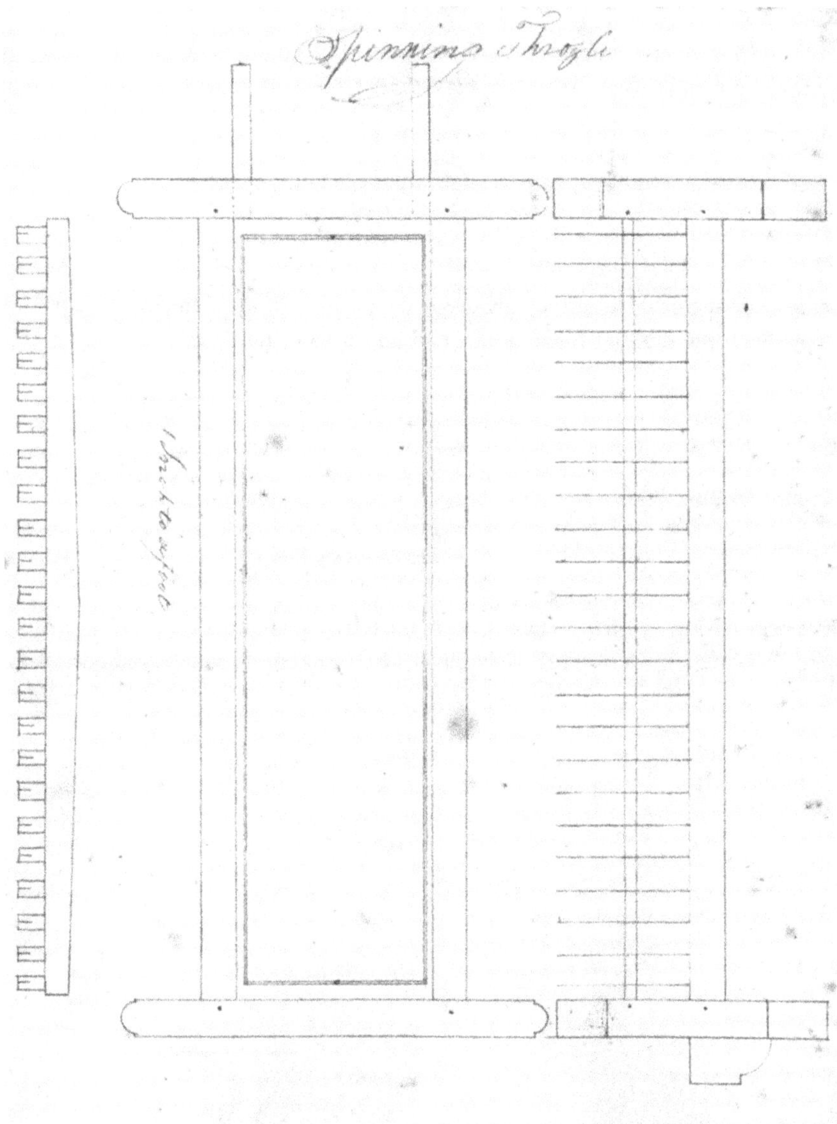

FIGURE 6.57 "Spinning Throzle." Scale drawings of three components.

Catalogue number: IEM 2.2

Document Description

Measurement: 26.9 x 22.2 cm (one page of a sheet 22.3 x 44.5 cm folded in half and sewn together with another sheet, making eight pages in all; see Figs. 6.38, 6.46–6.48, 6.52, 6.54, 6.55)

Medium: Ink

Condition: Pinholes in two corners; foxing, especially in fold; sewn; stitches on bottom are loose

Commentary

At left, (possibly) lifter rail for bobbins or for flyers. Shown are 24 inverted U-shapes, representing flyers. This identity is not wholly definite but it would fit with the depiction of the bobbin-lifter rail behind the flyers (when facing the front elevation of the machine) as seen in Rees's *Cyclopaedia* (see Fig. A3.9, p. 356, lower elevation).

At center, plan of lateral members for frame of throstle.

At right, elevation of spindle rail and part of frame, showing 23 gaps, presumably for a 24-spindle throstle.

Comparators

Drawings: Figure A3.9 (p. 356).

Artifacts:

American:

Somewhat later versions, with iron frames, are in the National Museum of American History, Smithsonian Institution, Washington, DC, and the Old Slater Mill Museum, Pawtucket, Rhode Island.

Spinning: Intermittent

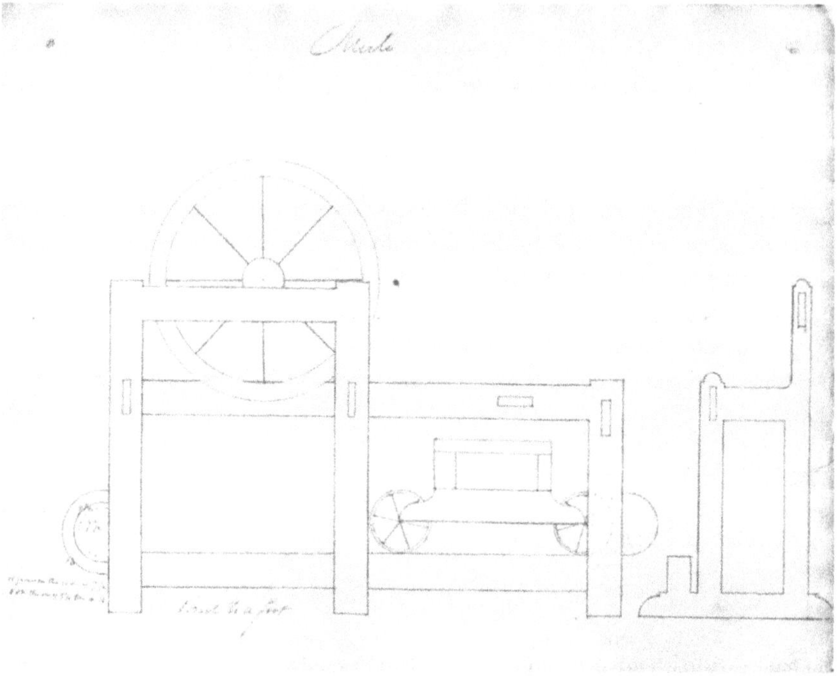

FIGURE 6.58 "Mule." Scale drawing, side elevation.

Catalogue number: IEM 17.3

Document Description

Measurement: 22.7 x 29 cm (half of folded sheet 29 x 45.6 cm that includes Figs. 6.58–6.60)

Medium: Ink

Condition: Pinholes in corners of page, minor foxing. minor tears on edges

Commentary

At the right is the beam supporting drafting rollers. At the upper left is the fly or large main driving wheel; in the center, the spindle carriage; and at the lower right and left, the pulleys for power, presumably for the wheeled track-guided spindle carriage.

Comparators

Drawings: Figure A3.10 (p. 357).

Artifacts:

British:

A section of an early spinning mule has been preserved in the Science Museum, London.

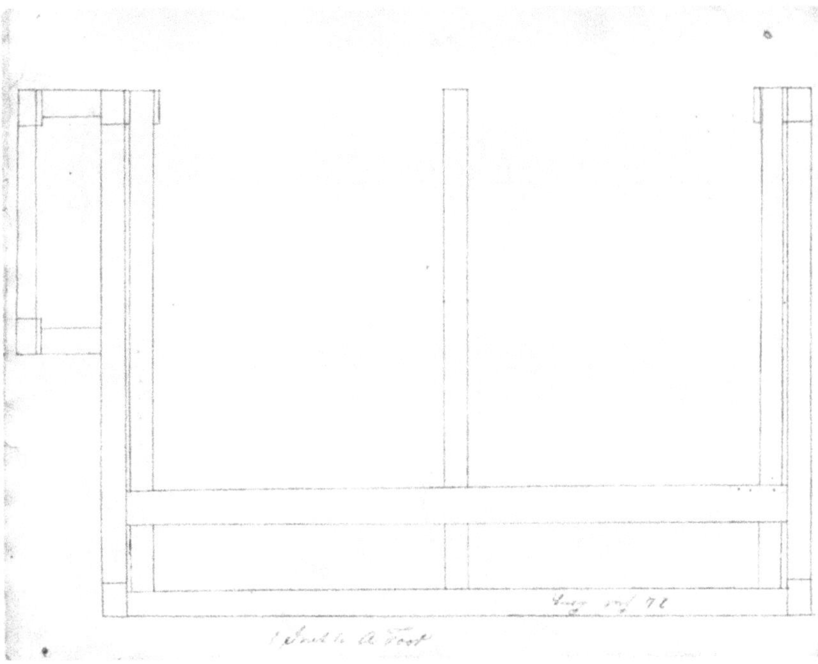

FIGURE 6.59 Mule frame, plan. Scale drawing.

Catalogue number: IEM 17.2

Document Description

Measurement: 22.8 x 29 cm (half of folded sheet that includes Figs. 6.58–6.60)

Medium: Ink

Condition: Pinholes in corners, bleed-through, minor foxing

Commentary

On the upper left is the frame housing the flywheel, a part of the machinery known as the headstock, built alongside the first of three parallel bases for the carriage track. At the bottom is the framing to support the drafting roller beam.

At 24 feet long, this mule probably had over 200 spindles. McConnell & Kennedy, the Manchester machine makers, in the mid-1790s quoted lengths of 21 feet, 4 inches for 180 spindles and 23 feet, 9 inches for 204 spindles.[54]

Comparators

Drawings: Figure A3.10 (p. 357).

Artifacts:

British:

A section of an early spinning mule has been preserved in the Science Museum, London.

[54]Lee, *Cotton Enterprise*, 19.

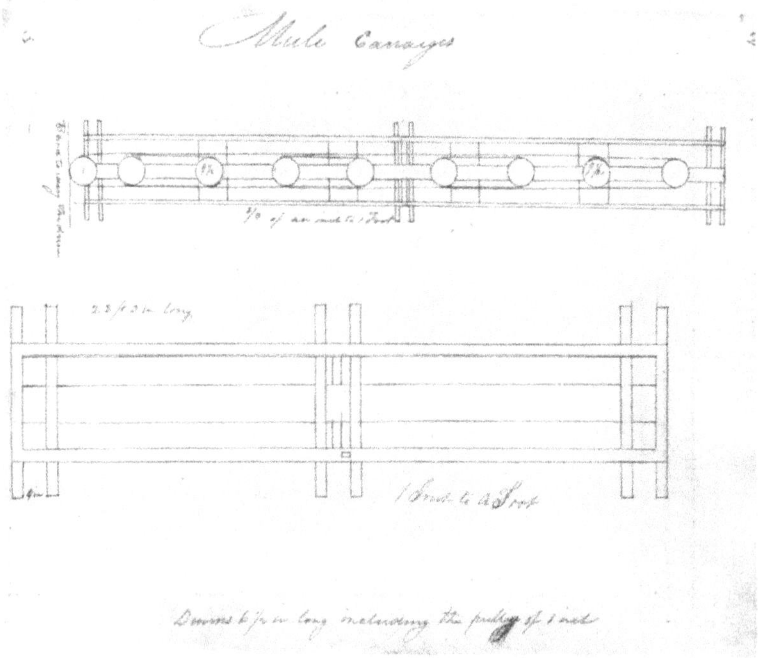

FIGURE 6.60 "Mule Carraiges" (*sic*). Scale drawings, plans.

Catalogue number: IEM 17.1

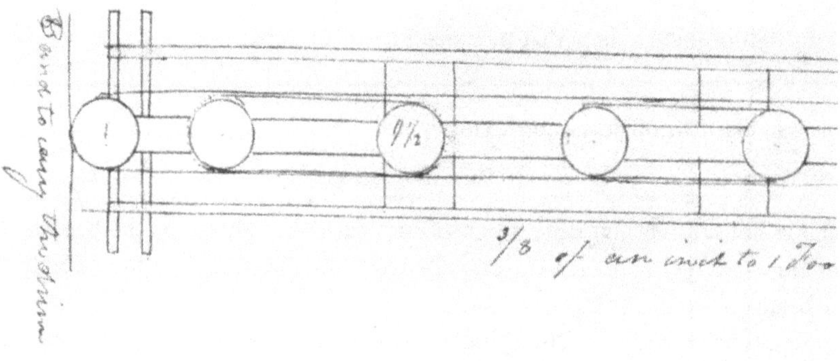

"Mule Carraiges" (*sic*), detail.

Document Description

Measurement: 22.8 x 29 cm (half of folded sheet that includes Figs. 6.58–6.60)

Medium: Ink

Condition: Pinholes in corners, bleed-through, minor tears on edges

Commentary

The upper plan shows the pulley arrangement for driving the spindles mounted on the mule carriage. The circles marked "9½" (inches, presumably) were the drums from which cords drove the spindles. If the arrangement was similar to the English mule reported by Thomson, the drums were powered by bands activated by a large horizontal pulley on the spindle carriage. That pulley took power from the main fly wheel by means of an endless rope extending between the roller beam and the headstock.[55]

The lower plan relates to a mule 22 feet, 3 inches long. However, as the paper is less than 12 inches square, the scale of 1 inch to the foot cannot be true unless it is not a plan but a side section less than 12 feet in length. The ⅜ inches to the foot scale of the upper plan fits the size of the paper on which it is drawn.

Comparators

Drawings: Figure A3.10 (p. 357).

Artifacts:

British:
A section of an early spinning mule has been preserved in the Science Museum, London.

[55]J. Thomson, "Manufacture of Cotton."

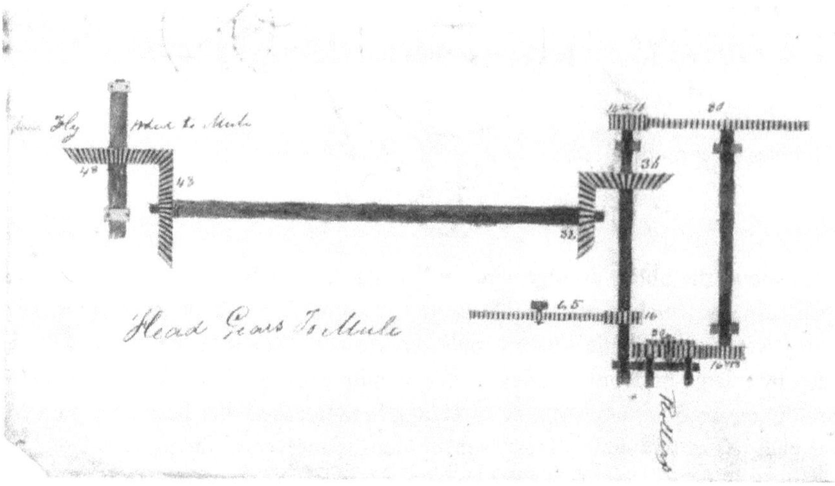

FIGURE 6.61 **"Head Gears to Mule," plan.**

Catalogue number: IEM 8.3

Document Description

Measurement: 22 x 26 cm (half of folded sheet 44 x 26 cm that includes Figs. 6.41A, 6.41B, 6.43, 6.61)

Medium: Ink, wash, and yellow watercolor

Condition: Pinholes in edges and corners, foxing, bleed-through, stains, tiny piece of red substance (wax?)

Commentary

Drawing showing gear sizes, in numbers of teeth per gear, for establishing appropriate speed ratios between the "mule fly" (*at left*) and the drafting rollers (*at right*).

The mule fly was on the primary driving shaft of the machine, taking its power from the mill water wheel by means of overhead shafting and belts. From the fly, power was distributed in three directions:

- by bevel gears and shafting, to the drafting roller beam (*at right*);
- (not shown) by endless ropes and pulleys, to the spindles on the carriage (see Fig. 6.60);
- (not shown) by endless rope, to the carriage itself, to power the drafting action of the carriage as it moved outward away from the drafting rollers toward the head stock where the mule spinner stood.

Comparators

Drawings: Figure A3.10 (p. 357).

Artifacts:

British:
A section of an early spinning mule has been preserved in the Science Museum, London.

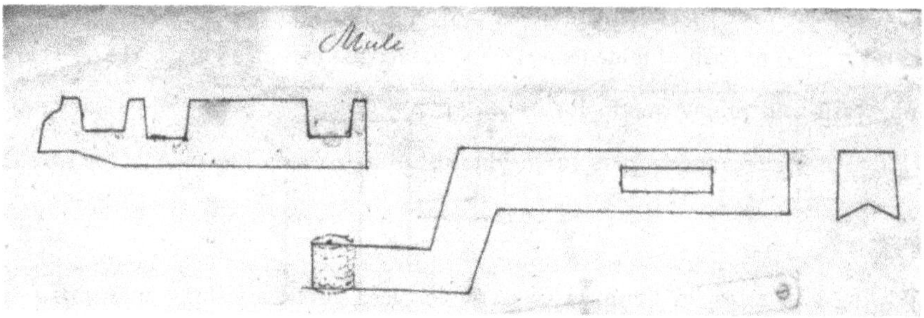

FIGURE 6.62 "Mule" elements. Three line drawings, without a scale.

Catalogue number: IEM 9.2

Document Description

Measurement: 27.5 x 27.3 cm

Medium: Ink

Condition: Pinholes in each corner, foxing, bleed-through, minor tears on edges; reverse side of Figure 6.44

Commentary

Components of unknown function or location in the spinning mule.

Comparators

Drawings: A3.10 (p. 357).

Artifacts:

British:
A section of an early spinning mule has been preserved in the Science Museum, London.

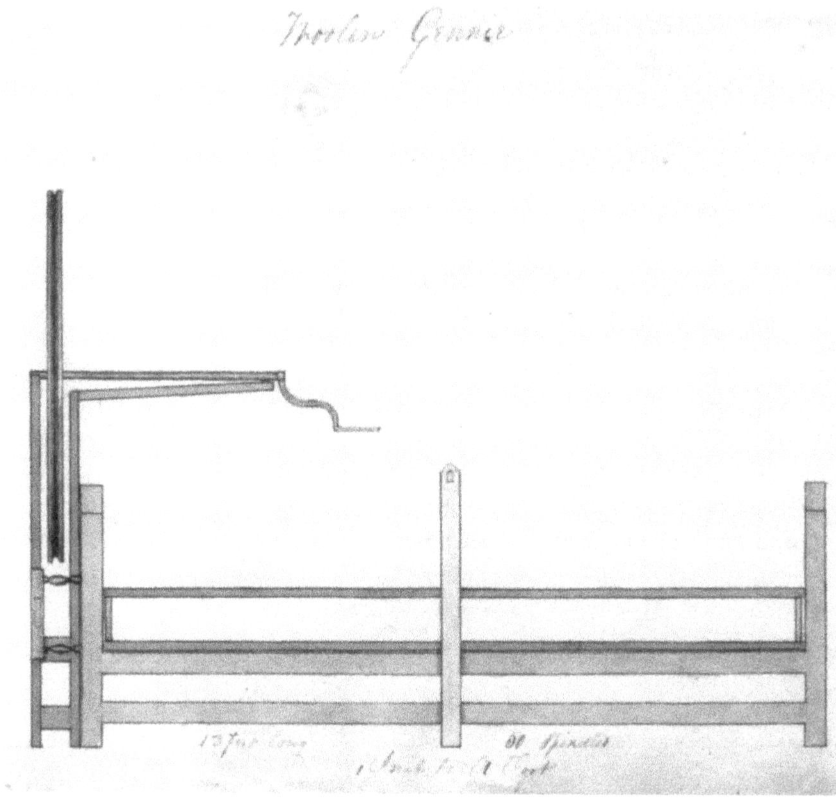

FIGURE 6.63 "Woolen Gennie." Scale elevation of a hand-powered woolen-spinning jenny, 13 feet long, for 50 spindles. If manual control of power to the fast-loose pulley was on the right, this is a rear elevation.

Catalogue number: IEM 16.2

Document Description

Measurement: 22.7 x 27 cm pages (half of folded sheet 27 x 45.4 cm that includes Fig. 6.64)

Medium: Ink and wash

Condition: Pinholes in corner, stains on edges, foxing, minor tears

Commentary

This appears to be a rear elevation of the jenny, when compared to the machine illustrated in Bakewell, "Woollen Manufacture" (see Figure A3.7, p. 354). Thus a right-handed spinner would turn the flywheel with his right hand.

Comparators

Drawings: American or Tully "jenny," Figure 3.4 (p. 52); French jenny, Figure 3.2 (p. 50); Figure A3.7 (p. 354).

Artifacts:

British:

Reconstructions of Hargreaves's jenny are in the Textile Museum at Helmshore, Lancashire, and the Science Museum in London. A later jenny is in the Science Museum in London.

American:

A reconstruction of Tully's jenny is in the American Textile History Museum at Lowell, Massachusetts.

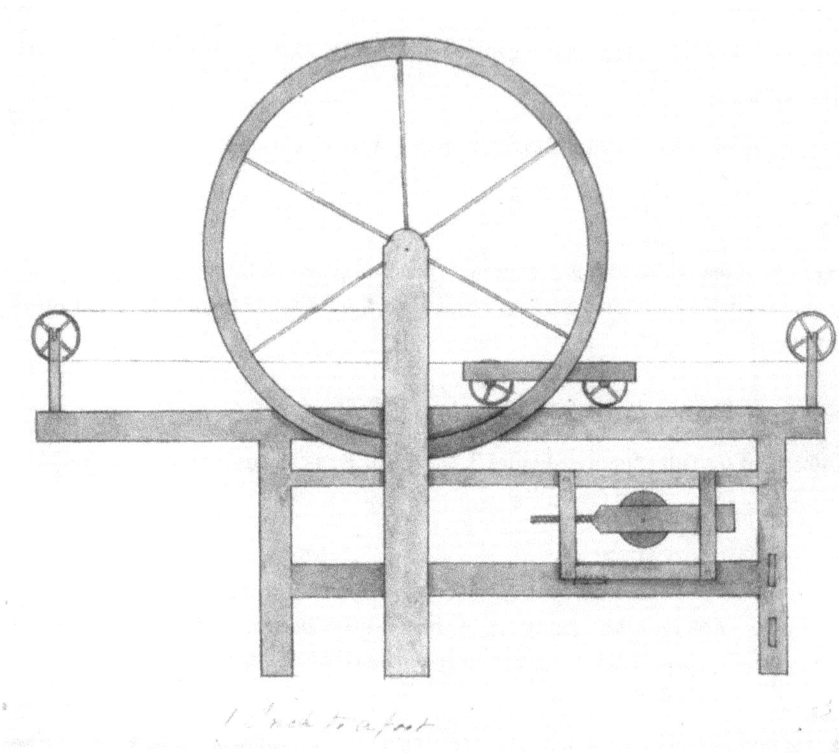

FIGURE 6.64 Woolen jenny, side elevation. Scale drawing, perhaps for the jenny seen at Figure 6.63.
Catalogue number: IEM 16.1

Document Description

Measurement: 27 x 22.7 cm pages (half of folded sheet 27 x 45.4 cm that includes Fig. 6.63)

Medium: Ink and wash

Condition: Pinholes in corners, foxing

Commentary

The large flywheel (*center*), probably turned manually, would have driven the long tin roller (*lower right*) from which individual bands powered the 50 spindles (*further to the right but not seen*). The screw to the left of the roller evidently allowed the position of the roller to be adjusted and thus the tension on the spindle pulleys to be loosened or tightened. The wheeled carriage supported a clasp through which woolen slubbings (rovings) passed. These came from a rack of slubbing cops (not seen but below and between the tin roller and the axis of the flywheel). From the wheeled clasp the slubbings passed to the spindles (to the right of the tin roller).

In operation the woolen jenny was intermittent like the cotton jenny or the cotton mule. First, more slubbing was drawn through the open clasp and into the drafting zone between the clasp and the spindles. Second, the wheeled clasp was shut, gripping the 50 slubbings, and as the carriage was drawn away from the spindles the flywheel was simultaneously and rapidly turned. This drafted and twisted the slubbings converting them into woolen yarns. Third, a cord-activated faller wire (triggered from the clasp) lowered the yarns ready for winding on the spindles. The flywheel was again rapidly turned and the carriage moved toward the spindles so that the yarns were wound onto them. The cycle was then repeated until the spindles were loaded with yarn (see Fig. A3.7, p. 354). This operation is described in Bakewell.[56]

Comparators

Drawings: American or Tully jenny, Fig. 3.4 (p. 52); French jenny, Fig 3.2 (p. 50); Figure A3.7 (p. 354).

Artifacts:

British:

Reconstructions of Hargreaves's jenny are in the Textile Museum at Helmshore, Lancashire, and the Science Museum in London. A later jenny is in the Science Museum in London.

American:

A reconstruction of Tully's jenny is in the American Textile History Museum at Lowell, Massachusetts.

[56]Bakewell, "Woollen Manufacture."

6. DRAWINGS: SPINNING—"FLY WHEEL SPINNING"

"Fly, in mechanics, is an apparatus employed to equalize and regulate the velocity and power of many machines which require such regulation from the inequality either of the moving power, or of the resistance it has to overcome."[57] Markham, in all likelihood, would have been familiar with this article, published in 1810, and the advice it gave on the use of flywheels in machines.

A more modern source confirms the utility of flywheels: "Flywheels are attached to a shaft and, because of the large mass involved, absorb and release energy to maintain a more uniform rotation."[58] Flywheels are most appropriate for use with reciprocating machines, like steam or petrol engines, in which impulses of energy occur cyclically.

Early-nineteenth-century American mills were invariably driven by waterpower. Compared to the steam engine, the water wheel produced a regular and uniform flow of energy. However, the transmission of power from the water wheel to production-line machinery required the use either of shafting (favored in England) or belting (preferred by American manufacturers). Shafting tended to "whip" (twist and uncoil) over long distances. Flywheels proved effective in checking this whipping effect and commonly accompanied shaft drives.[59]

In the first textile machines, a large hand-powered driving wheel, known as the "fly," was found in the jenny and the mule. Its function was twofold. As a large pulley driving a much smaller pulley, it multiplied the rotary movement conveyed to spindles. As a flywheel, it improved the uniform rotation imparted to the crank handle of the driving wheel by the partly reciprocal movement of the spinner's arm. This was well known from the Rees article: "A man, turning a winch by his hand, or a crank by his foot, can exert three times the power in many positions which he can in others."[60]

Apart from these two uses, this author knows of no other application of the flywheel in the early-nineteenth-century American textile industry. Markham's purpose in incorporating the flywheel in a spinning machine is open to some speculation.

Montgomery, in 1840, noting the prevalence of belting for power transmission in American cotton factories, observed that, at best, belts were "heavy, clumsy, and inconvenient, as well as more expensive" than shafting. Moreover belts were liable to stretch and when slack would slip on the drums driving them.[61] This suggests that belts for power transmission, like shafting, could cause irregularities in the power reaching individual production-line machines. Possibly, then, Markham employed a large flywheel to improve the uniformity of power transmitted to rollers and spindles in his spinning frame. The value of the modification is unclear. Certainly the device did not attract the attention of Montgomery, the most reliable commentator of the early American cotton industry.

---

[57]Rees, *Cyclopaedia*, sv "Fly."

[58]*Encyclopaedia Britannica*, 1969 edition, sv "Power Transmission."

[59]Hunter, *Water Power*, 61, 420, 441, 462n

[60]Rees, *Cyclopaedia*, sv "Fly."

[61]Montgomery, *Practical Detail*, 19–22; Jeremy, *Technology and Power*, 56–59, 231.

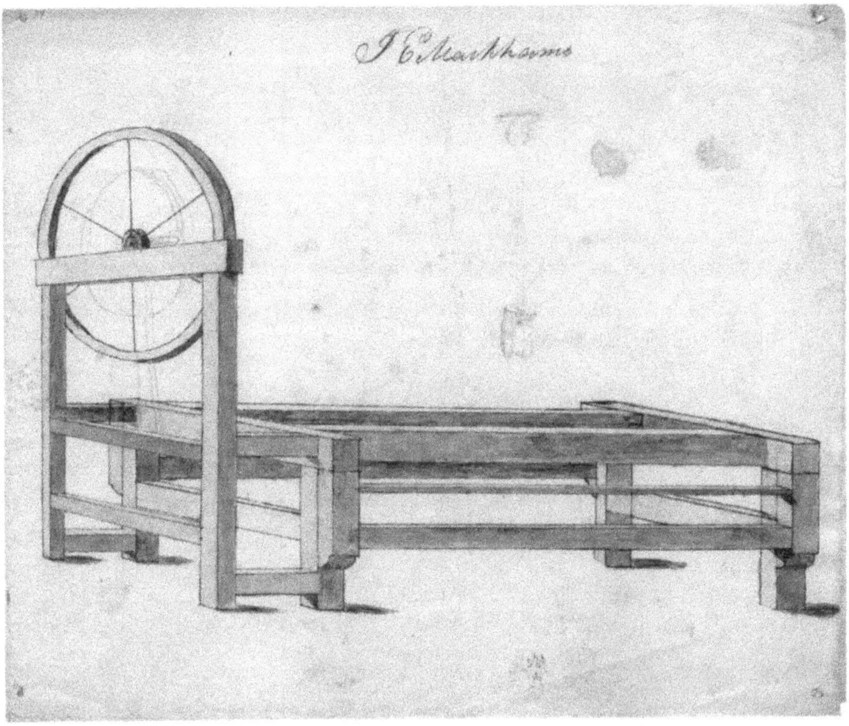

IGURE 6.65 "I E Markhams" Spinning machine, headstock, and driving wheel. Possibly the "fly wheel spinning" machine seen at Figure 69. Perspective sketch.

atalogue number: IEM 10.1

Document Description

Measurement: 22 x 26.8 cm (half of folded sheet 26.8 x 44 cm that includes Figs. 6.42, 6.45A, 6.45B, 6.65)

Medium: Pencil, ink, wash, blue watercolor

Condition: Pinholes in corners, stains, foxing, bleed-through

FIGURE 6.66 Spinning machine. Possibly the "fly wheel spinning" machine seen at Figure 6.69. Scale drawing, rear (?) elevation (manual control of power to the fast-loose pulley was on the right).

Catalogue number: IEM 18

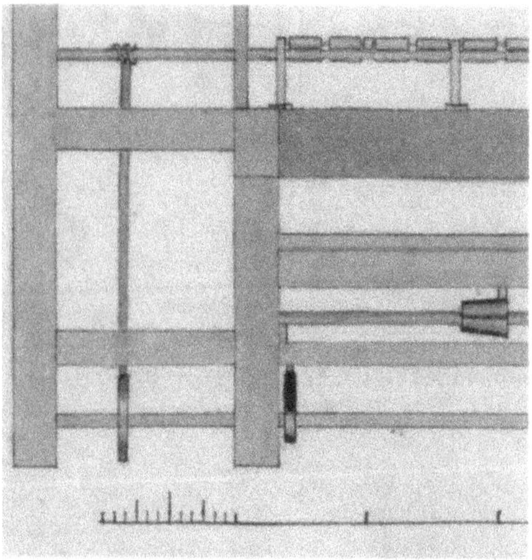

Worm gear and cone pulley, detail.

Document Description

Measurement: 26.5 x 42.6 cm, folded

Medium: Ink and wash

Condition: Pinholes in corners and fold, slight discoloration, foxing in fold

Commentary

Several features merit notice:

- this was a power-driven machine,
- it had a large flywheel (*left*) sharing the same axle as the fast-and-loose pulley (the usual clutch device for belt-driven machines),
- it contained opposed cones on a shared shaft,
- it also had a worm gear on a shaft behind the rollers.

The flywheel (see the introduction to this section) would have achieved a smoother transmission of power to the rollers and the spindles (or flyers). The same purpose would presumably have been served by the opposed conical pulleys. Usually such cone pulleys drove their twin cones, reversed. Assuming that they drove the tin roller, which, by means of bands, powered either the spindles or the flyers (see Fig. 6.71), belt slippage on each of the opposed cones would have been countered by its opposite cone and consequently might have increased efficiency in transmitting torque.

The worm gear in this position suggests the use of a guide bar holding small wire hoops (one per roving) to oscillate the rovings across their path into the drafting rollers. This was designed to prevent the leather on the rollers from becoming grooved and worn over time.[62]

[62]Ure, *Cotton Manufacture* II, 125–126.

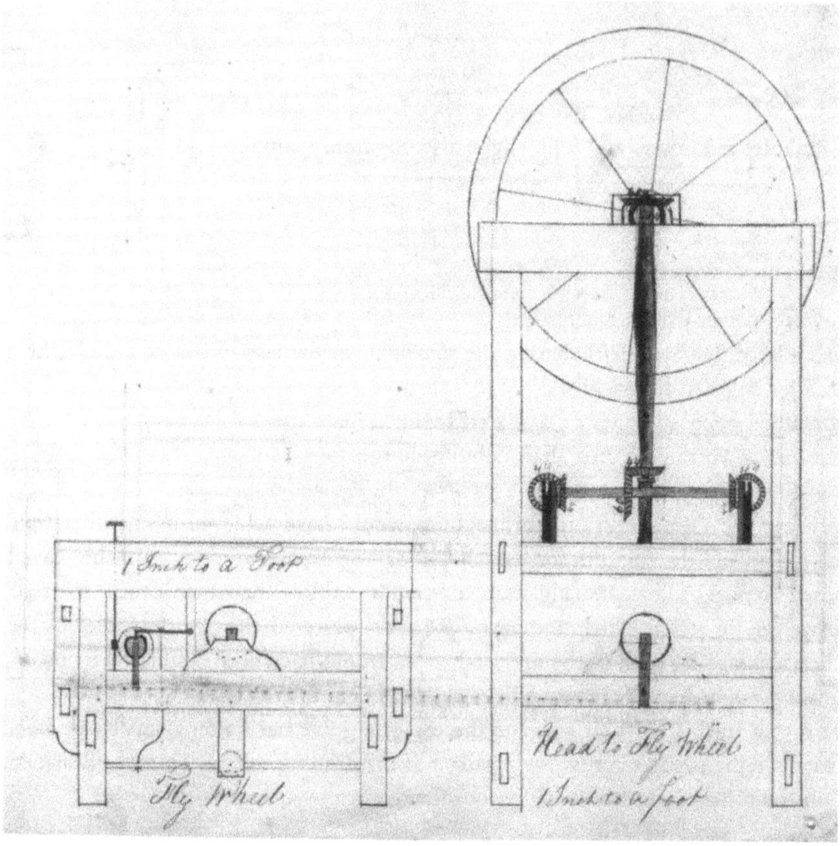

FIGURE 6.67 Flywheel spinning frame. Scale drawings, side elevations.

Catalogue number: IEM 4.1

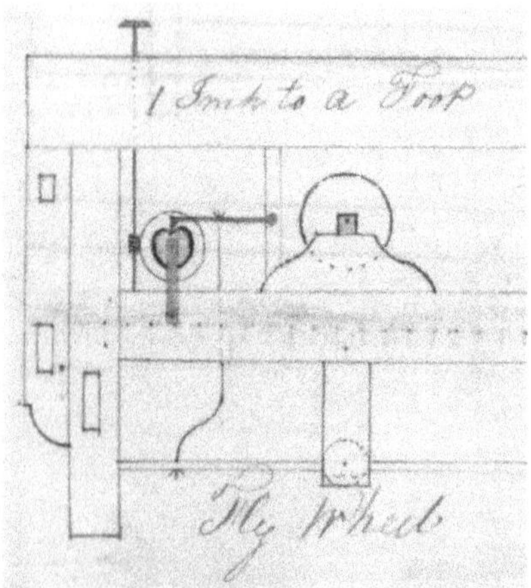

Flywheel spinning frame, detail.

Document Description

Measurement: 22.8 x 27.7 cm (one page of sheet 27.7 x 45.5 cm folded in half to make four pages in all; see Figs. 6.50, 6.67, 6.69, 6.70)

Medium: Ink, wash, blue and yellow watercolor

Condition: Pinholes in corners, bleed-through, minor foxing, remnants of red substance (wax?)

Commentary

The frame section (*left*) depicts several elements:

- a heart-shaped cam;
- the cam follower is a pivoted arm; midway along the arm, a connecting link is attached to a horizontal centrally pivoted lever that alternately raised and lowered the flyer (or spindle) lifter rails on each side of the machine;
- the tin roller (*center of frame*) running the length of the frame, which acted as a pulley to power the spindles (or the flyers, whichever was leading and was not traversing).

The head section (*right*) shows power from the main driving shaft taken by bevel gears to the drafting rollers on both sides of the spinning frame. Below is the tin roller, previously mentioned.

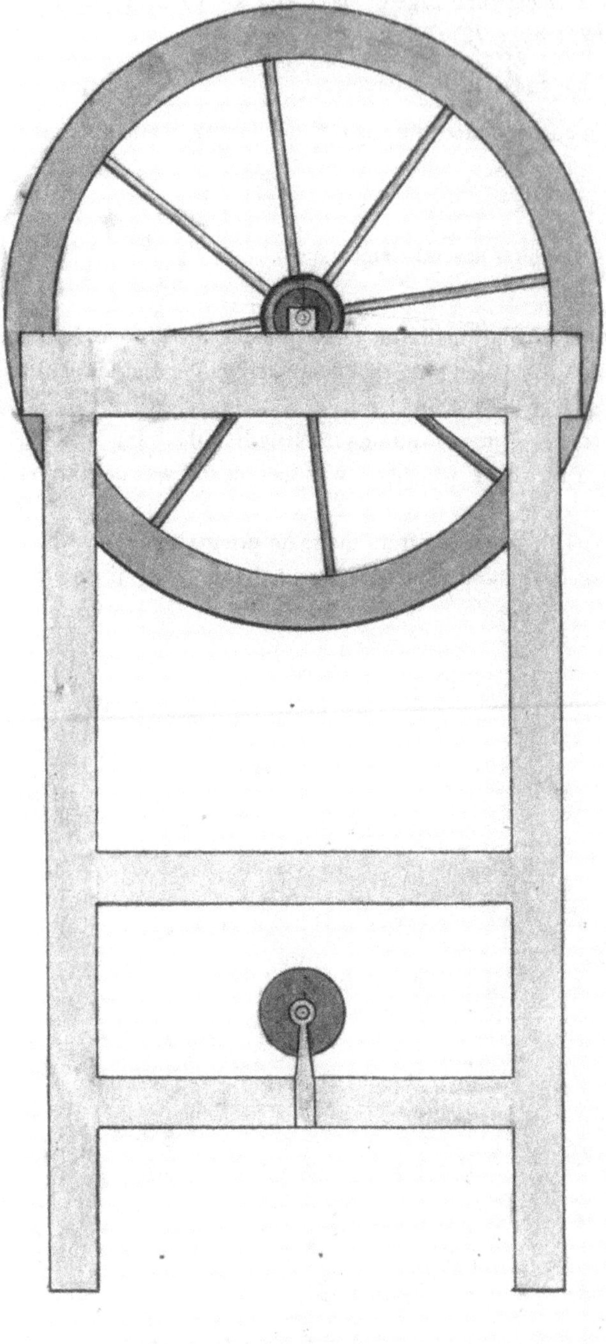

FIGURE 6.68 Flywheel spinning frame. Headstock. Drawing, side elevation.

Catalogue number: IEM 19.1

Document Description

Measurement: 26.5 x 21.2 cm (half of folded sheet 26.5 x 42.5 cm; including Figs. 6.68 and 6.71)

Medium: Pencil, ink, and wash

Condition: Pinholes in corners, foxing

Commentary

The drawing shows the mounting and the hub of the flywheel and (*below*) the tin roller driving spindles or flyers.

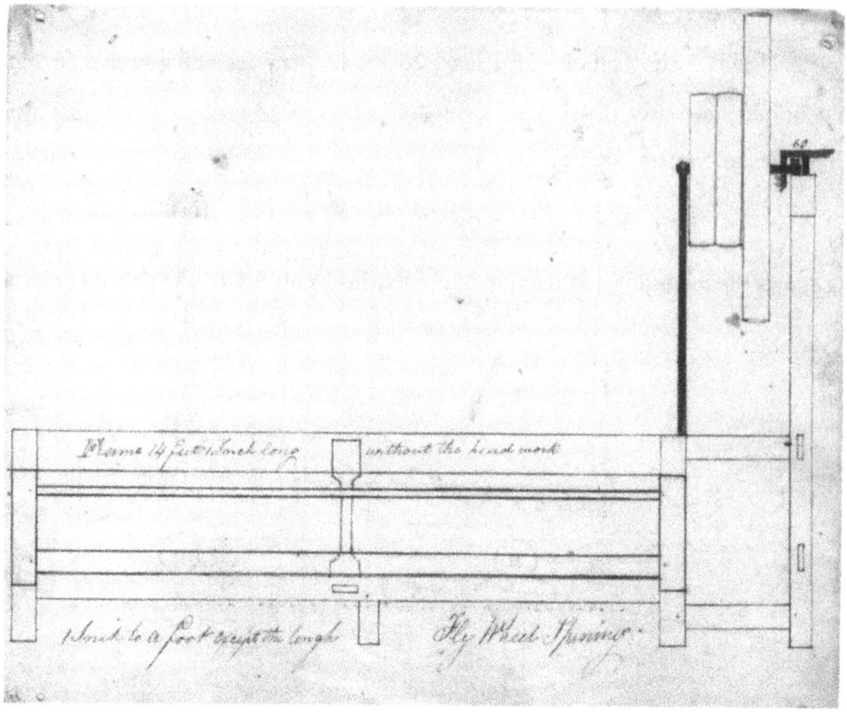

FIGURE 6.69 "Fly Wheel Spinning." Scale drawing, front elevation (if manual control of power to the fast-loose pulley was on the righ

Catalogue number: IEM 4.4

Document Description

Measurement: 22.8 x 27.7cm (one page of sheet 27.7 x 45.5 cm folded in half to make four pages in all; see Figs. 6.50, 6.67, 6.69, 6.70)

Medium: Ink, wash, and yellow watercolor

Condition: Pinholes in corners, minor edge damage, foxing, remnant of red substance (wax?)

Commentary

Drawing shows frame members and the head frame. At the upper right are the fast-and-loose pulley and the flywheel.

At 14 feet, 1 inch long, the frame might have held as many as 56 spindles on each side, a total of 112 spindles (judging by a contemporary Locks & Canals Co. spinning frame in the Old Slater Mill Museum, Pawtucket, Rhode Island).

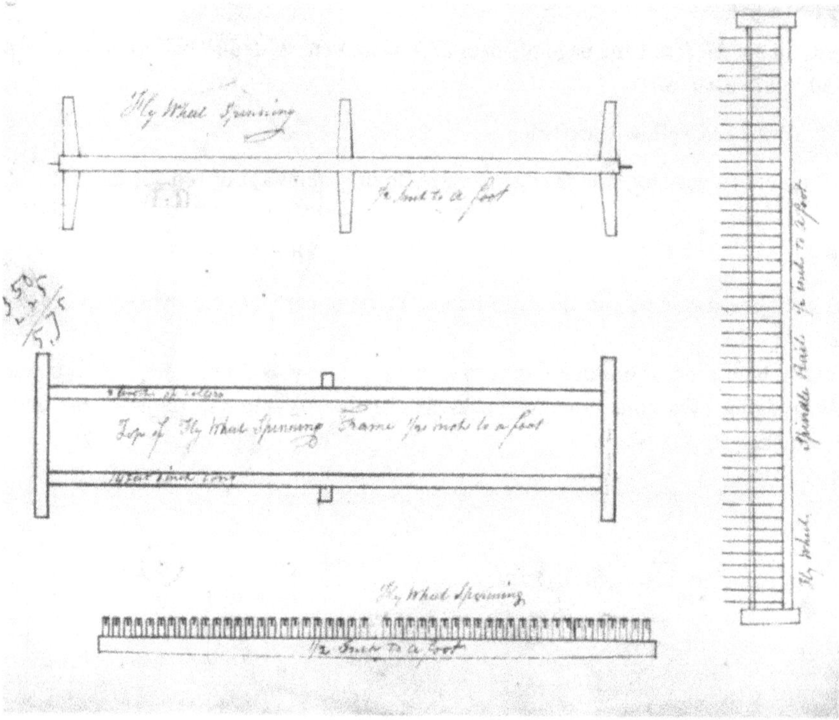

FIGURE 6.70 Flywheel spinning frame. Components. Scale drawings, plans and elevations.

Catalogue number: IEM 4.2

Document Description

Measurement: 22.8 x 27.7 cm (one page of sheet 27.7 x 45.5 cm folded in half to make four pages in all; see Figs. 6.50, 6.67, 6.69, 6.70)

Medium: Ink

Condition: Pinholes in corners, bleed-through, minor foxing, minor edge damage, remnants of (wax?)

Commentary

(*Top*), possibly part of creel holding roving bobbins; if so, this is a plan.
(*Middle*) top of frame, plan.
(*Bottom*) flyer rail, elevation.
(*Right*) spindle rail, elevation.

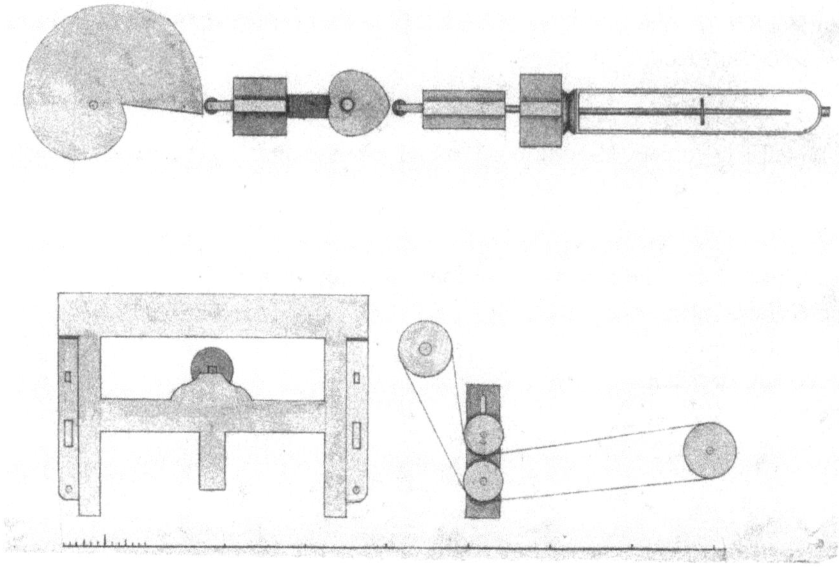

FIGURE 6.71 Flyer drives. Scale drawings.

Catalogue number: IEM 19.2

Document Description

Measurement: 21.2 x 26.5 cm (half of folded sheet 26.5 x 42.5 cm; including Figs. 6.68 and 6.71)

Medium: Pencil, ink, and wash

Condition: Pinholes in corners, minor foxing, bleed-through

Commentary

The upper drawing (which, to correspond with its machine positioning, should be rotated through ninety degrees) shows a spindle-and-flyer with a double-cam arrangement for effecting a lifter movement for winding the yarn onto the spindle.

This design of the flyer, with its legs fixed to a base plate and pulley, was described in England in the early 1830s as a "New Throstle."[63]

The double-cam arrangement is unusual. It appears to be a two-dimensional cam scheme. The upper follower element is the spindle (on which the bobbin sits on a plate, seen about halfway down its length), which traverses up and down within the flyer. Two frame sections below the flyer pulley is a small wheel at the lowest extremity of the spindle rod. This acts as a follower, moving up and down as the heart-shaped cam rotated. The axle of the upper cam turned on a vertical member, the lowest point of which carried another follower activated by the lower and larger of the two cams.

While the heart-shaped cam would produce a smooth up-and-down profile, the lower cam, with its sharp radial angle, would cause the spindle to drop suddenly to its lowest point. Imaginably this produced a crashing noise and considerable wear and tear, as the lower follower fell from the outer to the inner radius of the lower cam. One wonders if this was really as successful kinematically as was assumed on paper.

The lower left drawing is of a section of frame showing the tin roller for powering the flyers.

The lower right drawing is of a pulley arrangement. It is unclear whether this is a section or a plan. Whatever, there appears to be a slot to the left of the two center pulleys, possibly a tensioning feature.

[63]Montgomery, *Carding and Spinning Master's Assistant*, plate IX.

7. DRAWINGS: WARP PREPARATION—WARPING AND DRESSING

Warp Preparation

Before any weaving could commence a warp had to be prepared because the warp served, as it were, as the backbone of the cloth. The warp consisted of hundreds of yarns, each the length of the cloth (perhaps as much as 30 yards), laid in parallel. Obviously great care had to be taken that the yarn did not get entangled. Furthermore, the yarns needed to be treated with size, some form of starch (dressing), to give them increased strength and a smooth surface to withstand the tension applied to the warp in the loom and the abrasive actions of the heddle eyes as they were forced up and down in the shedding action (see the next section "Drawings: Weaving").

Preindustrial warping usually meant passing yarns to-and-fro between pegs set in a wall across a distance of 30 yards or less (which then meant doubling, as the clew would have to be reversed later to form the 30-yard length). By the late seventeenth century the system of wall pegs was being replaced by the use of a warping mill, a wooden framed reel about six feet in diameter and seven feet high, opposite which was a rack of bobbins loaded with the yarn. The yarns were passed through a separator or reed and taken in a clue onto the warping frame, which was rotated manually by an operator. In forming the warp, it was important to separate the yarns in a manner corresponding to the pattern that was to be woven. For this purpose separator rods were used to divide up the warp into two or more leases.[64]

After a warp was formed it had to be wound round a warp beam. This was then set in the loom and the warp yarns drawn through the heddle eyes of the shafts or harnesses (see next section "Weaving") and attached to the cloth beam. Set in the loom, the warp was tensioned by weights. The shedding action subjected the warp yarns to abrasive forces. To counter this wear and tear the warp was strengthened by being coated with size, starch derived from flour or potato. In the preindustrial period size was brushed onto the warp after it was installed in the loom.

Mechanization of warping accompanied mechanization of the handloom. Failure in the former presented a bottleneck to the productivity of the latter. In 1786 Edmund Cartwright, in his second power-loom patent, incorporated a sizing trough and brushes in the loom but the idea failed because the size took too long to dry.[65] At an unknown date the warping mill was replaced by the cradle warper, in which bobbins from the spinning frame were directly unwound to form a warp.[66] Sizing, or dressing, needed to be quite separate from weaving. This separation was first achieved by William Radcliffe of Stockport, Cheshire, in a patent taken out in the name of one of his workman, Thomas Johnson, in 1803.[67]

Radcliffe's dresser patent of 1803 showed both warping and dressing machines. In the first machine, yarn was taken from a creel of bobbins and wound onto a section beam. In a second machine the yarns from a dozen section beams were combined, passed over sizing brushes, and then combined on a section beam ready to be taken to a power loom. The Radcliffe-Johnson dresser of 1803 used a circular brushing motion. This proved unsatisfactory because it raised the fibers in the warp yarns. A year later they

[64]Duncan, "Weaving"; Wadsworth and Mann, *Cotton Trade*, 74.

[65]Edmund Cartwright, GB Patent of 30 October 1786 (1565).

[66]Montgomery, *Practical Detail*, 82–85.

[67]Thomas Johnson, GB Patent of 28 February 1803 (2684).

patented a second dresser in which the sizing brushes performed a linear brushing movement, much more satisfactorily smoothing down the fibers in the warp yarns (see Appendix 3 and Fig. A3.15). It was the Radcliffe-Johnson dressing machine that the Boston Manufacturing Co. at Waltham used as a model for the one built there and patented in the name of their engineer, Paul Moody.[68] Moody recorded, "I have seen the specification and drawings accompanying his patent, from which I constructed a machine resembling [Johnson's]."[69] Presumably Markham saw this when he worked at Waltham.

Insofar as the evidence of these drawings is concerned, Markham was clearly devoting much creative effort to the mechanization of warp dressing. Four of the drawings in this section record diverse arrangements for dresser designs. Two questions arise. Why did warp dressing attract so much of his attention? Why did he not patent any original and successful dresser design? The latter question is discussed in the introduction prefixing these drawings (see "Visual Aspects of the Generation of New Knowledge in Early-Anglo-American Industrial Engineering").

Technological and economic reasons seem to lie behind Markham's concentration on warp dressing. As noted earlier, it was essential to treat warp yarns with a size dressing so as to reduce the number of warp breakages during weaving. A well-sized warp might have allowed power looms to be run without a warp stop-motion, relying on operatives to piece breakages. But this was a practice dependent on the quality of the cotton and the coarseness of the yarns and, of course, on the supply and quality of the operatives. Among the mill girls at Waltham, tending the dresser attracted the third highest mean weekly pay rate among 11 grades of female operative. At the Dover Manufacturing Co. in 1827, females in the dressing room were the most highly paid in the works. That same year the agent of the Ware Manufacturing Co. complained, "There is no work in the factory considered so laborious for females as that of attending the Dresser." Presumably the Ware dresser was not dissimilar to the improved one used at Waltham and patented by Paul Moody, the company engineer, in 1818. Even so, tending the dresser demanded vigilance from the operative because there was no stop-motion to halt the machine when one of the sticky warp yarns broke. A willingness to tolerate the smell of size and imperviousness to the discomforts of a warm humid room were also needed. Not surprisingly, turnover rates for dresser tenders were high. For these reasons textile engineers in early-nineteenth-century New England sought to substitute capital for labor, machines for people, in the warp-preparation stage of cotton manufacturing.[70]

Markham evidently shared the desire of early-nineteenth-century American textile engineers to improve the throughput of their machines. An 1820 inventory of the Dover Cotton Factory, of Dover, New Hampshire, lists "2 warping machines with stop motions with 2 dressing of Adams and Markham's plan," all valued at $1,000.[71] (Adams may have been Asahel Adams: see Fig. 6.51A, p. 242.) The Waltham dresser of 1818 had drying fans and a measuring device coupled with a stop-motion, to halt the machine when a yarn end broke while the warp was being treated (see Table A3.2, p. 368). This made it the second most expensive machine built by the Boston Manufacturing Co., being sold for $1,000, compared to $2,000 for a speeder, $600 for a warper, and $125 for a power loom, prices reflected during the 1820s in other New England textile districts.[72] Warpers and dressers were large machines, necessarily so because space was needed to manipulate long clews of yarn, and this may further help to explain their high capital cost.

---

[68]Paul Moody, US Patent of 17 January 1818, US First Circuit Court, Case Files.

[69]Jeremy, *Transatlantic Industrial Revolution*, 99.

[70]Jeremy, *Transatlantic Industrial Revolution*, 173, 194–196, 199.

[71]Agreement between Williams & Wendell and A.& J. Wendell, 29 June 1820, Dover Cotton factory file, Williams & Wendell, 1815–1831, Case 11, Wendell Collection, Baker Library.

[72]Jeremy, *Transatlantic Industrial Revolution*, 209.

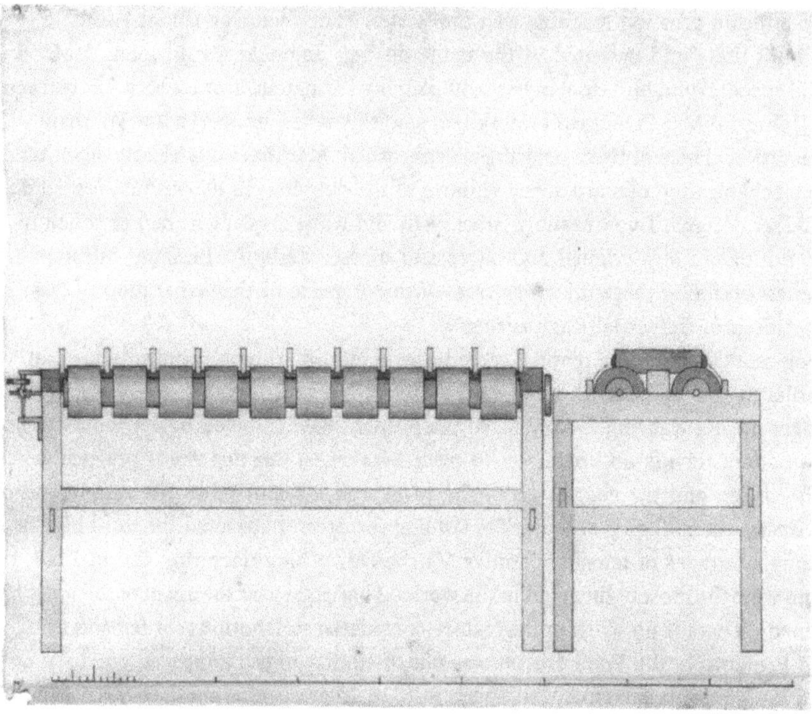

FIGURE 6.72 Spooling machine. Scale drawing, front and side elevations, aligned on a shared-base scale.
Catalogue number: IEM 23.1

Document Description

Measurement: 21 x 28.8 cm (one page of 42 x 26.8 cm sheet folded in half; see also Fig. 6.73)

Medium: Pencil, ink and wash

Condition: Pinholes in corners, stains, minor foxing, foxing in fold

Commentary

This was used for winding yarn from small bobbins or cops onto large bobbins (spools) in readiness for the spools to be set in a warping frame, for loading onto a section beam or warp beam. It was much less wasteful of cotton yarn than the earlier cradle warper. Montgomery, *Practical Detail*, Plate V, is almost identical to this drawing. But Montgomery had not seen this design in his native Scotland, or in Britain, and he was writing nearly twenty years after Markham made his drawing.[73]

The spools would sit on the drums and be driven by friction, thereby taking yarn from the small bobbins seated on spindles (not shown). The verticals between the drums are separator sheets of metal to keep the spools apart.

Comparators

Drawings: Montgomery, *Practical Detail*, Plate V; see Figure A3.19 (p. 366).

[73]Montgomery, *Practical Detail*, 82–85.

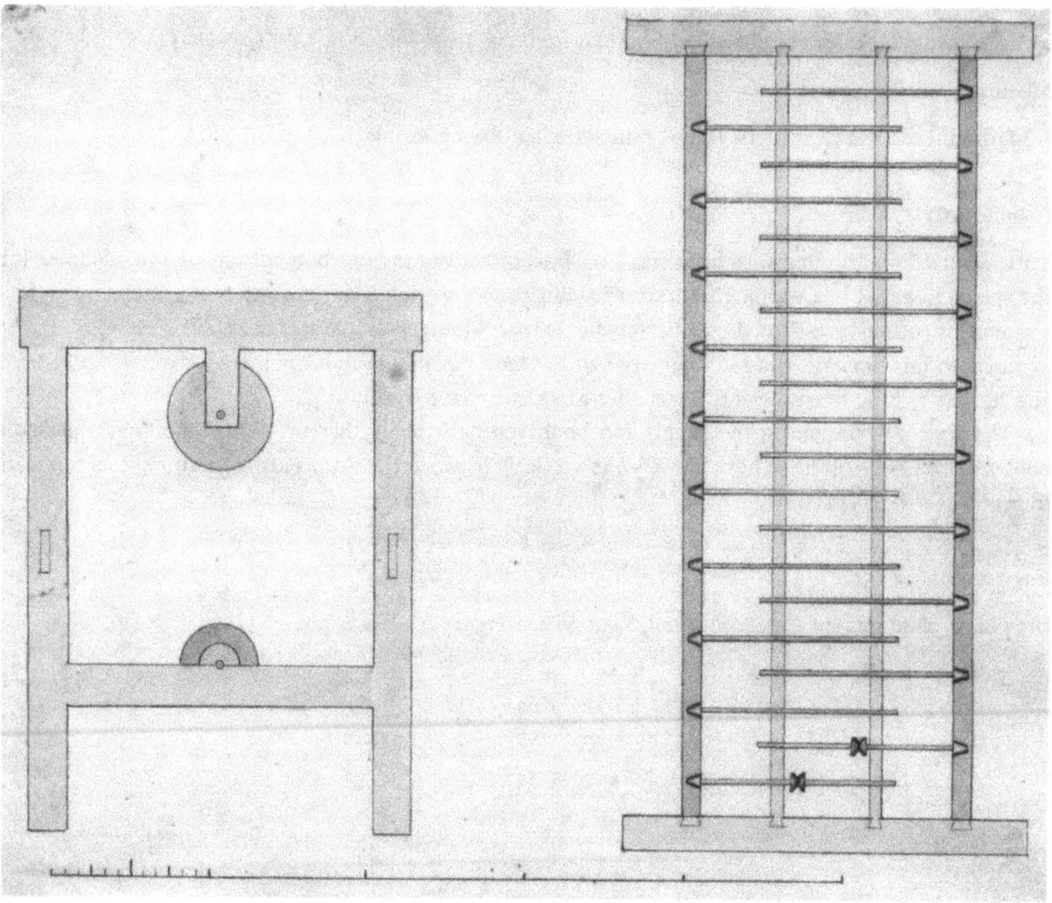

FIGURE 6.73 Spooling machine (?). Scale drawing, side elevation and plan.

Catalogue number: IEM 23.2

Document Description

Measurement: 21 x 28.8 cm (one page of 42 x 26.8 cm sheet folded in half; see also Fig. 6.72)

Medium: Ink and wash

Condition: Pinholes in corners, foxing, transfer from opposite page, minor corner damage

Commentary

The identification is tentative because the element on the right shows a frame in which the rods (or spindles?) would not readily have operated in a vertical plane. The "x"on two of the pulleys suggests that a crossed cord could have powered the rods from a drum at right angles, as in the elevation on the left.

If this was a spooling machine, then the cops from mule spindles would have been slid onto the rods and a clew of yarns formed to be wound onto a section beam (several of which would provide the warp for a the warp beam in a power loom). But this would require more equipment than shown here.

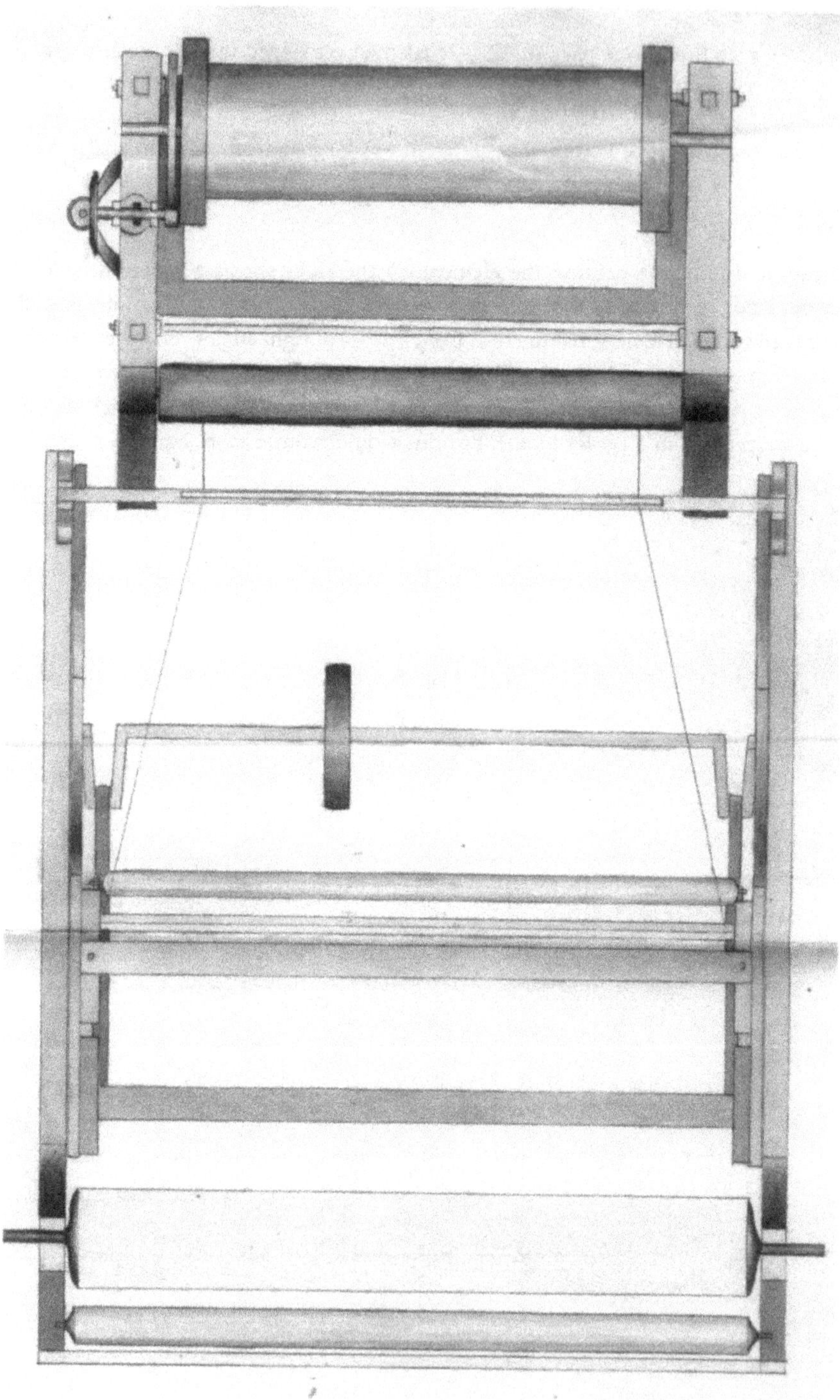

FIGURE 6.74 Intermediate warping frame (?). Perspective drawing (?).

Catalogue number: IEM 26

Document Description

Measurement: 47 x 28.6 cm; folded in half — drawing on full sheet

Medium: Ink and wash

Condition: Foxing

Commentary

The identification is tentative because, first, it is not clear what is being depicted: a perspective sketch or a plan-and-elevation. The shading and the two single lines linking the second to the third rollers (of five, in descending order) lend a three-dimensional appearance to the drawing. Nor is it clear whether there are one or two machines here.

If this is a warper, taking the yarns from a section beam (fourth roller) and passing them to a power-loom beam (at the top), where they are added to the warps from preceding section beams, then the cranked third roller would act as a tensioning device. This operation would have been preceded by the formation of a part-warp by taking yarns from throstle bobbins.

Comparator

This drawing has similarities to the plan of the mechanized warper in Johnson's patent of 28 February 1803 (GB Patent 2684; see Fig. A3.14, p. 361).

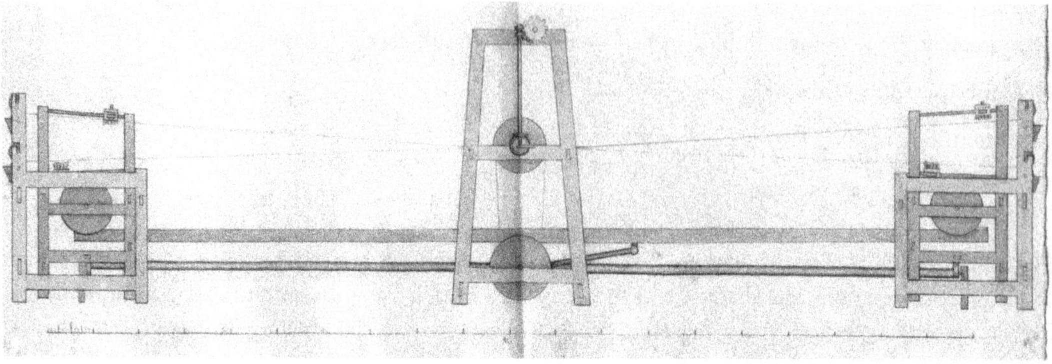

FIGURE 6.75 Dresser. Scale drawing, side elevation.

Catalogue number: IEM 24

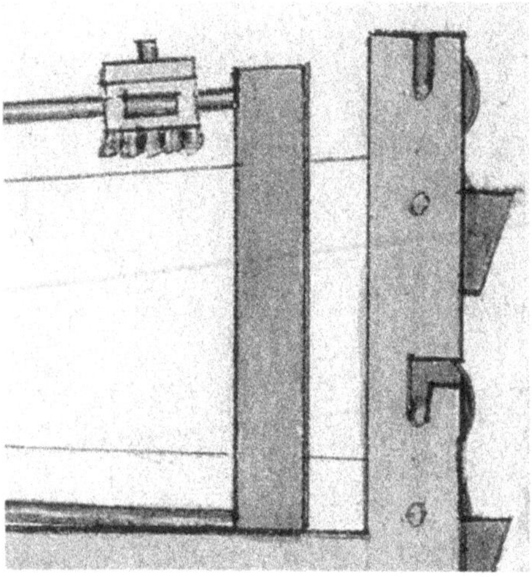

Dresser, detail.

Document Description

Measurement: 26.5 x 42.8 cm; folded in half—drawing on full sheet

Medium: Ink and wash

Condition: Pinholes in corners and middle of sides, minor foxing

Commentary

Assuming the scale was 1 foot per marked unit, this dresser was about 23 feet long. The principle of the machine can be gleaned by comparing it to the Lowell dressing machine of the 1830s at Figure A3.20.[74] Three features can be seen:

1. Warp yarns from section warp beams (circles, *extreme left and right*; see Figure A3.20, p. 367 and Table A3.2, p. 368) were combined into a single warp on a central loom warp beam (seated at the top of the central frame).

2. Size troughs (*extreme left and right*; see Figure A3.20, p. 367 and Table A3.2, p. 368) applied size as the section warp yarns passed through them. The lower of a pair of rollers pushed the yarns into the trough and passage between the two rollers then squeezed them before they reached the brushes.

3. A rocking "sweep,"[75] running the length of the machine (activated by a crank movement) imparted motion to the brushes so as to remove excess size fluid from the yarns.

The Lowell dresser incorporated a hot-air drying mechanism, not evident here.

Comparators

Drawings: Montgomery, *Practical Detail*, pp. 91–97, Plate VII. See Figure A3.20.

---

[74]Montgomery, *Practical Detail*, 91–97, Plate VII.
[75]Montgomery, *Practical Detail*, 94.

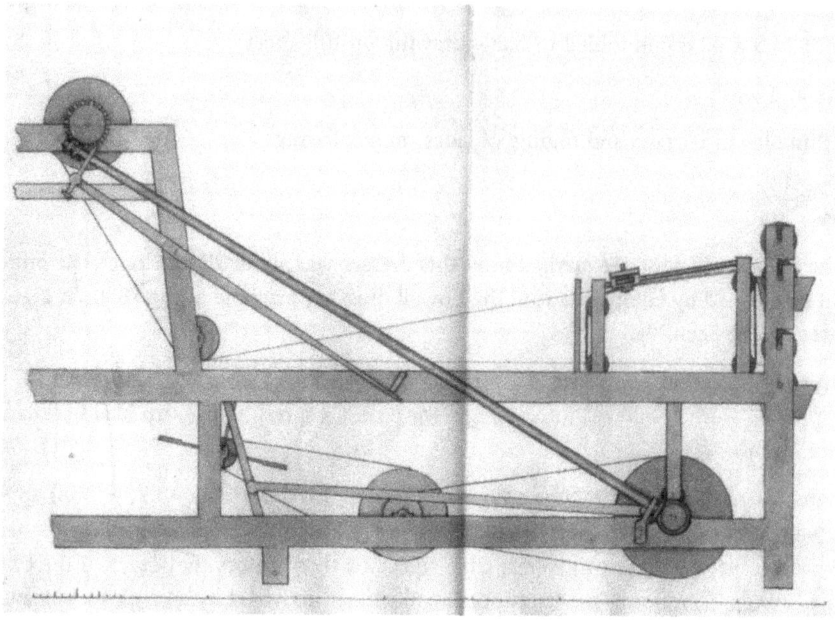

FIGURE 6.76 Dresser. Scale drawing, side elevation.

Catalogue number: IEM 25

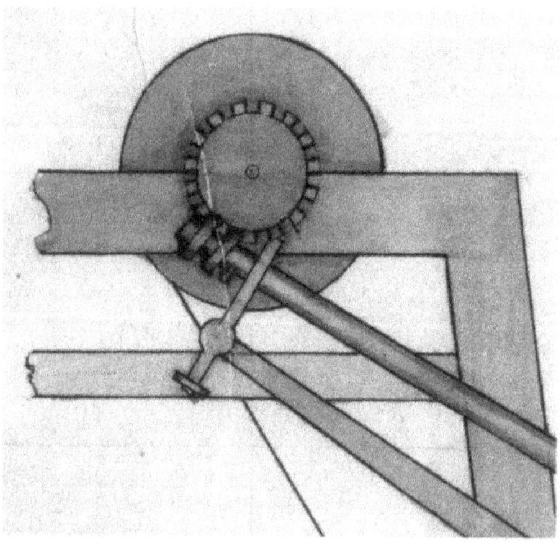

Dresser, worm gear, detail.

Document Description

Measurement: 28.8 x 41.8 cm; folded in half—drawing on full sheet

Medium: Pencil traces, ink, and wash

Condition: Pinholes in corners, center of top and bottom; minor stain; minor foxing

Commentary

The troughs on the far right hint that this is another design for a dressing frame. Here the section beams are positioned to the far right and a reciprocating brush arrangement is placed adjacent to the sizing troughs. The windlass at the lower left is a drying fan: Four arms characterize the fans of the Lowell dresser of the 1830s.[76]

A central warp beam (presumably) is seen at the upper left, where it could be stopped by using the pivoted arm to disengage the worm gear from the pinion.

[76]Montgomery, *Practical Detail*, Plate VII.

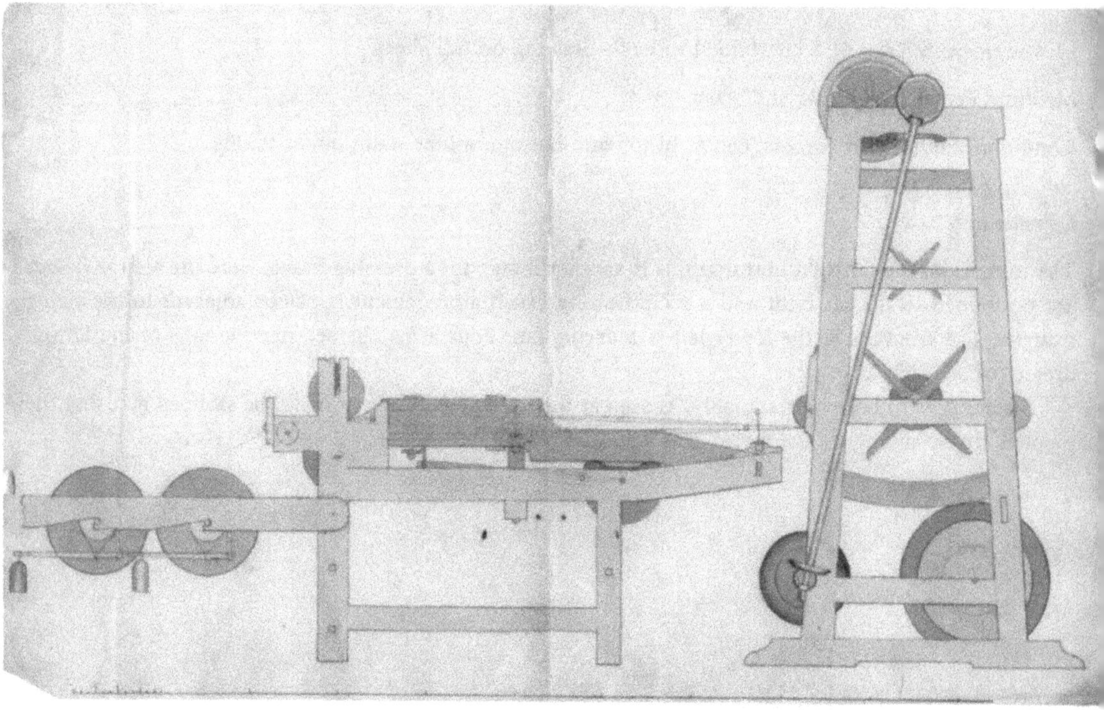

FIGURE 6.77 Dresser. Scale drawing, side elevation.
Catalogue number: IEM 27

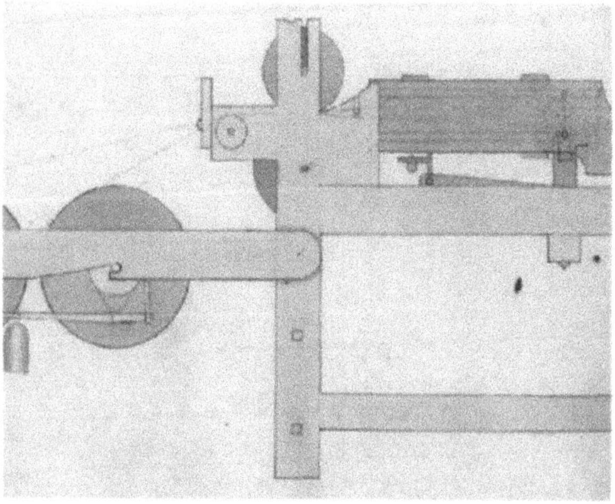

Dresser, detail.

Document Description

Measurement: 19.8 x 47.5 cm; folded in half—drawing on full sheet

Watermark: J. Whatman 1816

Medium: Traces of pencil, ink, and wash

Condition: Minor tears on edges, tear on fold, foxing

Commentary

In this design, warp from the weight-tensioned section beams (*far left*) passed upward and to the right over one roller and then between two rollers, the lower of which was partly submerged in a sizing trough. This is probably indicated by the shaded segment on the left side of the left-hand pillar of the central frame. The central frame supported a pivoted box-like structure that was possibly a brushing mechanism that used a vibrating or rocking motion.

Lease rods separated sections of the warp and two fans assisted drying before the warp was wound onto the loom warp beam. The latter is positioned at the top of the free-standing frame (*at right*). A geared drive to the loom warp beam appears to be able to be disengaged.

FIGURE 6.78 Dresser (?). Scale drawing, side elevation.

Catalogue number: IEM 22

Document Description

Measurement: 27 x 42.7 cm; folded in half—drawing on full sheet

Medium: Pencil, ink, and wash

Condition: Pinholes in corners and top and bottom center, torn upper corners, tearing at fold, foxing

Commentary

If this is another dresser, the large wheels (*right and left*) may have turned the axles supporting section beams and three rotating brushes across the top of the frame would have helped to spread size more evenly on the warp yarns. Yet the drawing is incomplete. There is no central frame to support a loom warp beam.

8. DRAWINGS: WEAVING

Weaving is the process by which one series of parallel yarns (the warp) is interlaced, at right angles and in the same plane, with a second series of parallel yarns (the weft or filling). Looms, hand or powered, made four basic types of cloth.

Simplest was plain weave. This fabric was characterized by a woven basket effect. To produce it, a minimum of two harnesses (or shafts; the frame holding the heddles, each of which controlled a warp yarn) was needed: one harness controlling even-numbered warp yarns and the other the odd-numbered warp yarns. The position of the harnesses was then reversed between each pick (the driving of the shuttle across the loom to insert one weft yarn into the weave of the fabric being woven).

Twill weave featured a diagonal line or variations of it, like herringbone, on the cloth surface. At least three harnesses were required for this weave, and at each pick the same number of harnesses was raised or lowered, the repeating sequence moving through the harnesses in turn.

Satin weaves were characterized by long floats of weft and required a minimum of five harnesses to make. At each pick only one harness was raised (or lowered), and a twill line was avoided by ensuring that harnesses containing adjacent warp yarns were not raised (or lowered) in succession.

Last, figured weaves required the raising at each pick of individual warp ends, rather than groups of them, so that intricate pictures could be woven. Figured weaving was first mechanized, though not immediately powered, by Jean Marie Jacquard of Lyons between 1801 and 1806.

Handbooks, aimed primarily at handloom weavers, were published in the early decades of the American Republic to show how these kinds of weaves, except figured weaves, were made. Earliest was *The Weavers Draft Book and Clothier's Assistant* (Baltimore, 1792) written by John Hargrove, an immigrant Irish weaver and then a Methodist minister in Baltimore.[77] Four more appeared before 1820.[78] There was therefore no shortage of information in New England about how to weave basic types of cloth. The challenge was to do this using the power loom.

Mechanization of the weaving process required two operations to be shifted from handicraft implements to powered machines. The first operation was warp preparation (see previous section); the second was weaving.

Power-Loom Weaving

In the handloom there were five different actions. All had to be a mechanized.

The movement of the warp through the loom involved three actions.

1. Let-off: releasing the warp (previously prepared and loaded on the warp beam).

2. Shedding: separating the warp yarns so that a shed or space was created for the shuttle carrying the weft yarn to be knocked through the shed from one side of the loom to the other. The shed was formed when one group of warp yarns was raised above the rest, giving a top shade (or shed) and a lower shade (or shed). Warp yarns were controlled by passing each warp yarn through a loop in a vertical

[77]Hargrove, *Weavers Draft Book.*

[78]Hargrove, *Weavers Draft Book* with new introduction by Rita Adrosko.

string (the heddle). Heddles were grouped according to the pattern being woven and each group was set in a separate harness frame (the harness is known as a shaft in England). The harnesses were then lifted and lowered by means of cords and pulleys activated by treadles.

3. Take-up: winding the woven cloth onto the cloth beam at a speed synchronized with that of the warp beam and enough to keep the warp taut. Equally important, the take-up motion controlled the number of picks per inch (i.e., weft yarns per inch) and hence the transverse density of the cloth (in contrast to the lateral density of the cloth, which was established at the warping stage).

The insertion of weft yarns involved two actions.

1. Picking: driving the shuttle carrying the weft through the warp shed, from side to side of the loom.
2. Beating-up: using the batten (the frame holding the reed that kept the warp yarns evenly spaced, known in England as a lay or lathe, along which the shuttle passed, knocked side to side of the loom) to pack the warp yarn into the fell (weft edge) of the cloth after it had been taken through the warp shed by the shuttle.

In sequence, the actions that needed to be mechanized were: let-off, shedding, picking, beating-up, and take-up.

Several supplementary devices were also needed for a power loom to operate smoothly. Most important was some sort of mechanism to stop the loom if the shuttle became jammed in the warp shed. Otherwise the batten, smashing against the shuttle, would drive it through the warp, leaving many ends broken, and might injure operatives. Besides this shuttle- or warp-protector motion, two other stop-motions were required, one to halt the loom whenever a warp yarn broke (warp stop-motion) and the other to halt the loom whenever the weft broke (filling stop-motion). A fourth device was necessary to keep the cloth taut across the loom. Grippers known as temples were used in the handloom. In 1816, a rotary version, suitable for a power loom, was patented in the United States by Ira Draper of Weston, Massachusetts.[79]

The search for an efficient and well-designed power loom, which occupied textile manufacturers and machinists during the first three decades of the nineteenth century, when Markham was busy in the cotton mills of Massachusetts and Vermont, was circumscribed by the mechanical technology available. In general, the goal was to use rotary rather than reciprocal power transmission arrangements—as in spinning, continuous, rather than intermittent mechanical movements.

The cloth (as opposed to the tape) power loom was developed in Britain ahead of other nations and numerous minds contributed to its mechanization. The Reverend Edmund Cartwright, an extraordinary gentleman of science, demonstrated what was possible when in middle age he turned to the challenge of mechanizing the handloom.[80] He took out four British patents between 1785 and 1788 moving from crude and excessively powered mechanical actions with a vertical warp loom (1785) to cams for shedding, picking, and beating-up (1786); spring- and cam-actuated picking (1787); and a crank-driven batten (1788).[81] Although his patents ambitiously described stop-motions for weft, warp, and warp protection, Cartwright's power looms evidently remained cumbersome and mechanically inefficient.

[79]Ira Draper, US Patent Drawing of 7 June 1816 (2608).
[80]O'Brien, "Micro Foundations."
[81]Edmund Cartwright, GB Patents of 4 April 1785 (1470), 30 October 1786 (1565), 1 August 1787 (1616), and 13 November 1788 (1676).

Robert Miller of Dumbarton, Scotland, patented in 1796 the improvements of cams to activate the treadles that raised and lowered the harnesses (or shafts) carrying the heddles that controlled the warp yarns. He also used cams, with pulleys and belts, for picking and beating-up (see Fig. A3.16, p. 363).[82]

A major fillip to the development of the power loom was given by William Horrocks, cotton manufacturer of Stockport, Cheshire (see Figs. A3.17 and A3.18, pp. 364, 365). His design of 1813, his third patent, introduced a much more efficient power loom, which more closely mimicked the actions of the handloom and the handloom weaver. By means of a compound lever in its driving crank, the batten made a swift stroke, beat up the weft, and then returned quickly to a stationary position, at which point there was more time for the harnesses to open and form a larger shed and more time for the shuttle to cross the loom. Numerous economic benefits followed. A larger shed permitted larger shuttles, which could take bobbins straight from the mule spindles, eliminating rewinding onto smaller bobbins; larger cops, in turn, meant fewer weft-piecing operations. Also, claimed Horrocks, "Waste from bottoms in the cops will be less." The short, sharp stroke of the batten reduced the tension required for the warp, which thereby diminished warp breakages (and hence knotting costs) and power for the shedding mechanism; it also packed in more weft per inch, making the goods "stronger, more uniform, and better wrought in all respects."[83] Britain's reservoir of power looms, by now made of iron, increased from 2,400 machines in 1814 to 14,000 in 1820.[84]

Like John Todd's power loom of 1803,[85] Horrocks's power loom of 1813 shows an under-picking motion, with a single, central picker stick (which by means of cords tugged the sliding hammers or pickers that drove the shuttle from side to side of the loom) positioned underneath the warp. In contrast, and similar to the design of Miller, Horrocks's power loom of 1803, Robert Bowman's pioneering design of 1820 for a twill loom, and Richard Roberts's patent of 1822, retained the picker stick pivoted above the warp.[86] Most of the power looms patented before 1825 took the side-to-side motion of the picking peg from the treadles or from parallel cam-activated levers. An early under-picking motion (pivoting the picker-stick mechanism under the warp) entailed stirrups linking the two extremities of a semicircular plate, at the base and pivot of the picker stick, to the treadles or levers. The latter alternately tugged the extremities of the plate downward and thereby the picker stick from side to side. By the early 1830s the under-pick motion as built by Sharp & Roberts prevailed.[87]

The big improvement embodied in Bowman's and in Roberts's power looms was a tappet (multicam) wheel by which six harnesses were controlled, allowing plain, twill, or satin weaves to be woven.

The Rees *Cyclopaedia* article on weaving by John Duncan appeared four years before Roberts's power loom was patented and the latest designs that he reported in any detail were those of Miller and Horrocks.[88] Nicholson mentioned Miller's loom, one in which the shafts (harnesses) were actuated by cams, and another power loom in which cranks powered the shafts.[89]

The latest British improvements to the power loom reached New England in 1815 when William Gilmour, a Scottish immigrant, arrived in Boston bringing a design quickly known as "the Scotch loom."

---

[82]Robert Miller, GB Patent of 28 June 1796 (2122).

[83]William Horrocks, GB Patent of 31 July 1813 (3728).

[84]Baines, *Cotton Manufacture*, 235.

[85]John Todd, GB Patent of 14 April 1803 (2698).

[86]William Horrocks, GB Patent of 20 April 1803 (2699); Robert Bowman, GB Patent of 20 July 1820 (4488); and Richard Roberts, GB Patent of 14 November 1822 (4726).

[87]Baines, *History*, 234–235; Ure, *Cotton Manufacture* II, 289–310; Ure, *Dictionary*, 1287.

[88]Duncan, "Weaving."

[89]Nicholson, *Operative Mechanic*, 421–422.

Other early-American experiments were made with power looms but the Scotch loom, incorporating the Horrocks variable-speed motion, had only one rival, namely, a machine also derived from English models built in 1813 under the direction of Francis Cabot Lowell in the works of the Boston Manufacturing Co. at Waltham, Massachusetts (see chapter 2 and Appendix 4).

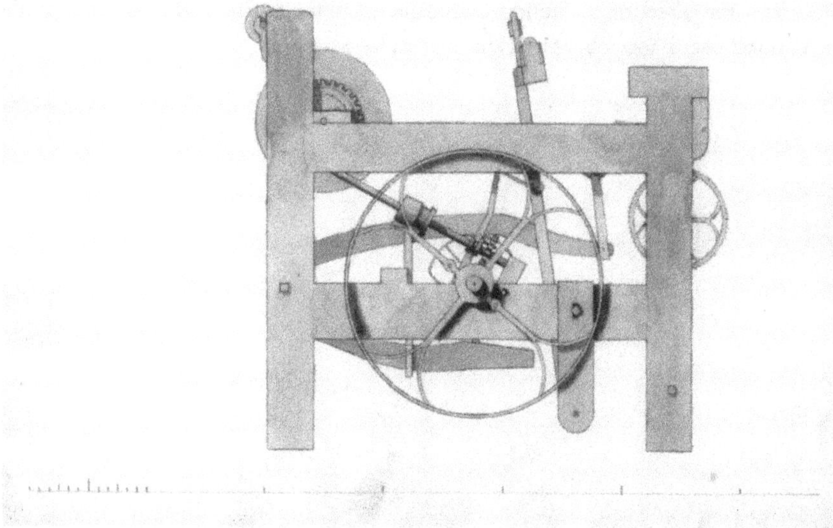

FIGURE 6.79 Power loom. Scale drawing, side elevation.

Catalogue number: IEM 11

Document Description

Measurement: 24 x 30 cm (60 x 47.5 cm sheet is folded in half then half again to make eight pages, only one of them used)

Medium: Pencil, ink, and wash

Condition: Pinholes along one edge, minor foxing, minor stains, bottom edge ragged

Commentary

Scale drawing of side elevation of power loom (warp beam on the left; cloth beam on the right).

The drawing shows shedding motion of power loom. On the axle of the large pulley, by which power entered the loom, can be seen one of two heart-shaped cams. Each of the cams activated a pivoted follower arm (the treadle) in an up-and-down movement that lifted and lowered the harnesses (here unseen but to the left of the batten frame, the vertical, bottom-pivoted arm, topped with the end view of one of its shuttle boxes) to make a shed for the shuttle to pass through.

The geared shaft angled from the main driving axle to the half-seen gear on the upper left would have powered the let-off motion by which the warp was fed under tension through the loom.

Noticeable are the materials from which the loom was built: wood for the framing; metal for the pulleys, gears, cams, and axles.

Comparators

Drawings: This section shows features that are similar to those in the wooden power loom made by the Lowell Machine Shop and advertised in 1848 (Fig. A4.2, p. 373).

Artifacts:

British or American: none known.

FIGURE 6.80 Power loom. Scale drawing, side elevation (*left*); front elevation (right, at 90 degrees to the vertical). Catalogue number: IEM 12.1–2

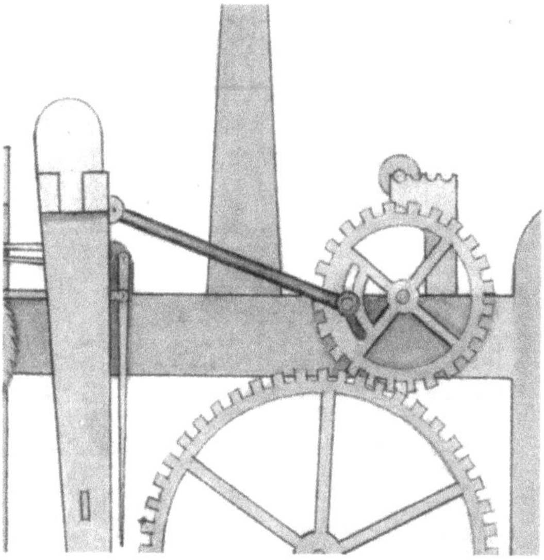

Beating-up device, detail.

Document Description

Measurement: 26 x 42.5 cm

Medium: Ink and wash

Condition: Folded, pinholes in each corner of full sheet, transfer from opposite page, stains, foxing in fold, tears at edges of fold

Commentary

Power Loom
Scale Drawing, Side Elevation (left)

This depicts the beating-up motion of the power loom. Like Horrocks's design of 1813 (Fig. A3.17 and A3.18, pp. 364, 365), this drawing shows a batten motion pivoted below the warp (to the left of the large driving wheel). Unlike Horrocks's design, this loom has a simple crank linked directly to the batten. That is, no attempt has been made to add the clever compound-lever arrangement by which Horrocks achieved a variable batten speed. This batten motion is positively driven by gears, rather than by pulleys and belts; in which case there would, most likely, have been a warp protector stop-motion to halt the loom automatically if the shuttle jammed in the shed. Alternatively, the large belt-driven pulley (seen in Fig. 6.79, p. 306) at the other end of the axle of the largest gear here may have been stopped, when there was a shuttle jam, by the combined strength of a warp of low-count, high-quality cotton yarns such as were made at Waltham.[90] Of great interest is the apparently unique mechanism powering the shedding movement. The drawing shows a connecting rod between the batten and the gear crank. The gear crank is held in place by a stud in a slide. The effect of this is discussed in Appendix 4.

On the far left, the cloth beam is turned slowly by the batten's reciprocal movement. The ratchet wheel on the axle of the cloth beam is pushed a notch or two (depending on the picks per inch) by the crank linked to the batten. At each beat up (as the batten moved to the right) the left pawl was released but the pawls on the right dropped down a notch or two. They held the cloth beam from unwinding when the left pawl, driven by the return of the batten, wound up the cloth beam a notch or two.

This loom is similar to the first power loom built by the Boston Manufacturing Co. at Waltham in that it has no variable batten speed. Yet it is dissimilar in that it employs gears instead of pulleys and belts in beating-up, one of the power loom's main motions. If this implied the presence of a warp-protector motion, then the design represents a definite improvement over its Waltham counterpart. Markham's loom, compared to other American looms ca. 1810–1812 (see Fig. A4.1, p. 370), appears to be a neat, well-fitted design for the limited purpose of weaving plain cloth made from strong, low-number cotton yarns. In contrast, contemporary British loom designs reveal more complexity of mechanism and more flexibility of output.[91]

Comparators

Drawings: A4.2 (p. 373), the power loom made by the Lowell Machine Shop and advertised in 1848 (see introduction to this section) there is a similar crank-driven batten.

[90]In 1820 the Boston Manufacturing Company made cotton shirtings and sheetings, of which there is one sample in the National Archives, RG 92, Coxe and Irvine Papers, 1794–1842, box 32 A, enclosure in Benjamin C. Ward & Co (BMC's selling agents) to Callender Irvine (Commissary General). The sample was "Chain 48 [ends] to the inch; Filling 52" (noted in Jeremy, *Transatlantic Industrial Revolution*, 324, n24).

[91]Jeremy, *Transatlantic Industrial Revolution*, 99.

Artifacts:

British or American: none known.

Commentary

Power Loom Scale Drawing, Front Elevation (at 90 Degrees to the Vertical)

This depicts the picking motion of this power loom. The design of the picking motion is very similar to that given by William Horrocks in his British patent of 1813 (GB patent 3728), in that it is under, rather than above, the warp. Like that English design, a central-picking peg extending from a circular-shaped element is linked to the pickers in each of the shuttle boxes, to the extreme left and the extreme right of the machine. Presumably the picking stick was linked to picking blocks by cords, as in a handloom.

If this power loom was similar to Miller's loom of 1796 (see Fig. A3.16, p. 363) and Horrocks's of 1813 (see Figs. A3.17 and A3.18, pp. 364, 365), and the other early British patented power looms, then the circular-shaped element would be a pivoted half-plate with flexible links (stirrups) at its extremities to levers or treadles below. Whether they were also treadles for pulling the harnesses up and down or whether they were simple levers, their cam-activated up-and-down movement would be converted into a side-to-side motion by the half-plate and stirrup arrangement—and hence a side-to-side motion for the picking peg. The five bars (bottom center) were probably guides (like later grid frames) to keep aligned the treadles activating the harnesses and the levers powering the picking peg.

The lever on the left is almost certainly a starting handle, to shift the power belt across from the loose to the fast pulley, and vice versa.

Other parts shown in this drawing are the cloth beam (shown across the central picking peg) and the frame and mountings for the harnesses (*top center*). Possibly the latter indicates a roller-top motion, whereby leather straps, contra-fixed, act to pull up the harnesses alternately.[92]

Comparators

Drawings: In the wooden-framed cotton power loom made by the Lowell Machine Shop and advertised in 1848 (Fig. A4.2, p. 373) there is a not wholly dissimilar picking movement. Two cam-driven treadles alternately pull a pivoted arm, the under- and center-mounted picking stick, from side to side.

Artifacts:

British or American: none known that are very similar.

---

[92]Suggestion of Ian Birtles, loom tackler at Quarry Bank Mill, Styal, Cheshire, 12 August 2008.

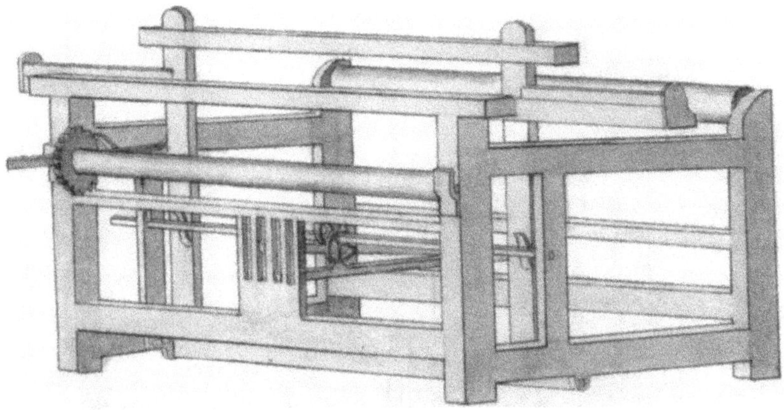

GURE 6.81 Power loom. Perspective drawing, front view.

talogue number: IEM 13

Document Description

Measurement: 22.5 x 24.5 cm

Medium: Ink and wash

Condition: Pinholes in corners

Commentary[93]

This illustration shows the take-up motion of a power loom. The ratchet wheel and pawl effecting the slow winding on of the cloth can be seen on the far left. In turn, the ratchet and pawl will have been moved, a notch at a time, probably by the reciprocal movement of the frame of the batten.

In the middle of the front part of the frame is a slotted block in which the treadles (activating the shedding motion) move up and down, following the heart-shaped cams, which are just visible.

Just above the frame side on the right can be seen one of the shuttle boxes.

The end of the warp beam appears where the frame pillar at the extreme right has a curved top.

Comparators

Drawings: Miller's 1796 patent (Fig. A3.16, p. 363) and Horrocks's 1813 patent (Figs. A.3.17, p. 364 and A.3.18, pp. 365).

Artifacts:

British or American: none known.

[93]My thanks to Ian Birtles for making some identifications here.

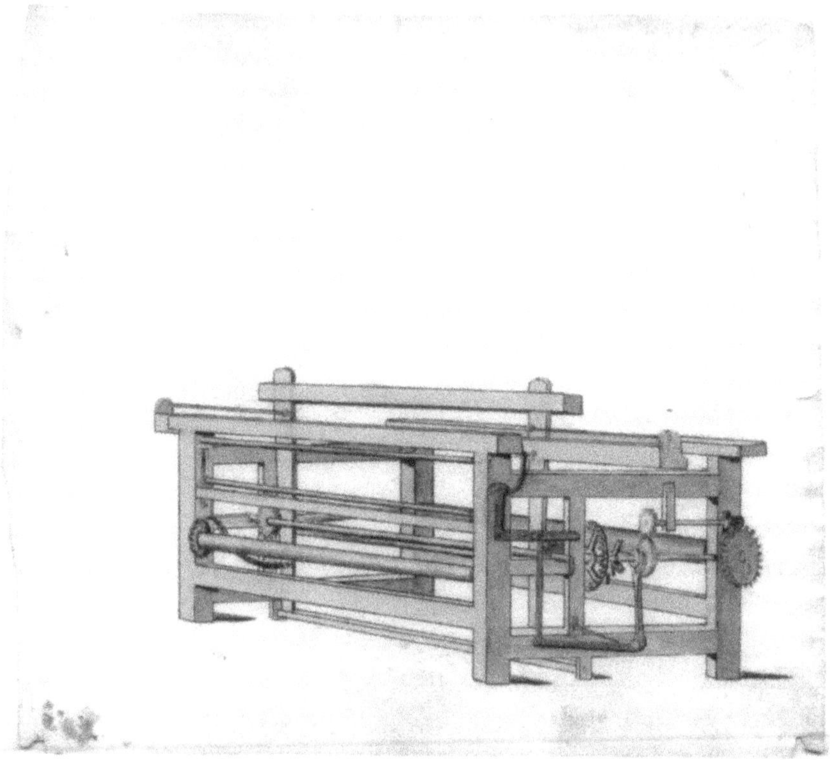

FIGURE 6.82 Power loom. Perspective drawing, front view.

Catalogue number: IEM 14

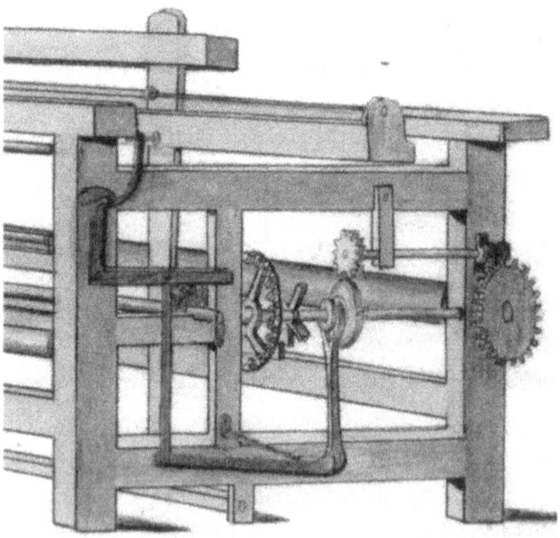

Power loom front view, detail.

Document Description

Measurement: 21.4 x 25 cm

Medium: Ink and wash

Condition: Unevenly cut edge, short tear on right edge, torn lower corners, discoloration of lower left corner

Commentary

Again showing the ratchet and pawl (*far left*) effecting the take-up motion, winding the cloth onto the cloth beam. This power loom is slightly different from the one at Figure 6.81 because the ratchet and pawl are situated lower down the loom frame and because, at its warp-beam end (*right*), vertical frame members are capped by horizontal members (in contrast to the curved pillar head in Fig. 6.81). Otherwise it is very similar to the loom frame at Figure 6.81.

The ironwork visible on the right-hand side of the loom frame might be evidence of a warp protector stop-motion mechanism, as shown in both Miller's 1796 patent (Fig. A3.16, p. 363), and Horrocks's 1813 patent (Figures A3.17 and A3.18, pp. 364, 365, respectively) and because the right-angled bell crank ends with a fork intended to act as a clutch, disengaging power from the loom.

On the other hand, why would there be a large gear wheel on the end of the warp beam? A worm gear appears to power that gear wheel, suggesting a let-off motion (in place of a traditional handloom system of tension weights to restrain the warp bam from unwinding).[94]

Near the clutch the gear with two or four teeth or arms, on the same shaft as the clutch, has an unclear purpose.

Again by means of cranks, the ironwork appears to support or be supported by a feeler pin inside the shuttle box. This is suggestive of a filling or weft stop-motion. The cumbersome nature of the mechanism suggests that it might not have been very efficient, but this is impossible to tell.

Comparators

Drawings: Miller's 1796 loom patent (Fig. A3.16, p. 363); Horrocks's 1813 loom patent (Figs. A3.17 and A 3.18, pp. 364, 365, respectively).

[94]My thanks to Ian Birtles for this suggestion.

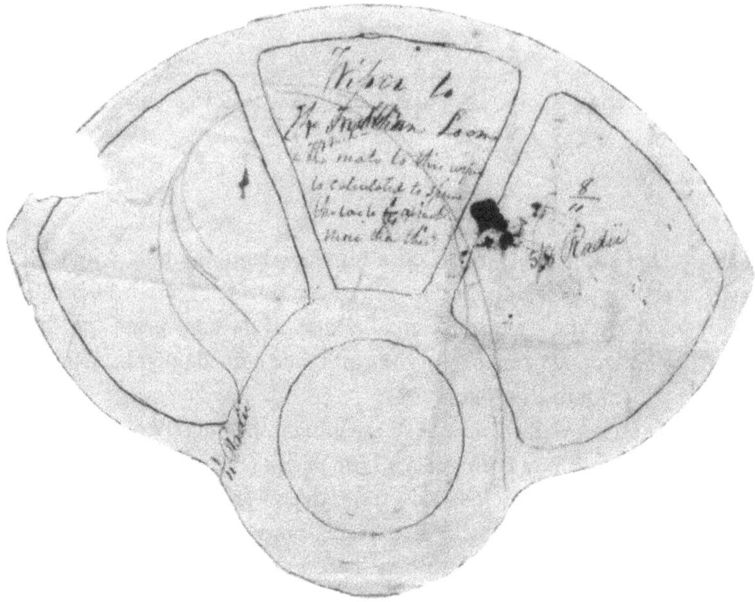

FIGURE 6.83A Paper pattern for "Wiper to the Waltham Loom."

Catalogue number: IEM 15.a

Document Description

Measurement: 16.9 x 20.5 cm

Medium: Pencil and ink

Condition: Ink smear, gap on edge, foxing; reverse side of Figure 6.83B

Commentary

The inscription reads "the mate to this wiper [cam] is calculated to spring the back?/lock?/cock? $1^{1}/_{4}$ inches more than this."

As already noted, cams were most frequently used in the power loom to activate the treadles controlling the harnesses and the shedding mechanism. However, cams could be used in other positions too, for example, the picking stick. The shape of this cam suggests that it might be unsuitable to drive harness treadles, but perhaps harsher movements were tolerated in early power looms. The picture of a weaving shed in Baines[95] shows a shape similar to this at the base of the picking stick (see Fig. 6.80, p. 308).

Comparators

See British loom patents, Figures A3.16–A3.18 (pp. 363, 364, 365, respectively); and the later Lowell loom (Fig. A4.2, p. 373).

[95]Baines, *Cotton Manufacture*, 239.

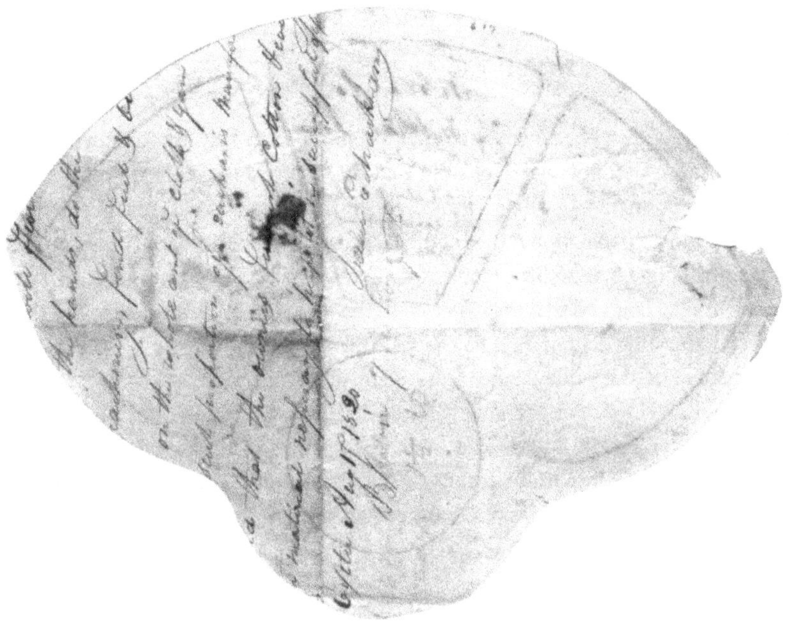

FIGURE 6.83B Reverse side of paper pattern for "Wiper to the Waltham Loom."

Catalogue number: IEM 15.b

Document Description

Measurement: 16.9 x 20.5 x cm

Medium: Pencil and ink

Condition: Bleed-through, gap on edge, stains; reverse side of Figure 6.83A

Fragment of document on back

Evidently the tracing was cut from an old letter, draft letter, or notes made by Isaac E. Markham because the lines of the text are cut as dictated by the shape of the cam. Fortuitously signed, the date and signature have been preserved: "Exeter Aug. 17 1820 Isaac E. Markham."

9. PORTFOLIO COVER: PULLEYS, CAMS

Carefully drawn on two sheets of heavyweight paper, each folded in half, these drawings formed a storage cover for Markham's machine drawings. Not only that, he seems to have used them as a bank of master drawings of critical components—pulleys, gears, rollers, and cams. The machines mentioned in the following plates were those, presumably, that could be made in the Middlebury Cotton Factory machine shop: carding machine (Figs. 6.38, 6.90); wool-carding machine (Figs. 6.86, 6.87); drawing frame (Figs. 6.85, 6.90); roving frame (Figs. 6.51B, 6.85); can frame (Figs. 6.50, 6.84); throstle (Figs. 6.52, 6.85, 6.90, 6.91); flywheel spinning frame (Figs. 6.67, 6.86); mule (Figs. 6.58, 6.87, 6.88); and power loom (Figs. 6.69, 6.89). Some of the pulleys and gears were to be made of iron (Figs. 6.90, 6.91), some of brass (Figs. 6.86, 6.90, 6.91). Some gears were beveled (Figs. 6.85, 6.90, 6.91). The exact dimensions supplied would have governed the speeds and interrelationships of the moving parts within each of the machines. As such these were important technical data, forming a body of reference materials against which components would have been designed and, after being made, could be checked.

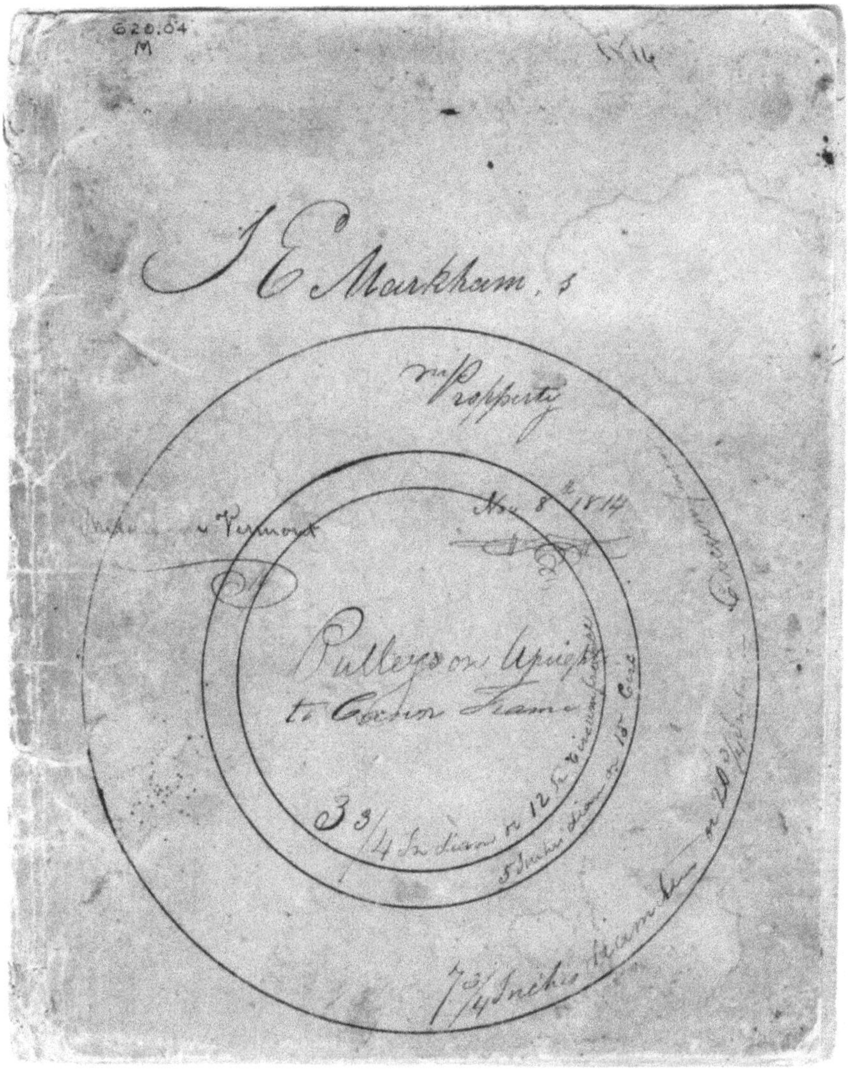

URE 6.84 Front cover to Markham's drawings: "I. E. Markham's Property, Middlebury Vermont Nov 8th 1814."

alogue number: IEM 1.1

Document Description

Measurement: 29.5 x 23.8 cm (one page of a 29.5 x 47.6 cm sheet folded in half that includes Figs. 6.84, 6.85, 6.90, 6.91)

Medium: Ink

Condition: Heavy (plied?) laid paper folded; holes in the fold; remnants of thread in the bottom holes of sheets; water (?) stains, fly specks

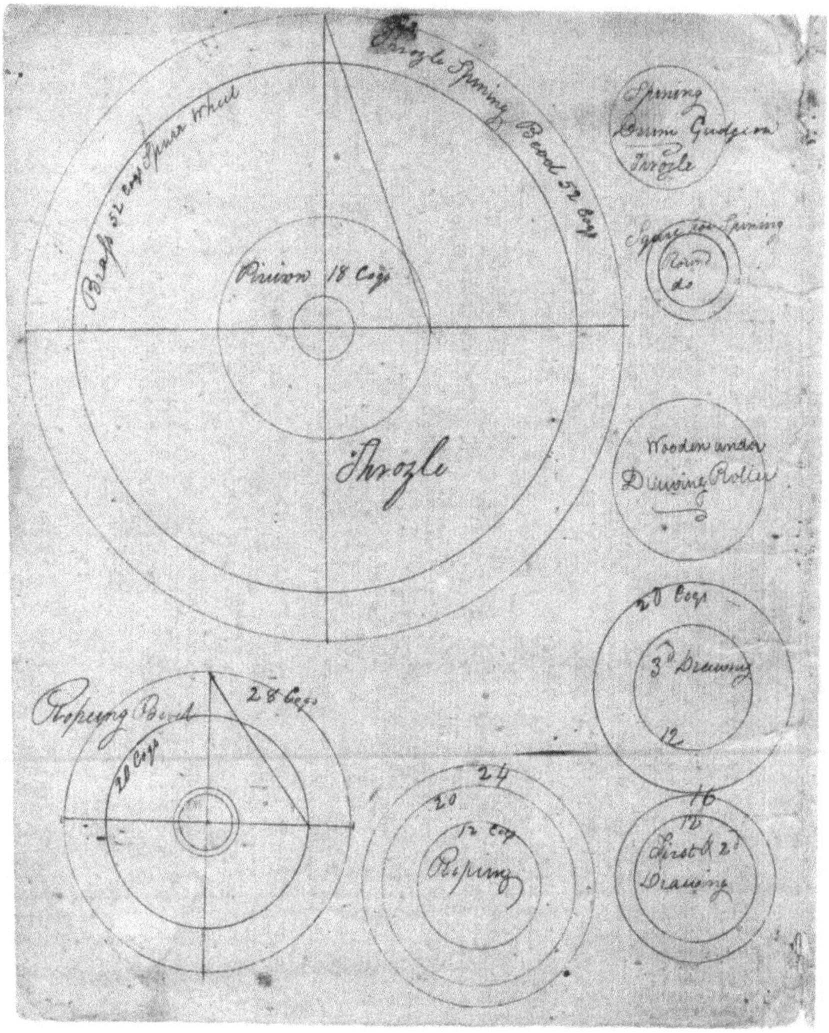

FIGURE 6.85 Front cover (verso): Gears, pulleys, rollers.

Catalogue number: IEM 1.2

Document Description

Measurement: 29.5 x 23.8 cm (one page of a 29.5 x 47.6 cm sheet folded in half that includes Figs. 6.84, 6.85, 6.90, 6.91)

Medium: Ink

Condition: Heavy (plied?) laid paper folded in half, holes in the fold, remnants of thread in the bottom holes of sheets, water (?) stains

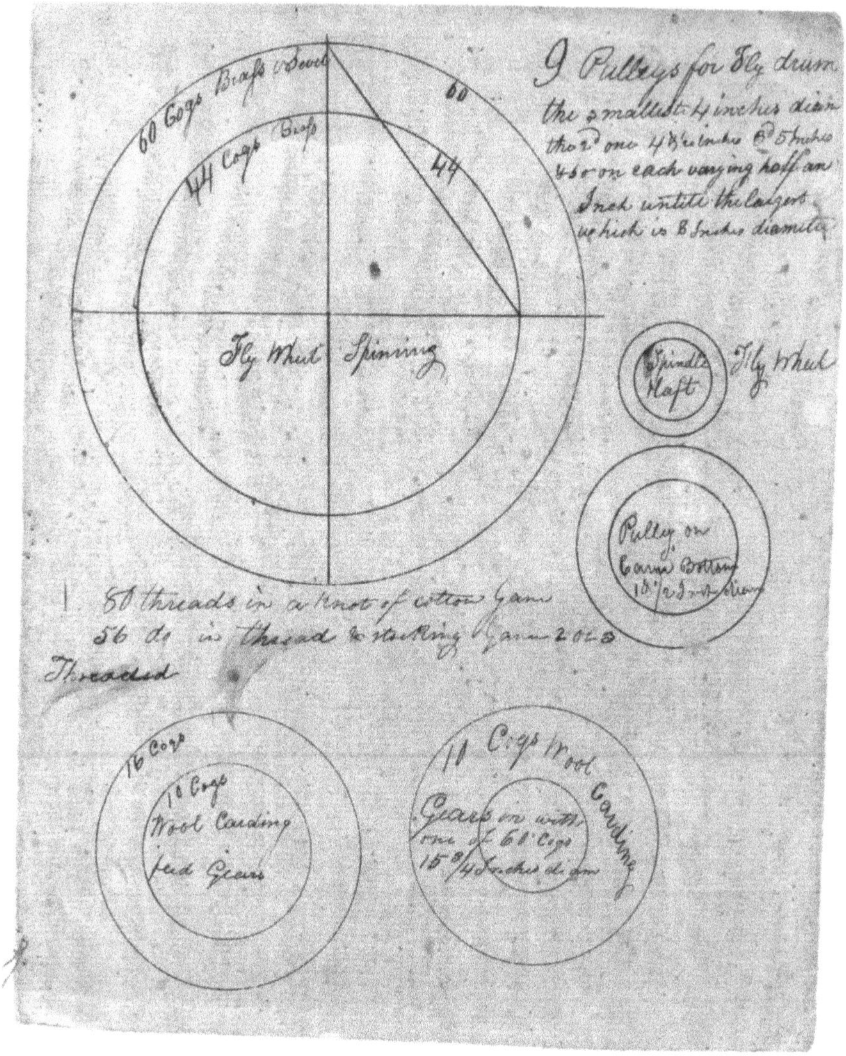

FIGURE 6.86 Front cover, inner sheet: Gears, pulleys.

Catalogue number: IEM 1.5

Document Description

Measurement: 29.5 x 24 cm (one page of a 29.5 x 48 cm sheet folded in half that includes Figs. 6.86–6.89)

Medium: Ink

Condition: Heavy (plied?) laid paper folded in half, holes in the fold, stains

Commentary

The formulae

"80 threads in a knot of cotton yarn
56 ditto in thread & stocking yarn per oz"

refer to methods of gauging cotton yarn and thread fineness. Eighty turns of yarn on a 1½ yard (54") reel produced a skein 120 yards long. Seven skeins, a length of 840 yards, comprised one hank of cotton yarn. The number of hanks in 1 pound avoirdupois indicated the number or fineness of a cotton yarn. In Britain before 1788, ounce thread had been made up in hanks 40 yards in length. Thereafter it was made up in hanks 30 yards long (30 turns on one-yard reel) but the weight unit was 1 ounce.[96]

[96]Jeremy, "British and American Yarn Count Systems."

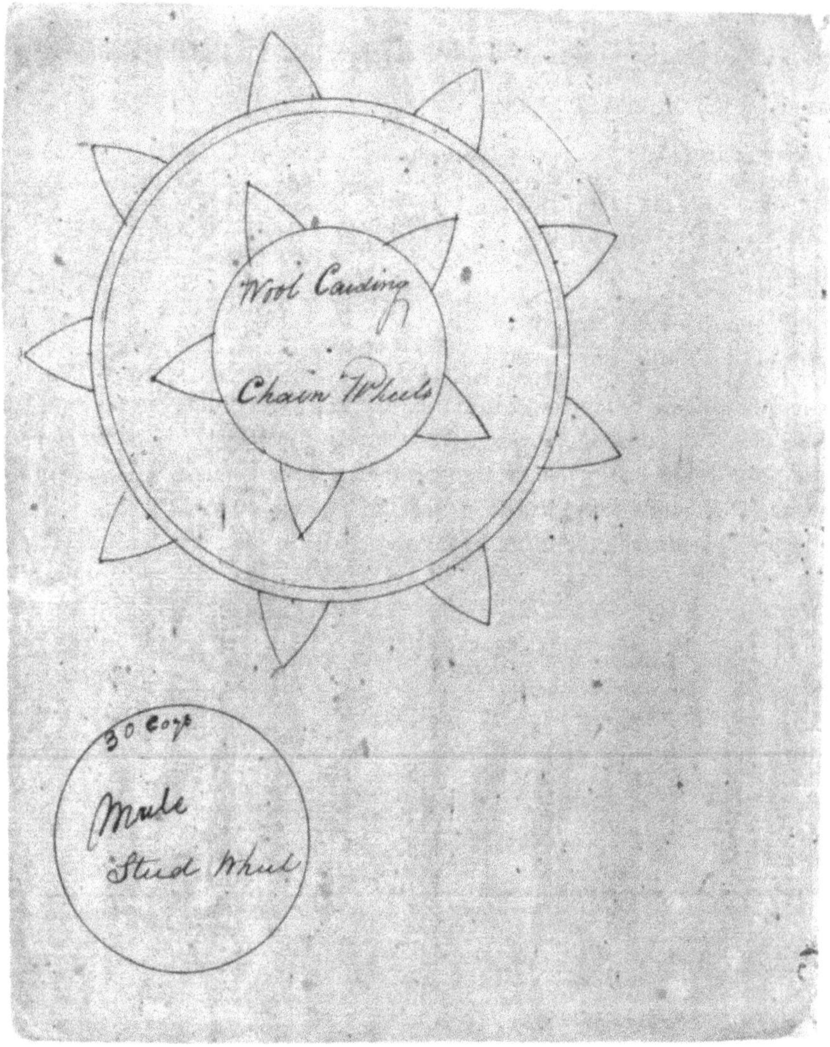

FIGURE 6.87 Inside front cover, inner sheet: Sprocket wheels and gear wheel.

Catalogue number: IEM 1.6

Document Description

Measurement: 29.5 x 24 cm (one page of a 29.5 x 48 cm sheet folded in half that includes Figs. 6.86–6.89)
Medium: Ink
Condition: Heavy (plied?) laid paper folded in half, holes in the fold

Commentary

Chain wheels (sprocket wheels) and chains drove worker cylinders in a wool-carding machine. Direct drive prevented slippage between the card teeth of the workers and those of the main drum, between which the wool was carded. Smaller, faster, belt-driven rollers, known as clearers or strippers, removed the wool from the workers and sent it forward over the main cylinder.[97] See chapter 6, section 3.

[97]Bakewell, "Woolen Manufacture."

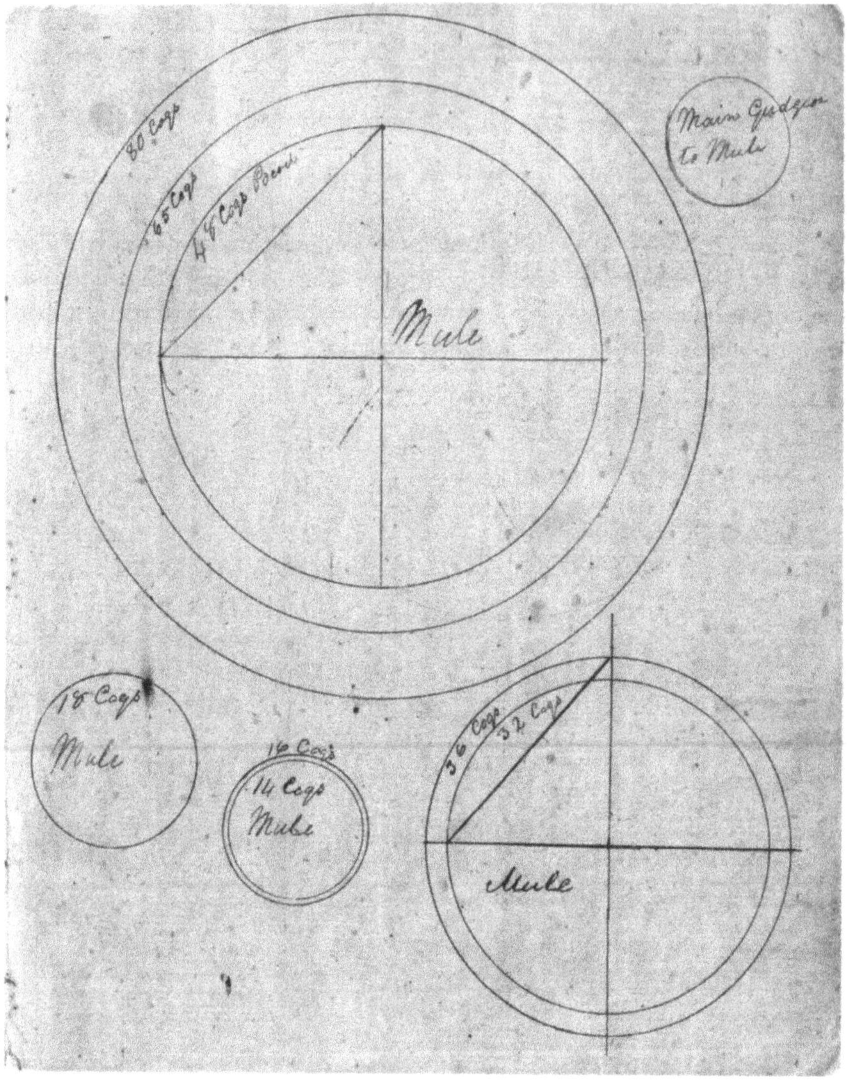

FIGURE 6.88 Inside back cover, inner sheet: Gear wheels for spinning mule.

Catalogue number: IEM 1.7

Document Description

Measurement: 29.5 x 24 cm (one page of a 29.5 x 48 cm sheet folded in half that includes Figs. 6.86–6.89)

Medium: Ink

Condition: Heavy (plied?) laid paper folded in half, holes in the fold

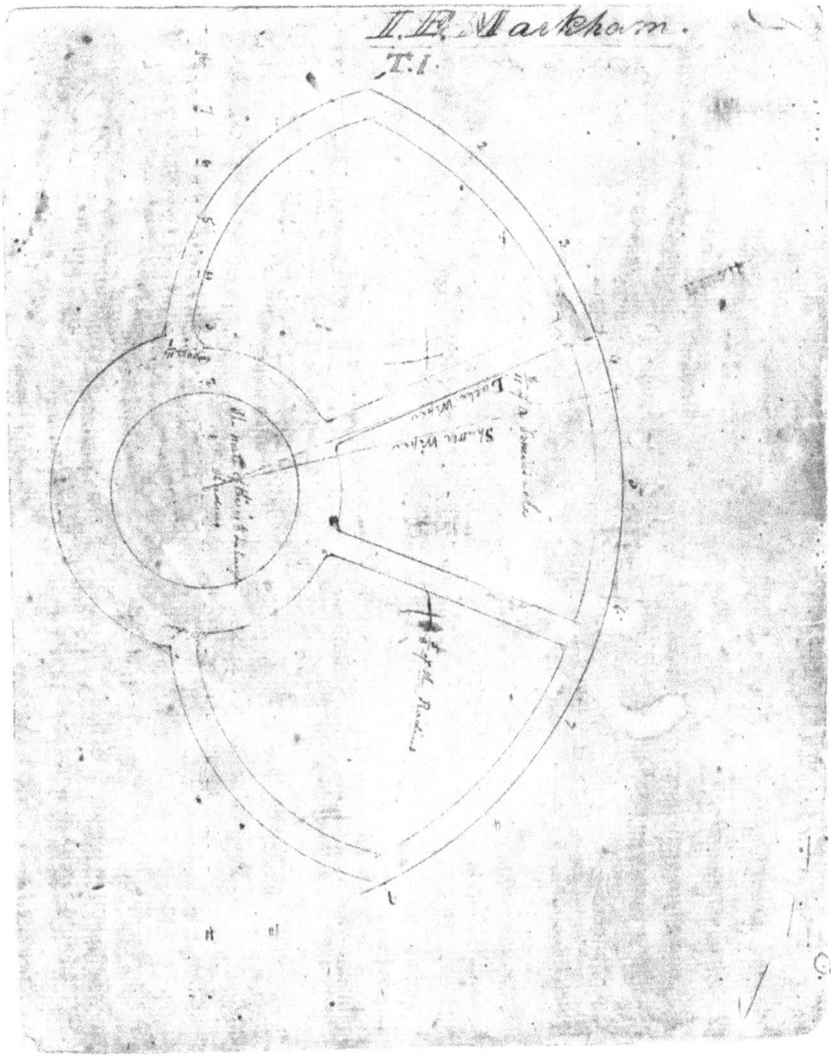

URE 6.89 Back cover, inner sheet: Loom cam.

alogue number: IEM 1.8

Document Description

Measurement: 29.5 x 24 cm (one page of a 29.5 x 48 cm sheet folded in half that includes Figs. 6.86–6.89)

Medium: Ink

Condition: Heavy (plied?) laid paper folded in half, holes in the fold, stains

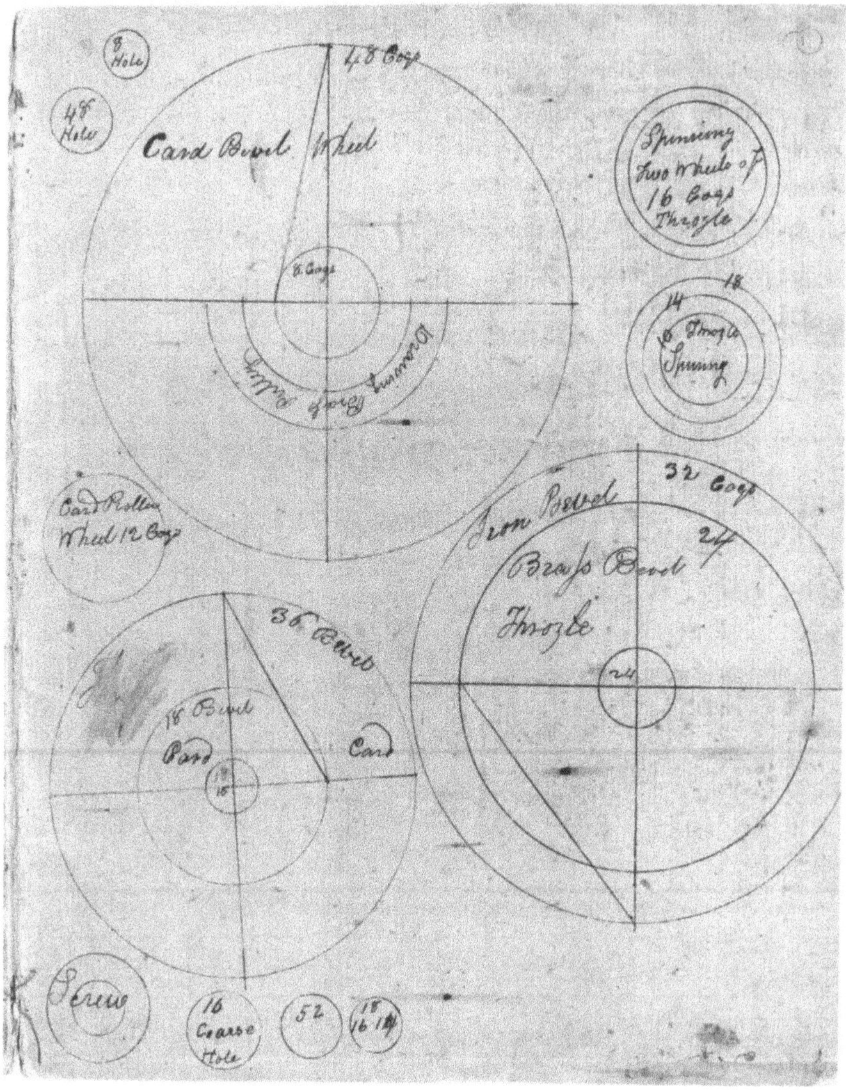

FIGURE 6.90 Inside back cover: Gear wheels, holes for shafts.

Catalogue Number: IEM 1.3

Document Description

Measurement: 29.5 x 23.8 cm (one page of a 29.5 x 47.6 cm sheet folded in half that includes Figs. 6.84, 6.85, 6.90, 6.91)

Medium: Ink

Condition: Heavy (plied?) laid paper folded in half, holes in the fold, remnants of thread in the bottom holes of sheets

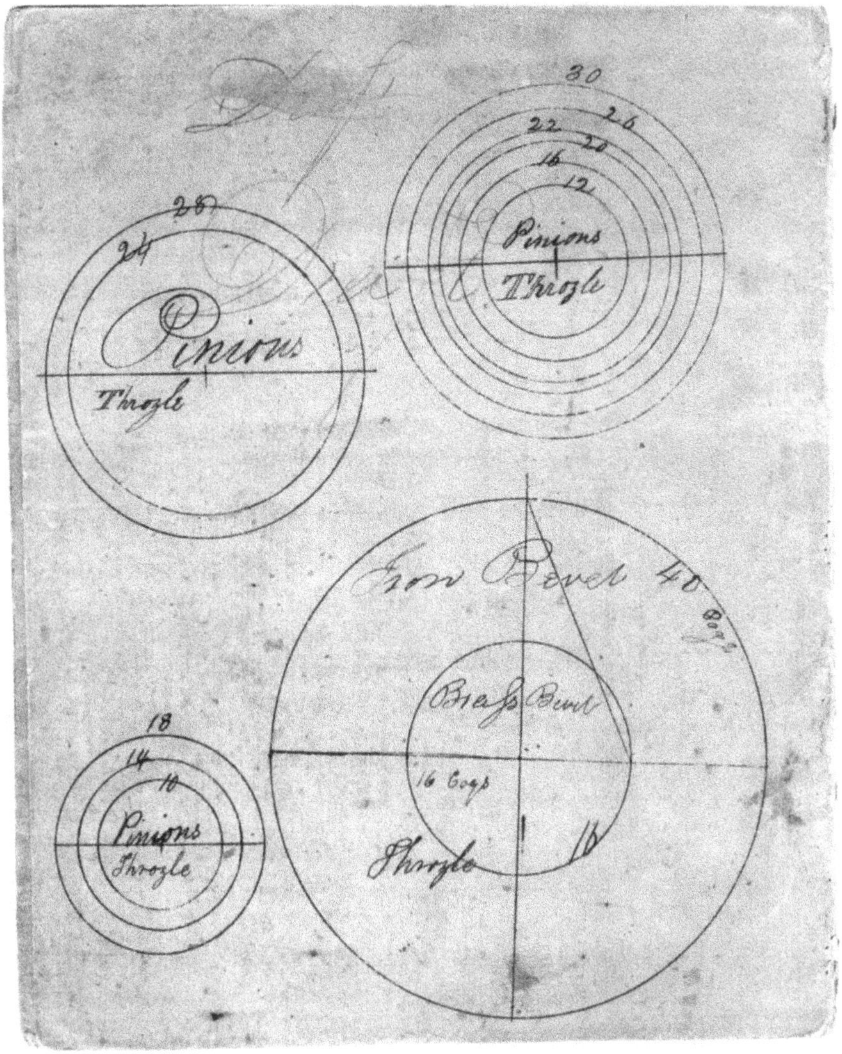

URE 6.91 Back cover: Gears.

talogue number: IEM 1.4

Document Description

Measurement: 29.5 x 23.8 cm (one page of a 29.5 x 47.6 cm sheet folded in half that includes Figs. 6.84, 6.85, 6.90, 6.91)

Medium: Ink

Condition: Heavy (plied?) laid paper folded in half, holes in the fold, remnants of thread in the bottom holes of sheets

Part III

APPENDICES: MICRO-STUDIES

Appendix 1

History and Provenance of the Papers

☆ ☆ ☆ ☆ ☆ ☆ ☆

IN 1882, THE SHELDON ART MUSEUM, Archaeological and Historical Society was incorporated in Middlebury, Vermont, by Henry L. Sheldon to preserve artifacts and documents of local history. For the next twenty-five years he collected personal papers, business records, and objects, often donated by townspeople, to commemorate the early inhabitants of Middlebury, influential citizens, earlier ways of life, and local institutions. He also collected all local publications, including newspapers, pamphlets, and books, to document the town of Middlebury and the surrounding area, creating one of the most comprehensive collections of local history materials in the Northeast.

When a new town hall was built in 1884, there was talk of installing a plaque to honor inventors from Middlebury. Isaac Markham's sister, Catherine Markham McLeod, supported the claim that Isaac Markham, at the age of ten, had invented the method of suspending gangs of saws in a frame from a fixed support using cords to give the frame an oscillating motion, which was used in Eben Judd's 1804 marble mill. She remembered the working model that he made and that her father showed the model to Dr. Judd, but a commission formed to evaluate the claims ruled against it, as sawing marble with ganged toothless saws and sand using waterpower was not a new idea.[1]

Perhaps because of the lack of recognition in those quarters, or perhaps because of his friendship with Henry Sheldon, Catherine McLeod's son, Thomas Henry McLeod, made several donations of family memorabilia to the museum, including in 1900, the pocket compass that Capt. Markham brought from Canada in 1783; in 1886 a portrait; and in 1896, "old letters, writs, licenses, and other papers"—probably including the drawings.

Sheldon scattered some of the papers through scrapbooks, such as *St. Stephen's Church with reminders of those by whom it was organized, 1810*, ca. 1880–1901. Others were kept together until a cataloging project in the 1930s separated the contents of all manuscript collections by type of document: account books were kept together, letters were interfiled in chronological order and individually cataloged, and so on. The drawings, with the addition of a few by other people, were simply identified as machinery drawings with no further description and shelved among the books.

They drew little or no attention until the early 1980s, when a manuscript-cataloging project was undertaken. Seeking more information about the drawings, museum staff put the word out on the industrial archaeology grapevine that the museum had a drawing tentatively identified as the mills in

[1] *Marble Border of Western New England.*

Waltham along with machine drawings. Michael Brewster Folsom, then director of the Charles River Museum of Industry, located in one of the Boston Manufacturing Co. mill buildings, made preliminary identifications, and circulated copies among scholars in the field, where they generated considerable interest. Markham's story and one of his drawings was included by Brooke Hindle and Steven Lubar in their *Engines of Change: The American Industrial Revolution, 1790–1860* (Washington, DC: Smithsonian Institution Press, 1986). The Markham sketch of the Boston Manufacturing Co. mills was used on the cover of Robert Weible (ed.), *Essays from the Lowell Conference on Industrial History, 1982–1983* (North Andover, MA: Museum of American Textile History, 1985) and discussed in one of the essays by Richard Candee.

Jeremy received copies in the autumn of 1981 when Michael Folsom, excited by the possibility of reconstructing the Waltham power loom from the drawings, sent him Xeroxes with the suggestion that the discovery would require a second edition of Jeremy's *Transatlantic Industrial Revolution*, which had just appeared.[2] Although he vouched for the importance of the collection, Jeremy was already involved in other projects at the Business History Unit in the London School of Economics, projects that preoccupied him for the next twenty or more years. Retirement allowed him to return to some of his earlier textile-history interests and in autumn 2004 he used the new technology of the World Wide Web to contact Polly Darnell, who as archivist, had compiled a profile of Markham that she had supplied to Folsom twenty-five years earlier. Meantime Folsom, sadly, had died in 1990. Although the drawings were referenced and discussed by scholars, they had never been the subject of the thorough scholarly review that they deserve. Jeremy was therefore delighted when Polly Darnell agreed to collaborate with him on an edition of the drawings and their efforts were facilitated by the ready cooperation of the Sheldon Museum and nearby Middlebury College, which undertook the photographic work necessary.

The history of the papers establishes their provenance back to the 1880s. Beyond that the main objective evidence of their authenticity must lie in their physical characteristics. Most readily detectable are the watermarks. These, as Table A1.1 demonstrates, prove that some of Markham's drawings were done on papers made by several American and one English papermaker in precisely the decade when Markham was most active, 1814–1825. The English papermaker, William Balston, supplied papers to the finest watercolorists of his day, including J.M.W. Turner, who used some paper with this watermark.[3]

[2]Michael Brewster Folsom to David J. Jeremy, 2 September 1981.

[3]Bower, *Turner's Papers, passim*; Bower, *Turner's Later Papers, passim*.

able A1.1 Makers of Papers Used by Markham for His Letters and Drawings

ocument	Watermark	Paper maker	Dates of use of watermark
etter from Waltham	C Baker & Co.[a]	Charles Baker & Company, Dorchester, MA	1819–1823
ig. 6.24	E. Burbank[b]	Abijah Burbank, of Sutton, MA, was joined at some point by his brother Elijah	1810–1823
ig. 6.40	T G & Co.[c]	Thomas Gilpin & Co., near Wilmington, DE	after 1800
ig. 6.49	T G & Co.	Thomas Gilpin & Co., near Wilmington, DE after 1800	
ig. 6.77	J. Whatman 1816[d]	James Whatman (retired 1794); succeeded by William Balston working the Springfield Mill, near Maidstone, Kent, England	1816
ketchbook, all pages igs. 6.27–6.29, 6.36–37)	Image of dove [e]	Charles Baker & Company, Dorchester, MA	1819–1823 (dove watermark)
ketchbook	C Baker & Co.[f]	Charles Baker & Company, Dorchester, MA	1819–1823
ketchbook, cover (Figs. 30–6.35)		William Hurd, Paper Manufactory, Newton, MA[g]	1816–1820

[a] Gravell and Miller, *American Watermarks*, 30, 237.
[b] Gravell and Miller, *American Watermarks*, 39, 246.
[c] Gravell and Miller, *American Watermarks*, 198, 265–266.
[d] Gravell and Miller, *Catalogue*; Bower, *Turner's Papers*, 37–38, 102, 103, 115; Bower, *Turner's Later Papers*, 47.
[e] Gravell and Miller, *American Watermarks*, 30, 237.
[f] Gravell and Miller, *American Watermarks*, 30, 237
[g] Gravell and Miller, *American Watermarks*, 30, 237.

Appendix 2

The Diffusion of Technology: British and American Published Sources of Industrial Textile Technology before 1825

☆ ☆ ☆ ☆ ☆ ☆ ☆

AMERICANS WERE AVID ACQUIRERS OF EUROPEAN BOOKS, as noted in chapter 2. Among these books were publications informing readers about new technological knowledge. Which of these books published before 1825 (when Markham died) were the most important and to which might Markham have had access?

Undoubtedly, the most important was edited by Abraham Rees (1743–1825), *The Cyclopaedia; or Universal Dictionary of Arts, Sciences and Literature* (39 volumes of text and six volumes of plates [London: Longman et al., 1802–1820]). This was a massive 39-million-word record in which Rees was assisted by more than one hundred contributors.[1] Before considering how Markham might have been able to consult the Rees volumes, several questions arise about the usefulness of the textile articles and drawings contained in these volumes. In a period of fast-moving technical change, to what points in time did its technological descriptions relate? How technically reliable were those descriptions? How comprehensive was the coverage of the technology in question and did it depict representative technology or simply the experimental?

The dating of the Rees articles was known to contemporaries from the list of contributors and a chronology of the articles published in *The Philosophical Magazine and Journal* in 1820. This was forgotten by late Victorian scholars and then rediscovered by Negley Harte.[2] The Rees articles were published alphabetically and grouped in parts, two parts comprising one volume. The parts appeared at irregular intervals over the course of eighteen years. When the end of the alphabet was reached the parts were bound up in volumes with the accompanying plates collected into separate volumes. In this form, rather than as unbound parts, almost all known copies of the book survived. The sequence and authors listed in Table A2.1 come from the definitive lists in *The Philosophical Magazine and Journal*.

Much less certain is the publishing history of the plates of drawings. At first sight it might be assumed that the plates accompanied the articles to which they related, as they seem to do in the bound version of Rees. This arrangement is listed in Table A2.2. However, the dates of the plates do not always coincide with or even precede the dates of the articles to which they relate.[3] This can be seen in Table A2.3. It

[1]Woolrich, "Abraham Rees."

[2]*Philosophical Magazine* 56 (1820) 218–224; Jackson, *Attempt*; Harte, "On Rees's *Cyclopaedia*."

[3]At the foot of each plate, besides the names of the artist and the engraver, was the formula "Published as the Act directs, 1811 [or whenever], by Longman, Hirst, Rees, Orme & Brown, Paternoster Row." This was part of the information recorded on each plate in order to gain copyright protection of twenty-eight years under the Copyright Act of 1767.

Table A2.1 Authors and Dates of the Textile Articles in Abraham Rees (ed.), *The Cyclopaedia* (45 vols., London, 1802–1820)

Title of publication	Date	Author	Trade/profession
"Cotton" in Rees (ed), *Cyclopaedia*	2 May 1808 (London)	James Thomson (1779–1850)	calico printer
"Cotton Manufacture" in Rees (ed), *Cyclopaedia*	2 May 1808 (London)	Thomson	calico printer
"Manufacture of Cotton" in Rees (ed), *Cyclopaedia*	4 Nov 1812 (London)	Thomson	calico printer
"Weaving" in Rees (ed), *Cyclopaedia*	29 May 1818 (London)	John Duncan	author of textbook on weaving
"Wool" in Rees (ed), *Cyclopaedia*	30 July 1818 (London)	Robert Bakewell (1768–1843)	geologist
"Woolen Manufacture" in Rees (ed), *Cyclopaedia*	30 July 1818 (London)	Bakewell	geologist

Sources: *Philosophical Magazine* 56 (1820) 218–224; Harte, "On Rees's *Cyclopaedia*"; Aspin, "James Thomson"; Torrens, "Robert Bakewell."

Table A2.2 Textile Articles and Accompanying Plates, as Published in Abraham Rees (ed.), *The Cyclopaedia* (45 vols., London, 1802–1820)

Plate number	Title of drawing or article
	Manufacture of cotton*
1	Calico printing
2	Batting machine
3	Devilling
4	Carding
5	Drawing frame
6	Roving can frame
7	Roving frame termed double speeder
8	Roving frame
9	Water spinning frame
10	Throstle spinning frame
11	Mule spinning
12	Reeling
13	Doubling machine and twisting machine
14	Messrs Strutt's Cotton mills at Belper, Derbyshire
	Weaving
1	Weaving: Mr Austin's engine loom
2	Weaving: Duff's draw boy
	Woolen manufacture
1	Slubbing machine or billy
2	Spinning jenny
3	Shearing machine
4	Carding engine
5	Gig mill

*Article title appears in bold type.

would have been perverse if the plates postdated their related articles. In contrast to Thomson's "Cotton Manufacture" (1808) article, which had no detailed machine descriptions, his "Manufacture of Cotton" (1812) article was devoted to lengthy descriptions of the various machines and how they worked, all carefully keyed to the plates' drawings. It is known that the London plates had to be engraved twice over and it may be that there were earlier and later impressions of the *Cyclopaedia* with the later impressions of the plates appearing in the completed and bound edition.

Table A2.3 implies that Longman, the publisher, had a bank of plates prepared ahead of, and on one occasion, after (apparently) the articles were written, and sometimes for related articles. This inference appears to be confirmed by Thomson, who noted in his "Manufacture of Cotton" article, "For the explanation of these [machines], we have appropriated 13 of our plates, which are entitled *Cotton Manufacture*."[4]

How technically reliable were the drawings and technical descriptions in Rees? Technical reliability depends on the technical competence of the authors and draftsman who produced the Rees articles and plates. What do we know about them? Harte has identified the contributors to Rees's *Cyclopaedia*.[5] Most respected was James Thomson (1779–1850) because he was intimate with the cotton-manufacturing technology that led technical change in the textile industries and which he was describing. Born in Blackburn, he studied in Glasgow and then joined the London branch of the Peels (to whom he was related), pioneering cotton manufacturers in Lancashire. From the London branch of the firm Thomson was moved in the 1800s to manage their print works at Church, near Accrington, becoming a partner of the Peels. In 1810 he set up his own calico-printing firm at Primrose, near Clitheroe. For the remaining forty years of his life he was an acknowledged leader of the calico-printing trade.[6] This kind of experience lent great weight and authority to the articles that he contributed to Rees.

Robert Bakewell (1767–1843), not to be confused with the improver of British livestock, was born at Nottingham and, following the family trade, for twenty years was a wool stapler at Wakefield, Yorkshire. Bankrupt in 1810, he fled to London and devoted the rest of his life to studying and writing about geology. John Duncan, from Glasgow, was the inventor of the tambouring process, a weaving technique, and the author of *Practical and Descriptive Essays on the Art of Weaving* (Glasgow, 1808). He also contributed articles on textile subjects to David Brewster's *Edinburgh Encyclopedia* (ca. 1808–1830).[7]

Thomson, Bakewell, and Duncan were all "men whose knowledge derived from direct personal experience and from earnest inquiry among other practical men of a scientific bent, as well as from extensive reading among the growing body of scientific periodicals and other contemporary literature diffusing improving information."[8]

As for the accuracy of the plates, there is one indication that they were carefully and competently executed: the training and talent of the lead draughtsman, John Farey (1791–1851). Schooled in drafting patent specifications under William Nicholson, by the age of fourteen he was making drawings for Rees's *Cyclopaedia*. Highly talented, he contributed articles as well as engravings to technical journals and books and he designed draughtsman's instruments, becoming an engineer of repute, an expert in patent law, and author of a highly regarded treatise on the steam engine.[9] It may be assumed also that the drawings engraved for Rees would have been first inspected and approved by the authors of the articles to which

[4]Col. 1 of article.
[5]Harte, "On Rees's *Cyclopaedia*."
[6]Aspin, "James Thomson."
[7]Harte, "On Rees's *Cyclopaedia*"; Torrens, "Robert Bakewell."
[8]Harte, "On Rees's *Cyclopaedia*."
[9]Woolrich, "John Farey."

they related. Farey, whose drawings and engravings established his reputation, was most probably the author of *The Operative Mechanic* (1st ed. London, 1825; Philadelphia 1826), a wide-ranging description of new technologies, almost a mini-Rees. The author's name is given on the title page as "John Nicholson, Esq, Civil Engineer." No such individual is listed in London directories of the 1820s and is otherwise unknown. The evidence, not least the combination of Farey's first name with that of his old master's surname, points to Farey as author, though the drawing style is skeletal compared to that of Farey's drawings in Rees.[10]

Another indicator of the quality of the engraving, as opposed to the drawing, of the Rees plates is that they were wholly executed by Wilson Lowry (1760–1824) (Table A2.4), who gained a high reputation in the 1790s as an engraver of architectural prints. His work engraving drawings of mechanisms for Rees was a huge task: Some 6,000 copies of the *Cyclopaedia* were made and the plates had to be engraved twice over, to meet demand for copies.[11]

Were Rees's textile articles comprehensive in the coverage of their subjects and did they describe representative technology or the more experimental? Thomson, in his 1808 article "Cotton Manufacture," having recounted the history of Hargreaves's jenny, Arkwright's water frame, and Crompton's mule, observed, "This compleats [*sic*] the series of machines now in use."[12] For his "Manufacture of Cotton" article of 1812, Thomson opened his 56-column article by noting:

> *Cotton*, one of the leading and most important branches of our national industry and commerce. The history of its astonishing progress in the last century, the successive improvements in the machinery, which have been made by various inventors, and the extent of the trade, with other curious important facts, are detailed under the article COTTON: it is needless, therefore, to recapitulate these circumstances, and we shall proceed at once to describe this extensive manufacture, as conducted on the most improved system in some of our largest cotton mills. Many of our readers may have viewed a cotton mill with wonder, but not with intelligence, nor with leisure to trace the steps by which the wool from the bag ultimately assumes the form of a very fine thread. ...Such readers will not think a page or two unemployed, in giving a brief account of the different operations the cotton passes through, from the raw cotton or cotton wool, as imported, to the finished thread; and we shall afterwards enlarge upon each subject, and describe the machinery by which these operations are effected in the most expeditious and perfect manner. For the explanation of these, we have appropriated 13 of our plates, which are entitled *Cotton Manufacture.*[13]

Robert Bakewell, in a 46-column article describing "WOOLLEN MANUFACTURE, *Progress of the*," noted of the West Riding of Yorkshire and of Lancashire:

> These districts were the first to introduce mechanical improvements into the woollen manufacture, and thus gained a decided advantage over the ancient seats of the woollen trade. At present of all kinds of machinery that have hitherto been applied to wool are extensively employed in the West of England, and the manufacture of superfine cloth is in a flourishing state in the counties of Gloucestershire, Somersetshire, and Wiltshire, all ancient seats of the clothing trade.[14]

Concluding his 37-column article on "WOOLLEN MANUFACTURE, *Process of*," Bakewell remarked, "There has been a great number of other projects and patents for the improvement of different branches of the woolen manufacture; but as we have already noticed most of those which have come into use, we shall not enumerate any more of the unsuccessful attempts."[15]

---

[10] Woolrich, "John Farey, Jr., Technical Author"; my thanks to Tony Woolrich for conversations and materials about Farey.
[11] Guyatt, "Wilson Lowry."
[12] Thomson, "Cotton Manufacture," col. 16.
[13] Thomson, "Manufacture of Cotton," col. 1.
[14] Robert Bakewell, "Woollen Manufacture, Progress of the," col. 38.
[15] Robert Bakewell, "Woollen Manufacture, Process of the," col. 37.

On the basis of this evidence, there is no doubt but that the Rees articles described the most commonly used (as it happened, improved) machinery at the time of writing; that is, they depicted Britain's representative textile technologies prior to 1812. This was the new technical knowledge that was irregularly but progressively moving across the Atlantic to the infant United States.

There were problems for users of Rees, however. The *Cyclopaedia* was not regularly updated after completion in 1820. This meant that improvements in cotton-manufacturing technology appearing after 1812, and those appearing in woolen-manufacturing technology after 1818, were not in Rees. It was suggested that at least one supplementary volume be added, to record "such additional discoveries and improvements in the sciences and arts as have been made at various times since the work commenced."[16] This was never done. Until another magisterial survey appeared, watchers of the changing technologies available to British manufacturers would have to rely on the somewhat incomplete indexing and reprinting of the latest British patents. In cotton manufacturing, where technical change was ahead of change in woolen manufacturing, the innovations that before 1825 established power-loom weaving and more or less perfected mule spinning, could only be adequately tracked through the single patent publication of the time, the *Repertory of Arts*.

It should be added that in Rees's *Cyclopaedia*, for the first time, the descriptive text is accompanied by drawings of the new, improved machinery. This was especially useful because the two-dimensional drawing, far more than words, has the power to create or enhance in the viewer's mind the three-dimensional image of an object, even one as moderately complicated as a spinning mule or a power loom pre-1820 (see chapter 3, "Seeing in a Different Way" [p. 42]; see also Appendix 5). The drawings most likely to interest Markham, because he built machines of their type, are presented in Appendices 3 and 4.

If the Rees articles and plates offered the English-speaking world the first comprehensive glimpse of the new technologies transforming numbers of British industries and trades, how and when did they become accessible to American readers, not least those in the remote communities of a distant New England state like Vermont?

Rees was initially published in London in parts, 79 parts in all between 1802 and 1820, with an enormous total of 6,000 copies printed. The original English price, however, was high. When the last volume of text, volume 39, and six volumes of plates were published in 1820, one review reported the price per volume as £3.00.[17] If each of the volumes sold at the same price, then the *Cyclopaedia* sold for a total of £135. However, *The Philosophical Magazine and Journal* cited prices of "85 parts at £1 in Bds [boards] royal 36s," totaling a minimum of £85 or a maximum of £153 for the whole set.[18] These were huge sums in the early 1820s when male spinners on hand mules, commanding some of the highest wages in the cotton industry, earned 35s a week in the Manchester district.[19] In other words, to buy a copy of Rees's *Cyclopaedia* would take all the wages of a highly paid skilled worker for between 49 and 87 weeks—perhaps twice as long in reality.

Translated into American circumstances, the English edition of Rees, *Cyclopaedia*, would have cost (at an exchange rate of $4.85 to the £1 sterling) between $412 and $742 in the United States, without taking into account transatlantic transportation costs. In Samuel Slater's New England mills, a weaving room overseer earned an average of about $1 a day, whereas a machinist received $15 to $20 a month.[20]

---

[16] *Philosophical Magazine* 54 (1819) 379.
[17] *Edinburgh Review* 34 (1820), 507.
[18] *Philosophical Magazine* 56 (1820) 218.
[19] Wood, *History of Wages in the Cotton Trade*, 22; Burnett, *History of the Cost of Living*, 251–252.
[20] Tucker, *Samuel Slater*, 151.

This made the English edition of Rees's *Cyclopaedia* in the United States equal to between 82 and 148 weeks' wages, again of a well-paid worker—absorbing the savings of a skilled American over the course of three to six years. What chance was there for Markham, an itinerant mill mechanic when the *Cyclopaedia* articles were being published, of seeing the work's textile articles and drawings?

As already noted, there existed a strong American market for European books. Some copies of *Cyclopaedia* parts may have been imported into Boston from England, though the American publisher had a nonimportation agreement diminishing this possibility. If copies of the English edition reached educated men of capital in New England's rural communities, commercially minded men could have let their mechanically minded associates read them. In this way Markham might have seen articles relating to cotton manufacturing in their alphabetically appearing part-form within months of their London appearance. But the evidence connecting Markham to Rees's *Cyclopaedia* is more definite than this.

The wide-ranging utility of Rees, coupled with its high English price, excited demand for an American edition and Samuel F. Bradford, a Philadelphia publisher, seized the opportunity, contracting for the printing of 2,500 copies. It proved a Herculean task that eventually bankrupted him and his successor, who completed the work with a reduced print run of 2,000. The publishing history of the American edition of Rees's *Cyclopaedia* has been thoroughly researched by Professor Robert D. Arner of the University of Cincinnati.[21] Published in parts (two comprising a volume) starting in May 1806 and sold at $4 per part or $8 per volume,[22] it was completed in a bound set of 41 volumes of text and six volumes of plates sold at $400 (bound set) in December 1820. Dates of the appearance of the textile articles in the American edition can be seen in Table A2.3.

The table establishes a remarkable, hitherto unknown fact: within the space of six to twelve months of its London appearance the American edition of Rees was being distributed to American purchasers. The text of the American edition (insofar as the cotton and woolen articles were concerned) closely followed the text of the London edition. So too did the plates, as Table A2.4 suggests.

The strong likelihood of Markham seeing the American edition of Rees's *Cyclopaedia* sharply emerges from an examination of the edition's 1,851 subscribers. Of these, 430 lived in northern New England, including six in Vermont. Of these six, one lived at Brattleboro (John Holbrook), two at Middlebury (Joshua Bates and Daniel Chipman), one at Randolph (William Nutting), and two at Windsor (Isaac Green and Erastus Torrey). Two more were at Dover, New Hampshire (Oliver Crosby and Jabez Dow, MD).[23] These individuals emerge as professionals and community leaders.

Joshua Bates (1776–1854) was the pastor of the Dedham, Massachusetts, First Church, 1803–1818, and then the third president of Middlebury College, 1818–1839.[24] He had no known connection to Markham or his family, but it was a small community and if they were not personally acquainted, they would certainly have known of each other after Bates arrived in 1818.

Daniel Chipman (1765–1850), brother of Nathaniel Chipman (1752–1843) United States Senator and federal judge, was an attorney who served in both the state legislature and Congress. He was involved in some land transactions relating to the property on which David Page built the cotton factory. Isaac Markham's brother, David, also an attorney, studied law in Chipman's office and conveyed an interest in

---

[21]We thank Professor Arner for most generously sharing his knowledge with the authors.

[22]This was the original price and was still the price in 1818: *North American Review* 7 (1818) 436.

[23] Rees, *Cyclopaedia* American edition, vol. 41, which lists all subscribers. Regional analysis is given in Jeremy, *Transatlantic Industrial Revolution*, 70.

[24]Raum, "Joshua Bates."

ble A2.3 Rees's *Cyclopaedia*, Textile Articles and Accompanying Plates, as Separately Published, London and Philadelphia

		London edition		Philadelphia edition	
ate umber	Title of plate or article	Volume and part containing article	Date of article or plate	Volume and part containing article	Sale date of article or plate
plates?	**Cotton***	**X, part 1 (19)**	**2 May 1808**	**X, part 1**	**17 Jan 1809**
plates?	**Cotton manufacture**	**X, part 1 (19)**	**2 May 1808**	**X, part 1**	**17 Jan 1809**
4 plates	**Manufacture of cotton**	**XXII, part 2 (44)**	**4 Nov 1812**	**XXIII**	**8 July 1813**
	Calico printing		1808 (1809)		
2	Batting machine		1809		
3	Devilling		omitted		
4	Carding		1809 (1810)		
5	Drawing frame		1808		
	Roving can frame		1810		
7	Roving frame termed "double speeder"		1811		
8	Roving frame		1811		
9	Water-spinning frame		1813		
)	Throstle-spinning frame		1813		
	Mule spinning		1814		
2	Reeling		1812		
3	Doubling machine and twisting machine		1814		
4	Messrs Strutt's Cotton mills at Belper, Derbyshire		1814		
plates	**Weaving**	**XXXVIII, part 1 (75)**	**29 May 1818**	**XXXIX, part 1**	**25 Sept 1818**
	Weaving: Mr Austin's engine loom		1814 (1815)		
	Weaving: Duff's draw boy		1819		
o plates?	**Wool**	**XXXVIII, part 2 (76)**	**30 July 1818**	**XXXIX, part 2**	**24 Nov 1818**
plates	**Woolen manufacture**	**XXXVIII, part 2 (76)**	**30 July 1818**	**XXXIX, part 2**	**24 Nov 1818**
	Slubbing machine or billy		1811		
	Spinning jenny		1811		
	Shearing machine		1811		
	Carding engine		1809		
	Gig mill		1815		

urces: *Philosophical Magazine* 56 (1820) 218–224; Harte, "On Rees's *Cyclopaedia*." For the American edition, dates of advertisements for sale come from e *Salem (MA) Gazette*, except for vol. XXIII, announced in the *Boston Daily Advertiser* of 8 July 1813. Information on the publishing history of the nerican edition of Rees comes from Professor Robert Arner of the University of Cincinnati.

Note: Two sets of engravings were made for the plates of the London edition. The dates in parentheses, for the set held by Manchester Central Library, ow differences between those from which Cossons made a modern selection.[a]

[a]Cossons (comp.), *Rees's Manufacturing Industry*.

**Article information appears in bold type.*

Table A.2.4 London and American Artists and Engravers of the Textile Plates in Rees (ed.), *Cyclopaedia*, Both Editions

Plate number	Title of drawing	English date	English edition		American edition (all plates undated)	
			Artist	Engraver	Artist	Engraver
	Manufacture of cotton*					
1	Calico printing	1808 (1809)	William Thomson	Wilson Lowry	W. Thomson	Benjamin Tanner
2	Batting machine	1809	James Burton	Wilson Lowry	J. Burton	Benjamin Tanner
3	Devilling	omitted	James Burton	Wilson Lowry	omitted	H. Anderson
4	Carding	1809 (1810)	James Burton (Brown)	Wilson Lowry	omitted	omitted
5	Drawing frame	1808	omitted	Wilson Lowry	omitted	Benjamin Tanner
6	Roving can frame	1810	omitted	Wilson Lowry	omitted	omitted
7	Roving frame termed double speeder; horizontal plan	1811	J. Burton (omitted)	Wilson Lowry	J.Burton	H.Anderson
8	Roving frame; elevation in front	1811	omitted (J. Burton)	Wilson Lowry	omitted	omitted
9	Water spinning frame	1813	J. Farey Jr	W. Lowry	omitted	H.Anderson
10	Throstle spinning frame	1813	J. Farey Jr	W. Lowry	omitted	omitted
11	Mule spinning	1814	J. Farey	W. Lowry	omitted	H.Anderson
12	Reeling	1812	J. Farey	Lowry	omitted	H.Anderson
13	Doubling machine and twisting machine	1814	J. Farey, Jr	Lowry	omitted	C.G.E
14	Messrs Strutt's Cotton mills at Belper, Derbyshire	1814	J. Farey, Jr	W. Lowry	omitted	G.M.
	Weaving					
1	Weaving: Mr Austin's engine loom	1815	J. Farey	Lowry	omitted	Christian F. Gobrecht
2	Weaving: Duff's draw boy	1819	Farey	Lowry	omitted	H.Anderson
	Woolen manufacture					
1	Slubbing machine or billy	1811	J. Herbert	Wilson Lowry	omitted	omitted
2	Spinning jenny	1811	omitted	Lowry	omitted	D. Haines
3	Shearing machine	1811	omitted	Wilson Lowry	omitted	H.Anderson
4	Carding engine	1809	J. D. Herbert	Wilson Lowry	omitted	H.Anderson
5	Gig mill	1815	omitted	Lowry	omitted	Christian Gobrecht

Note: Two sets of engravings were made for the plates of the London edition. The information in parentheses, for the set held by Manchester Central Library, show differences with those from which Cossons made a modern selection.[a]

[a]Cossons (comp.), *Rees's Manufacturing Industry*.

**Article title appears in bold type.*

a piece of property to Chipman.[25] In addition, both Chipman and Isaac Markham belonged to the same Masonic lodge.

John Holbrook (1761–1838) of Brattleboro was a storekeeper, then

> was partner from 1805 to 1810 in the Brattleboro publishing firm of Holbrook, Fessenden & Porter...in January 1815, Holbrook returned from Connecticut to carry on Fessenden's thriving printing, publishing, papermaking, and bookselling activities. ...The firm of Holbrook & Fessenden eventually developed into the Brattleboro Typographic Company, which they incorporated in 1836.[26]

William Nutting (1779–1863) of Randolph graduated from Dartmouth College in 1807; was principal of the Orange County Grammar School in Randolph from 1807–1813; Town Clerk, 1811–1830; an attorney; he also served in the state legislature. He was also offered, and declined, a chair of mathematics and natural philosophy at the University of Vermont, of which he was a long-time trustee.[27]

Isaac Green (born 1759) was recalled as: "Dr. Green...one of the first directors of the Ascutney Bank, a director of the Cornish Bridge Company, was instrumental in getting the [state] prison located here [Windsor], was a merchant here for a long time and also practiced medicine."[28]

Of these individuals, Daniel Chipman's circumstances made him most likely to have shared his copy of Rees's *Cyclopaedia* with Markham. David, Isaac's brother, after attending Addison County Grammar School, studied law with Chipman, was admitted to the Bar in 1813, and practiced until 1823.[29] So Chipman had a close connection to the Markham family. As a lawyer, Chipman had professional gains to be made out of local economic development. That his attitude was Hamiltonian, proindustrial growth, rather than Jeffersonian, proagricultural expansionism, was suggested by his sale of land to Page for the Middlebury Cotton Mill, of which Isaac Markham became agent and superintendent. It was also suggested by his particular and publicized interest in the law of contracts, which governed men in business in Vermont.[30] In a remote rural community, of course, economic development benefited nearly everyone.

At an emotional level there were other potentially strong obligations that impelled Chipman to act in the interests of Isaac Markham. Chipman, thirty years older than Markham, could remember the American War of Independence. His economic interests suggested that this gave him a patriotic psychic investment in the future of Middlebury. One expression of this, and one that he shared with Isaac Markham's father, was his involvement in freemasonry and its fraternal and charitable activities.

Chipman belonged to the Union Lodge of Middlebury, which, originally chartered by the Grand Lodge of Connecticut, affiliated with the Grand Lodge of Vermont when the latter was established in 1794 (under the institution of Free and Accepted Masons). One of the early meeting places of the Union Lodge was the tavern kept by Ebenezer Markham, Isaac's father. He died in 1813 and, given the Freemasons' obligation of mutual assistance, members of his Lodge would have rallied round Markham's widow and children. Because Isaac started at the Middlebury Cotton Factory the previous year, Chipman would surely have shared any relevant articles in the *Cyclopaedia* with him.

Isaac Markham was an officer of Union Lodge at the time of his death in 1825, so Chipman would have met Isaac at lodge meetings and doubtless discussed the latest intelligence relating to the cotton

---

[25]Swift, *Middlebury; Middlebury Land Records* VI, 478, 480; VIII, 445; IX, 179; Selleck, *Bicentennial History of Union Lodge # 2*, 5; Sheldon, Scrapbook vol.156, "Twentieth Century Memorial of Middlebury, Vt. Copies of All Known Pictures of Persons Living in Town at the Beginning of the Nineteenth Century."

[26]McCorison, "John Holbrook."

[27]Hemenway et al. (eds.), *Vermont Historical Gazetteer* II, 995, 996, 1925.

[28]Child (comp.), *Gazetteer and Business Directory of Windsor County, Vermont for 1883-84.*

[29]Henry L. Sheldon Scrapbook vol. 156, 59.

[30] Chipman, *Essay on the Law of Contracts*, reviewed in the *North American Review* 17 (1823), 375-83.

mill and machine shop that Isaac managed and of which Chipman strongly approved, perhaps even had a financial interest in. This likelihood was strengthened by Isaac's standing among Vermont freemasons, a standing reflected in the presence at his funeral in 1825 of representatives from Bridport of the Morning Sun Lodge (also affiliated with the Grand Lodge of Vermont), as well as, from Vergennes, of the Jerusalem No. 2 Chapter of Royal Arch Masons.[31] Given all this circumstantial evidence, it is nigh certain that Chipman would have shown the textile articles in Rees' *Cyclopaedia* to Isaac Markham.

The complete omission from Rees's *Cyclopaedia* of an account of any of the American innovations that transformed the United States's cotton industry before the late 1820s is partly explained by the accident of alphabetization: the article "Manufacture of Cotton" appeared in 1812, just before Americans invented their modifications to British textile technology (as well as before the important Horrocks power loom and Roberts spinning-mule improvements were patented in Britain). The Rees article by Duncan on weaving, appearing in 1819, briefly described the Horrocks power loom (see Figs A3.17 and A3.18 [pp. 364, 365]) and, at greater length, how the Johnson (and Radcliffe) dressing machine worked (see Figs. A 3.14 and A3.15 [pp. 361, 362]), but illustrated neither in the *Cyclopaedia* plates. The *Circle of the Mechanical Arts* edited by Thomas Martin and published in London in 1813 likewise missed the appearance of the later weaving innovations.

One other contemporary source could well have informed Markham. This was the *Repertory of Arts, Manufacture, and Agriculture* (London, 1794–1862). About a quarter of 300 or so British textile patents issued in the period 1790–1830 were published here.[32] In view of his wide repertoire in textile machine building, those patents that were published with their drawings and would have most interested Markham, surely included Miller's power loom of 1796 (see Fig. A3.16, p. 363),[33] Rutt's card-making machine of 1810,[34] Raynor's roving frame (double speeder) of 1813 (see Fig. A3.13, p. 360),[35] and Horrocks's power loom of 1813.[36]

From the point of view of a stage-diffusion theory, Markham's experience straddled all four stages of early industrial diffusion. Transatlantic transfer of new knowledge of the power loom, via Joseph Gordon and (for best British practice) Rees's *Cyclopaedia*, exemplified diffusion Stage 1 (bringing demonstrable knowledge of a new technology to a potential host economy). The power loom production line Markham and Gordon set up at Middlebury, Vermont, illustrated diffusion Stage 2 (with the establishment of pilot production units in a host society). Markham's machine building, especially of his power loom, fell under diffusion Stage 3 (the replication of pilot production units across a region or economy). His improvements, for example, to his dressing frame, belonged to diffusion Stage 4 (local modifications to an imported technology).[37]

---

[31] See chapter 1, this volume, and Smith (ed.), *History of Addison County, Vermont*, chapter 13.
[32] Jeremy, *Transatlantic Industrial Revolution*, 47.
[33] *RAMA* 8 (1798) 148–154.
[34] *RAMA* 2nd ser. 26 (1814) 3–8.
[35] *RAMA* 2nd ser. 23 (1813) 321–330.
[36] *RAMA* 2nd ser. 25 (1814) 1–5.
[37] Jeremy, *Artisans*, Introduction.

Appendix 3

How Inventive Was Markham? Drawings of Exemplars of British Textile Machinery Available to Markham and Selected American Comparators

☆ ☆ ☆ ☆ ☆ ☆ ☆

PLAINLY, THE BIG QUESTION POSED BY MARKHAM'S WORKSHOP DRAWINGS and papers relates to the originality of the designs depicted. Which ones derived from British models, which from the technology pioneered by the Boston Manufacturing Co. at Waltham, and which were entirely of Markham's devising? The most difficult aspect of this question has to do with the design of his power looms, which has been separately treated in Appendix 4. How much of his technology came from Britain can be estimated by comparing the machine forms presented in Rees (ed.), *Cyclopaedia* (as seen in the previous section, almost certainly available to Markham in Middlebury), and in British patent designs published in the *Repertory of Arts, Manufactures and Agriculture*. Unpublished British patent designs were not necessarily denied to American competitors; thus the proprietors of the Boston Manufacturing Co. succeeded in obtaining the dresser design, which its inventor, William Radcliffe, thought he had secured from foreign eyes by having it patented in the name of one of his workmen, Thomas Johnson.[1] Table A3.1 shows what can be attributed to these British sources and what to other American sources.

Most detail is known about Markham's modifications to the Moody dresser of 1818, which deserves some amplified commentary. Mechanizing warp dressing presented early textile mechanics and engineers with a series of problems. Size had to be applied evenly to the warp. Then the warp had to be dried. Because the warping machine produced only part of a complete warp at a time (on a section beam), a multiple of section warp beams had to be combined in the dresser on a loom warp beam. The relative speeds of section beams and the loom warp beam would need to be adjusted to allow for the increasing diameter of the loop warp beam. A measuring device could signal the conclusion of the dressing operation. A stop-motion device would halt the machine when the warp was completed or when a piece of yarn broke.

Table A.3.2 summarizes how these challenges were resolved. In principle, most were solved in England by Thomas Johnson's (and William Radcliffe's) two dresser patents of 1803 (see Fig. A 3.14), which also included a warping frame, and 1804 (see Fig. A3.15). In the 1803 design the loom warp beam collected warps from the several section beams at one end of the machine. In the 1804 design, the centrally located

[1]Jeremy, *Transatlantic Industrial Revolution*, 48, 64, 99, 100.

loom warp beam collected warps from section beams arranged symmetrically on each side of the central frame. The warps absorbed size from a pair of rollers, the lower one of which rested on a third roller partly submerged in the size trough in the 1803 patent. In the 1804 design, just two rollers took up size, with the lower one half submerged in a size trough.

The size was then evenly spread by brushes. Johnson's 1803 patent shows two brushes on an endless belt. In his 1804 patent there are two sets of brushes, one mobile, the other static; the mobile one laid the size on the warp's topside, the static brush collected surplus size from the warp's underside. The moving brush, instead of following a circular motion as in the 1803 patent, followed a linear reciprocating track. Circular motion, it was found, raised fibers in the yarn instead of laying them flat. Changes in section beam and loom warp beam speeds were attained by attaching friction weights on the axles of the section beams, relying on the positive drive to the centrally supported loom warp beam to drag the warps through the dresser (both patents show this). Drying the warp was achieved by locating the section beam frames and the sizing zones far from the warp beam (the design in both patents). For this reason, dressers were large, elongated machines, about 30 feet long. Last, both of Johnson's patents show single drying fans.

Engineers at the Boston Manufacturing Co.'s cotton-manufacturing factories at Waltham, Massachusetts, got hold of a copy of the specification and drawings of Johnson's dresser patent of 1804 (see Fig. A3.15). None of them saw the machine itself but they proceeded to build a copy and made modifications, which enabled Paul Moody, BMC's chief engineer, to take out an American patent on the machine in 1818.

Moody's specifications listed changes as simple as positioning the section beams above one another to achieve better thread separation and thus operative access, instead of having them behind each other in a line, as was Johnson's preference; incorporating a gear-driven clock for measuring warp length; installing two drying fans instead of one; and arranging size rollers, troughs, and brushes so that broken threads could be more easily mended. Drying was assisted by the stop-motion mentioned in Moody's patent but not specifically claimed as his invention. Moody also noted that he had recently added a hot-air pipe to assist drying. All of this allowed Moody to claim that the Waltham dresser experienced fewer breakages, improved sizing, and faster drying than Johnson's dresser and that, consequently, it reached double the throughput of English dressers.

As seen in Table A3.2 (p. 368), Markham's drawings reveal strong similarities to the Waltham design for the dresser patented by Paul Moody in 1818. His major omissions are a measuring clock and a hot-air heating system. His major addition was the use of belt-driven twin cones (one a driver, the other a follower) to retard the speed of a positively driven loom warp beam. Intriguingly, the twin-cone pulley improvement survived in the dressing machine used at Lowell in the 1830s and described by Montgomery in 1840.[2] Was it original to Markham? Given his preoccupation with the possibilities of the twin-cone pulley, it would not be surprising if it were.

The chief difference between the American (northern New England?) and British dresser designs as developed in the decade or so after Markham's death lay in the American reliance on the direct drive to the loom warp beam to drag the section warps through the machine. This reflected a technical preference for simpler rather than more complex technology, permitted by the American use of superior cottons in strong, coarse yarns at the spinning stage. The use of cones, evident in Markham, suggests that eventually he might have attempted to power the section beams and the loom warp beam with changing differential speeds. English textile engineers, dealing with much finer yarn than their American counterparts, powered

[2]Montgomery, *Practical Detail*, 91–101, Plate VII.

able A3.1 British and American Sources of Markham's Designs

A) In the American Edition of Rees (ed.), *Cyclopaedia* (1810–1824)

ources	Machine	Date on English edition plate	Markham drawing figure number
1	Cotton-carding machine (Fig. A3.1)	1809	6.38, 6.39, 6.40 (but Markham's ground-to-rollers delivery system is novel)
2	Woolen-carding machine (Fig. A3.2)	1809	6.41A, 6.41B, 6.42
3	Cotton-drawing frame (Fig. A3.3)	1808	6.46, 6.47, 6.48, 6.49
4	Cotton can roving frame (Fig. A3.4)	1810	6.46, 6.47, 6.48, 6.50
5	Cotton double speeder, front elevation (Fig. A3.5)	1811	6.51 (A, B), with addition of rack and pawl to slow bobbin rail
6	Cotton double speeder, plan (Fig. A3.6)	1811	same
7	Woolen spinning jenny (Fig. A3.7)	1811	6.63, 6.64
8	Cotton water (powered) spinning frame (Fig. A3.8)	1813	6.46 (saddle and stand), 6.47 (drawing head)
9	Throstle-spinning frame (Fig. A3.9)	1813	6.52, 6.53, 6.54, 6.55, 6.56, 6.57, 6.71
0	Cotton-spinning mule (Fig. A3.10)	1814	6.58, 6.59, 6.60, 6.61
1	Cotton-reeling frame (Fig. A3.11)	1812	6.50

Notes: For their dates of publication in the London edition, see Tables A2.3 and A2.4.
There are no dates on the plates of the American edition.

B) From British Patent Sources

ources	Machine	Patent & date	Publication	Markham drawing figure number
2	Aaron Garlick's double speeder (Fig. A3.12)	2166 (1797)	unpublished	6.51A, 6.51B
3	Joseph Raynor's double speeder (Fig. A3.13)	3633 (1813)	*RAMA* 2nd ser. 23 (Nov. 1813), 320–330	6.51A, 6.51B
.4	Thomas Johnson's dresser and warper (Fig. A3.14)	2684 (1803)	unpublished	6.75, 6.76, 6.77, 6.78
.5	Thomas Johnson's dresser (Fig. A3.15)	2771 (1804)	unpublished	6.74, 6.75, 6.76, 6.77, 6.78
.6	Robert Miller's power loom (Scotch loom)—cams for shedding, picking, beating up; warp protector stop-motion (Fig. A3.16)	2122 (1796)	*RAMA* 8 (1798), 148–154	6.79, 6.80, 6.81, 6.82 (warp protector stop motion possibly); 6.83A, 6.83B (paper tracings for cam patterns)
.7 and .8	William Horrocks power loom (for variable batten speed motion, not in Markham) but shows underpick movement for picking stick (under warp) moved by cam-operated treadles (Figs. A3.17, A3.18)	3728 (1813)	*RAMA* 2nd ser. 25 (June 1814), 1–5	6.80 for underpick movement

(continued)

Table A3.1 *(continued)*

(C) Earliest American Drawings or Descriptions of Other Markham Machine Designs

Sources	Machine	Source	Date	Markham drawing figur number
19	Spooling machine (Fig. A3.19	Montgomery, *Practical Detail*, 82–83, Plate V	1840	6.72
20	Dresser (Fig. A3.20)	Paul Moody (Boston Mfg Co) patent of 1818; Montgomery, *Practical Detail*, 91–101, Plate VII	United States patent 1818; illustrated 1840	6.75, 6.76, 6.77, 6.78
21 (and 5 and 6)	Double speeder	Paul Moody (Boston Mfg Co) patents	United States patents 1819, 1821	6.51A, 6.51B
22	Fly-wheel spinning	None known	none	6.65, 6.66, 6.67, 6.68, 6. 6.70, 6.71

both the section and loom warp beams in their dressers by the late 1840s, if not earlier: By 1846 they had abandoned "the old drag system."[3]

Markham's papers and drawings testify to the extent to which the Waltham system of labor-saving factory production, building on a platform of English technology, spread through northern New England. Markham himself played a part both as conduit and as modifier. Perhaps nothing epitomized this labor-saving technology so much as the stop-motion. In the hands of the first generation of Waltham engineers, it was incorporated into the twisting frame, the warping frame, and the dresser, in addition to its English uses in the power loom and, more experimentally perhaps, in the roving frame, the dresser, and the spinning frame. Markham built a stop-motion into his warping frame where it would compensate for the limitations, in quantity and quality, of a workforce of teenagers, often farm girls (Fig. 6.16, p. 158). The presence of drop wires in his list of wrought iron components for the device suggests that it was of the drop wire and a rocker type (as opposed later to the latch type invented by Samuel Batchelder for the drawing frame in 1832). Its number of drop wires hints that it may not have been dissimilar to the warping frame built at Waltham in 1819.

The conclusions drawn from this comparison are that Markham was probably responsible for three, or possibly four, elements in textile machines modified elsewhere in New England. No other illustration of the ground-to-rollers feeding (delivery?) system in the carding machine appearing earlier than Markham's is known. Nor is there any earlier illustration or description of flywheel spinning. The spooling-machine drawing is certainly the earliest to survive, though Markham's failure to give it any significance suggests that it was already known. The distinctive element of Markham's dresser seems to have been the use of the double-cone drive to achieve a smooth and automatic increase in the external diameter of the warp beam as the warp was wound onto it. His drawings, it should be noted, provided several permutations of the form of the dresser.

The presumed movements of technical knowledge represented by the comparisons in these tables are affirmed by the available illustrations of the source technologies in Britain and source or receptor technologies in the United States seen in the plates included here.

[3]White, *Practical Treatise on Weaving*, 241.

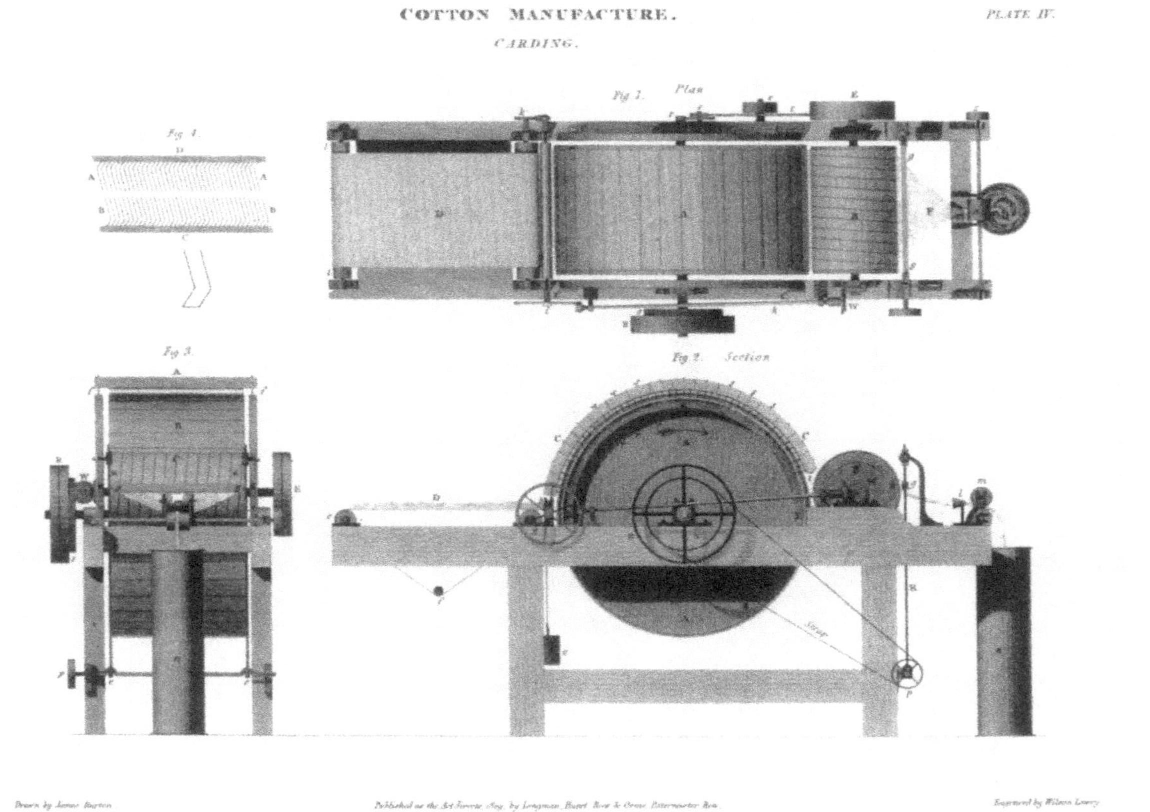

FIGURE A3.1 Cotton-carding machine, from Rees (ed.), *Cyclopaedia* (1812), Plates, vol. 2. *Above*, plan; *below*, elevations, delivery at left, side at right. For the operating principles of this machine, see chapter 6, section 3, pp. 210–211.

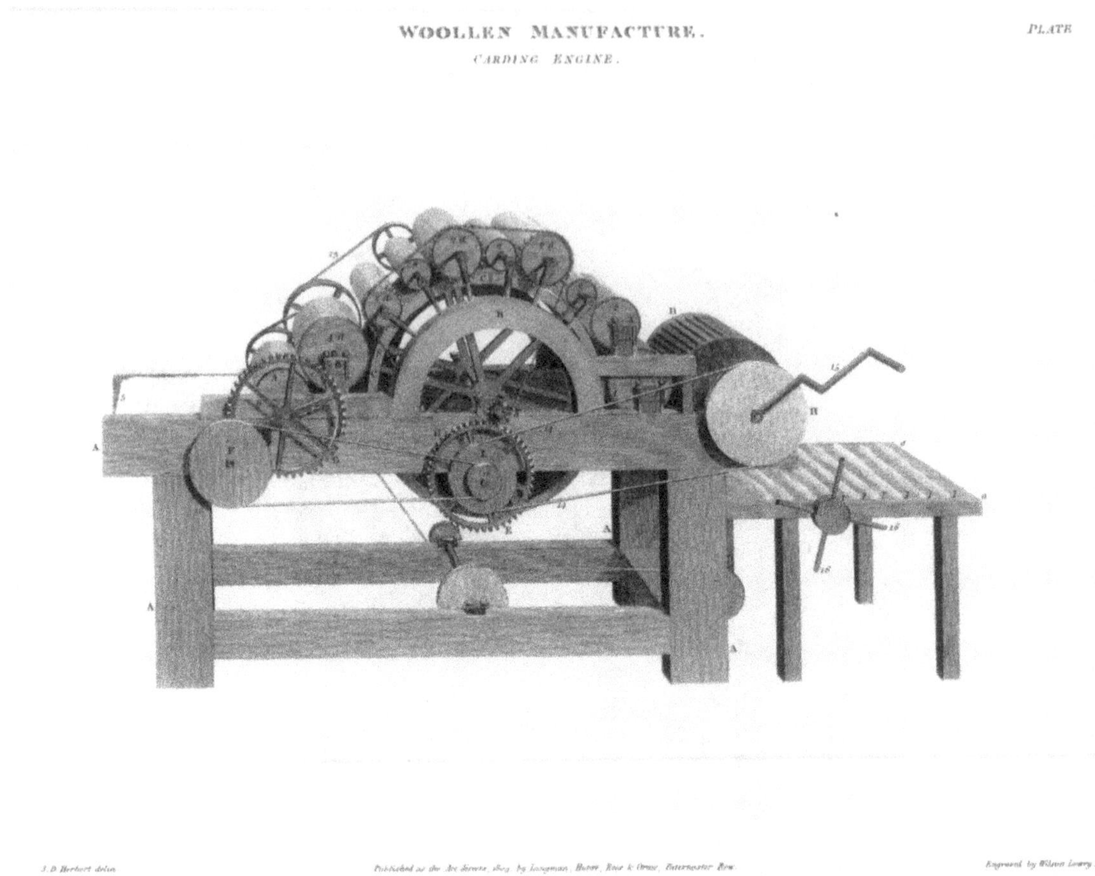

FIGURE A3.2 Woolen-carding machine, from Rees (ed.), *Cyclopaedia* (1818), Plates, vol. 6. For the operating principles of this machine, see chapter 6, section 3, pp. 210–211.

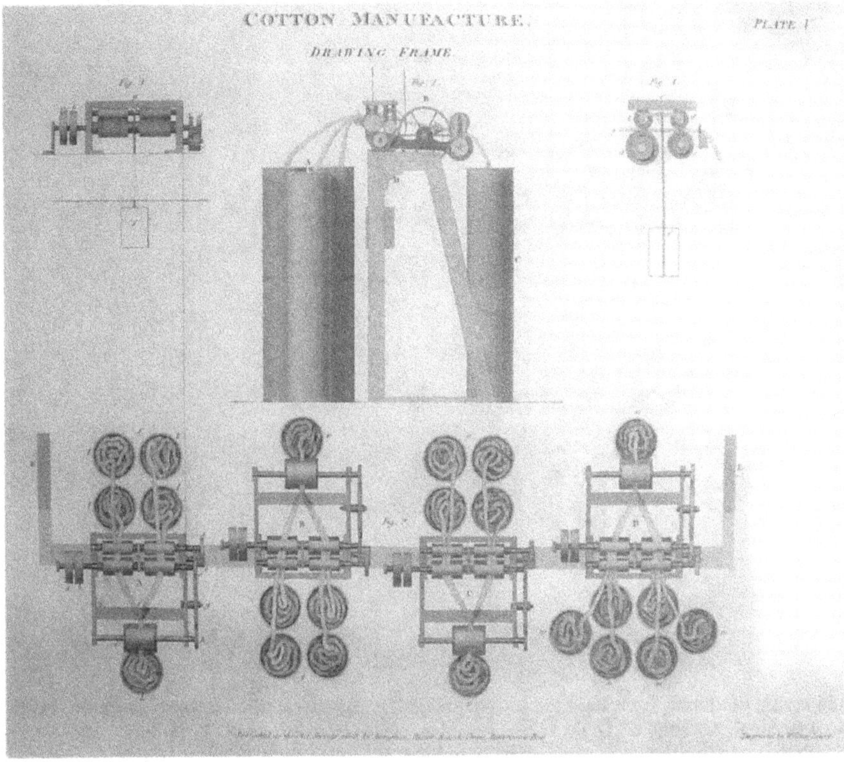

GURE A3.3 Cotton-drawing frame, from Rees (ed.), *Cyclopaedia* (1812), Plates, vol. 2. Plan of four drawing frames, along the ttom of the picture. *Above*, elevations: whole side, center; side of drafting rollers, right; front of drafting roller head, left. For the erating principles of this machine, see chapter 6, section 4, p. 228.

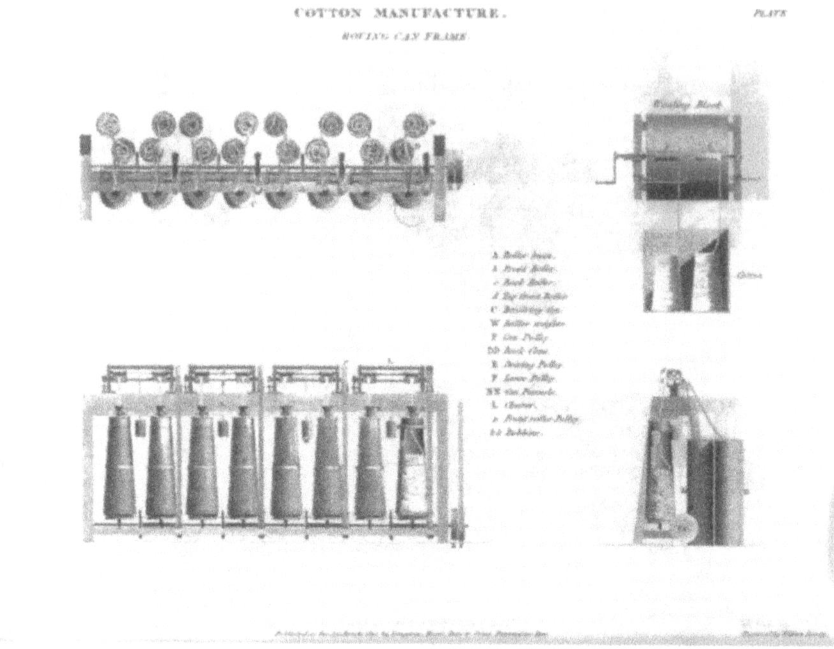

FIGURE A3.4 Cotton roving can frame, from Rees (ed.), *Cyclopaedia* (1812), Plates, vol. 2. Described in the text as "old roving can frame, first introduced by Sir R. Arkwright." *To the left, above*, plan; *below*, front elevation. *To the right, above*, front elevation of winding block by which rovings from the cans were wound onto bobbins. *To the right, below*, side elevation of roving can frame (Thomson, "Manufacture of Cotton"). For the operating principles of this machine, see chapter 6, section 4, pp. 228–231.

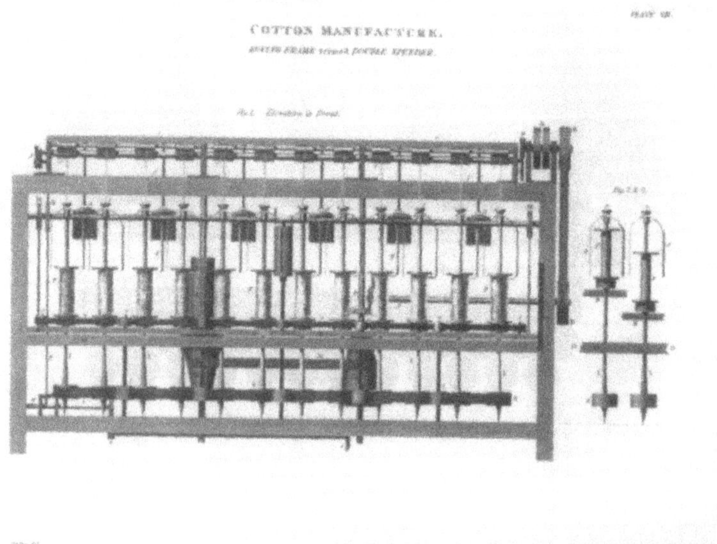

FIGURE A3.5 Cotton double speeder, from Rees (ed.), *Cyclopaedia* (1812), Plates, vol. 2. Plan. For the operating principles of this machine, see chapter 6, section 4, pp. 228–231.

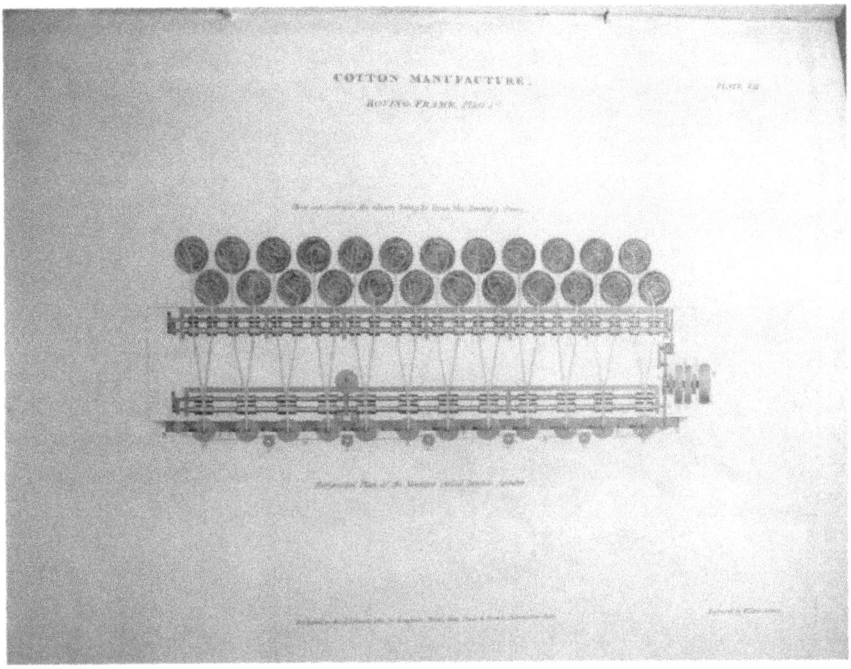

FIGURE A3.6 Cotton double speeder, plan, from Rees (ed.), *Cyclopaedia* (1812), Plates, vol. 2. Front elevation. For the operating principles of this machine, see chapter 6, section 4, pp. 228–231.

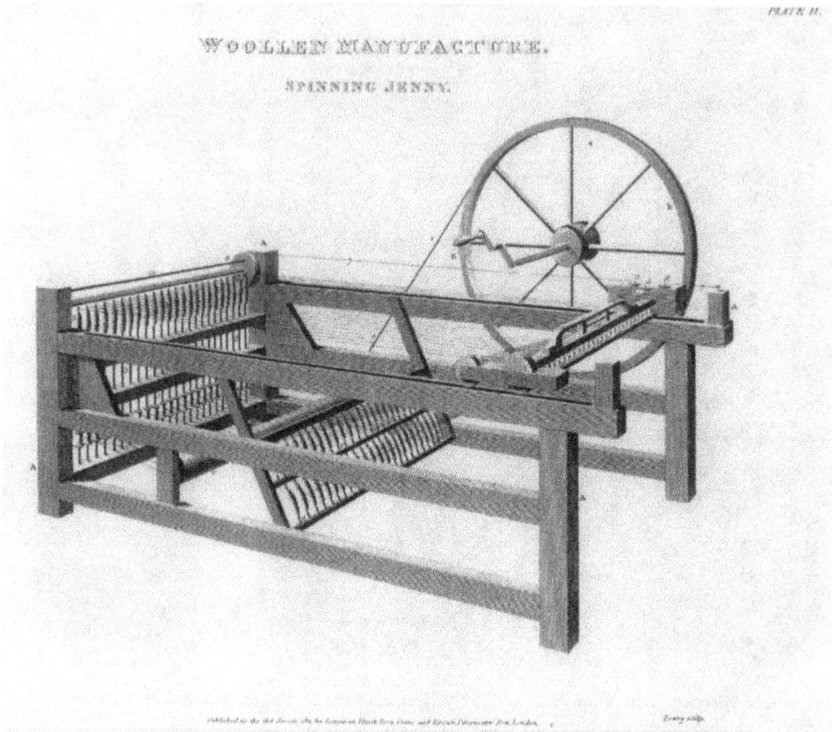

FIGURE A3.7 Woolen-spinning jenny, from Rees (ed.), *Cyclopaedia* (1818), Plates, vol. 6. Perspective view. For the operating principles of this machine, see chapter 6, section 5, pp. 246–247.

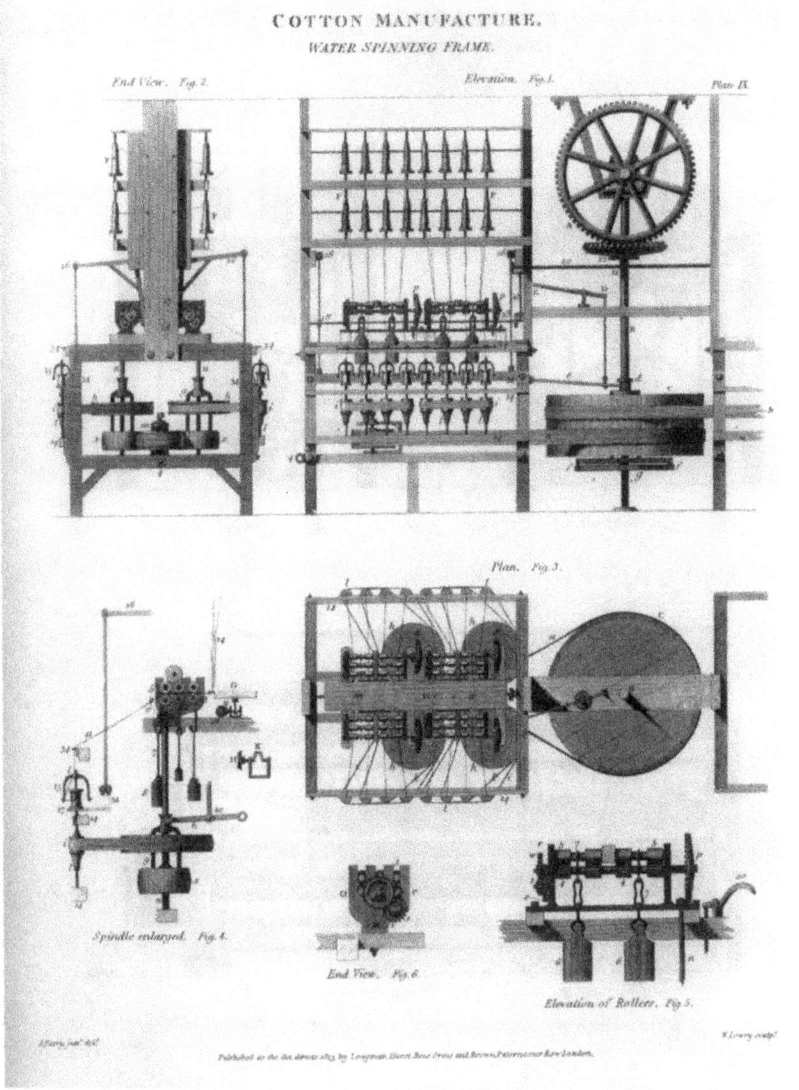

GURE A3.8 Cotton-water-spinning frame, from Rees (ed.), *Cyclopaedia* (1812), Plates, vol. 2. *Above*, elevations, side to the left, nt to the right. *Below*, plan (figure 3); elevations: two of the side (*bottom left* and *center*) one of the front. For the operating nciples of this machine, see chapter 6, section 5, pp. 246–247.

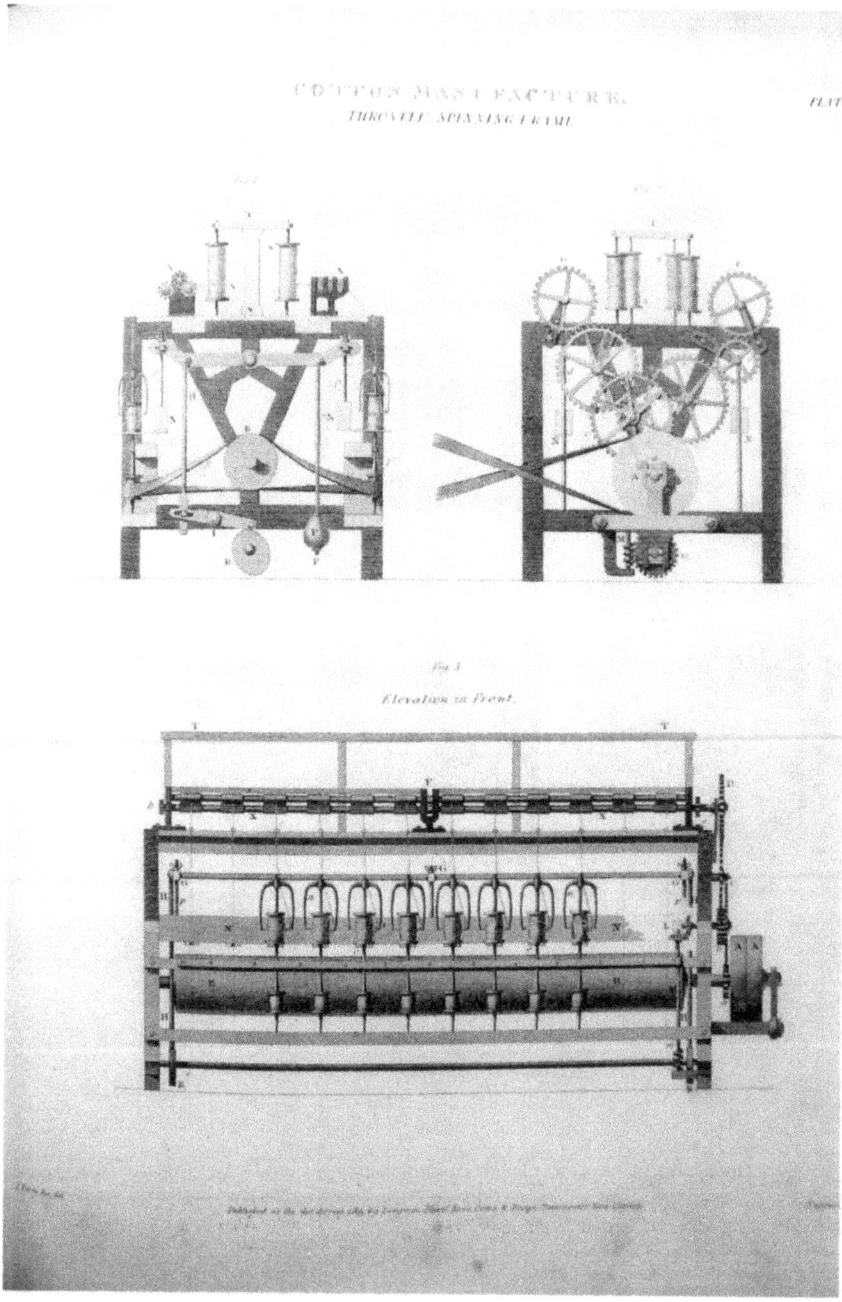

FIGURE A3.9 Cotton-throstle-spinning frame, from Rees (ed.), *Cyclopaedia* (1812), Plates, vol. 2. Elevations: the two sides above, t front below. For the operating principles of this machine, see chapter 6, section 5, pp. 246–247.

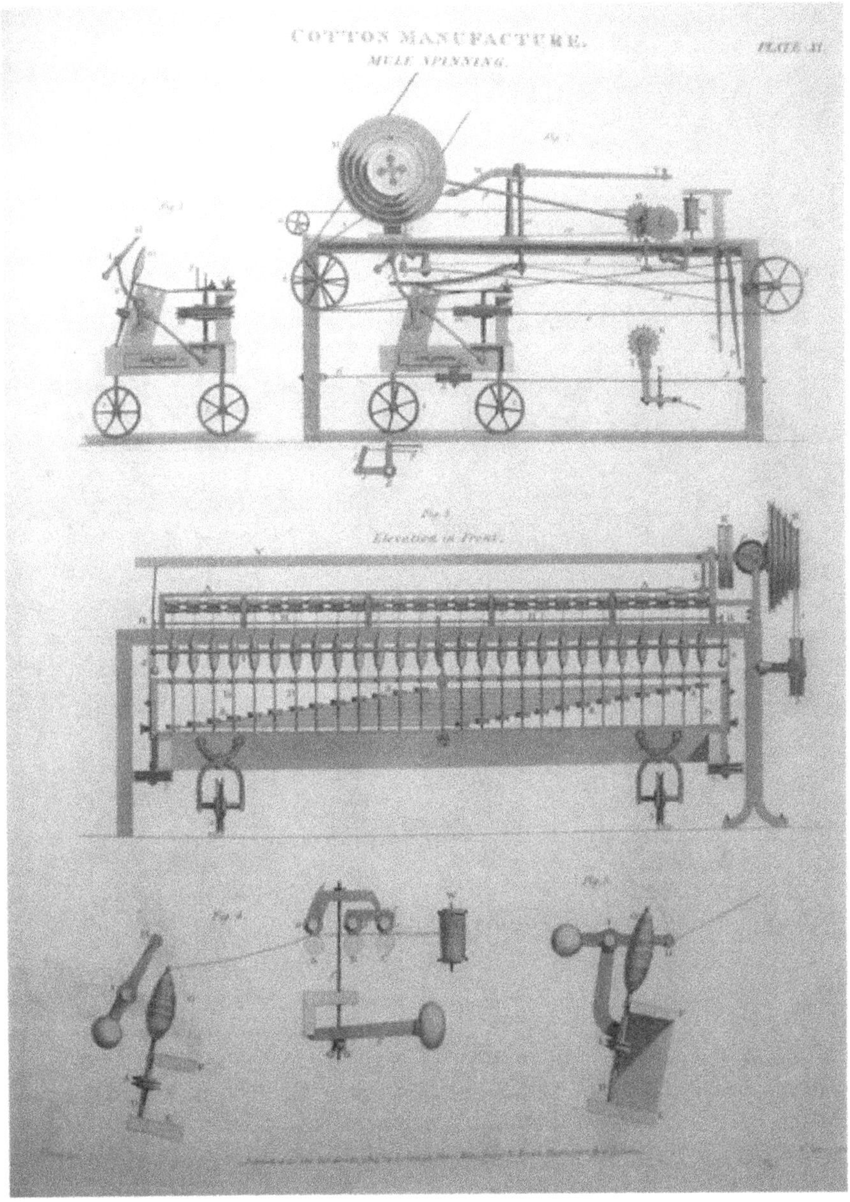

GURE A3.10 Cotton-spinning mule, from Rees (ed.), *Cyclopaedia* (1812), Plates, vol. 2. *Above*, side elevations of the machine and s carriage. *Middle*, front elevation. *Below*, side elevations of the action of the faller rail. For the operating principles of this machine, e chapter 6, section 5, pp. 246–247.

FIGURE A3.11 Cotton-reeling frame, from Rees (ed.), *Cyclopaedia* (1812), Plates, vol. 2. Here (upper drawings only) the reeling frame's side elevation, at the left; and front elevation, at the right, are shown. Reeling produced hanks of yarn, each 840 yards long. By this means the fineness of a yarn could be relatively gauged: whatever number of hanks weighed a pound avoirdupois was known as its "number" and the higher the number the finer the yarn (at this time No. 20s to 30s were middling; a No.100 was fine, achievable with a mule; and a No. 300 was exceedingly fine, reached by Indian hand spinners) (Jeremy, "British and American Yarn Count Systems"). For Markham's reel and the operating principles of this machine, see Figures 6.47 and 6.50, pp. 234–235 and 240–241.

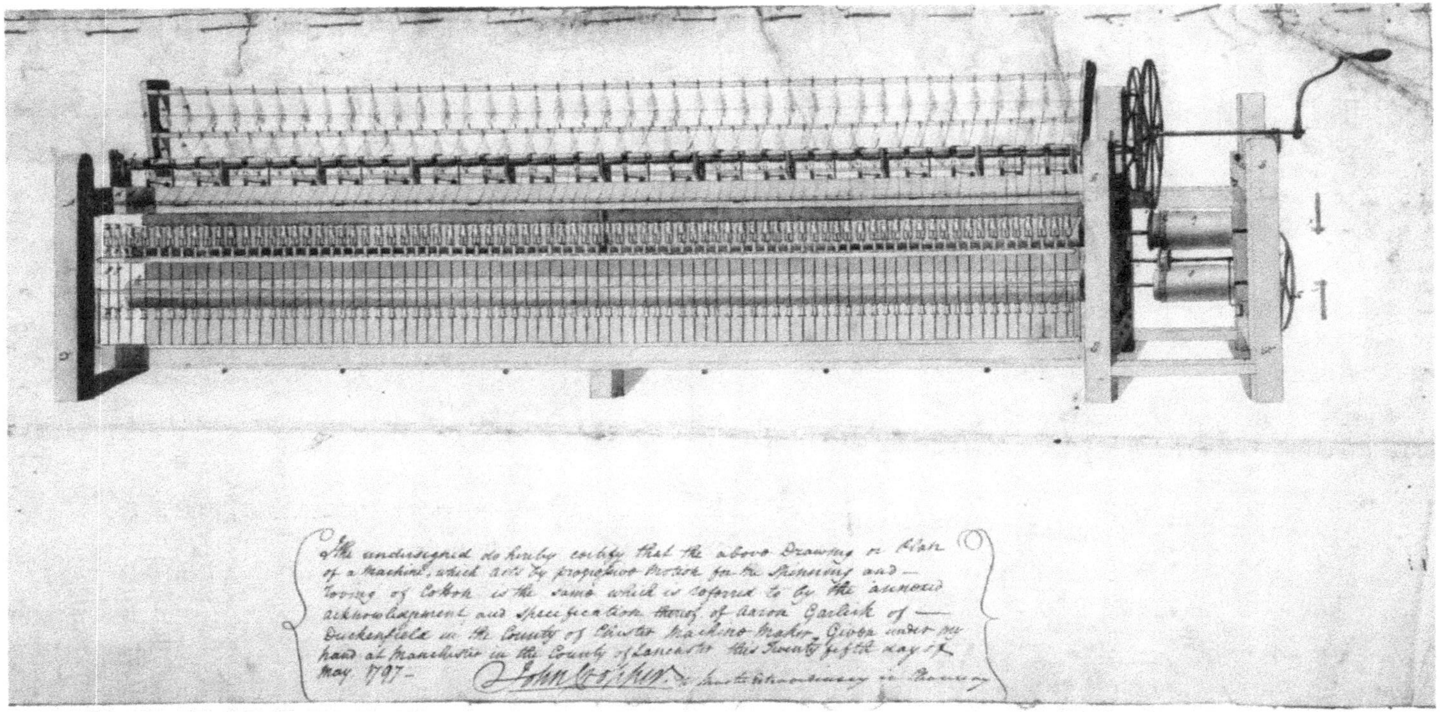

FIGURE A3.12 Aaron Garlick, English patent for the double speeder, 7 February 1797; perspective view; pen and wash on linen (?) (33.5 x 67.5 cm). National Archives C 210/54. GB Patent no. 2166. Not published in *RAMA*. See chapter 6, section 4, pp. 228–231, for technical details.

Courtesy of the National Archives, London.

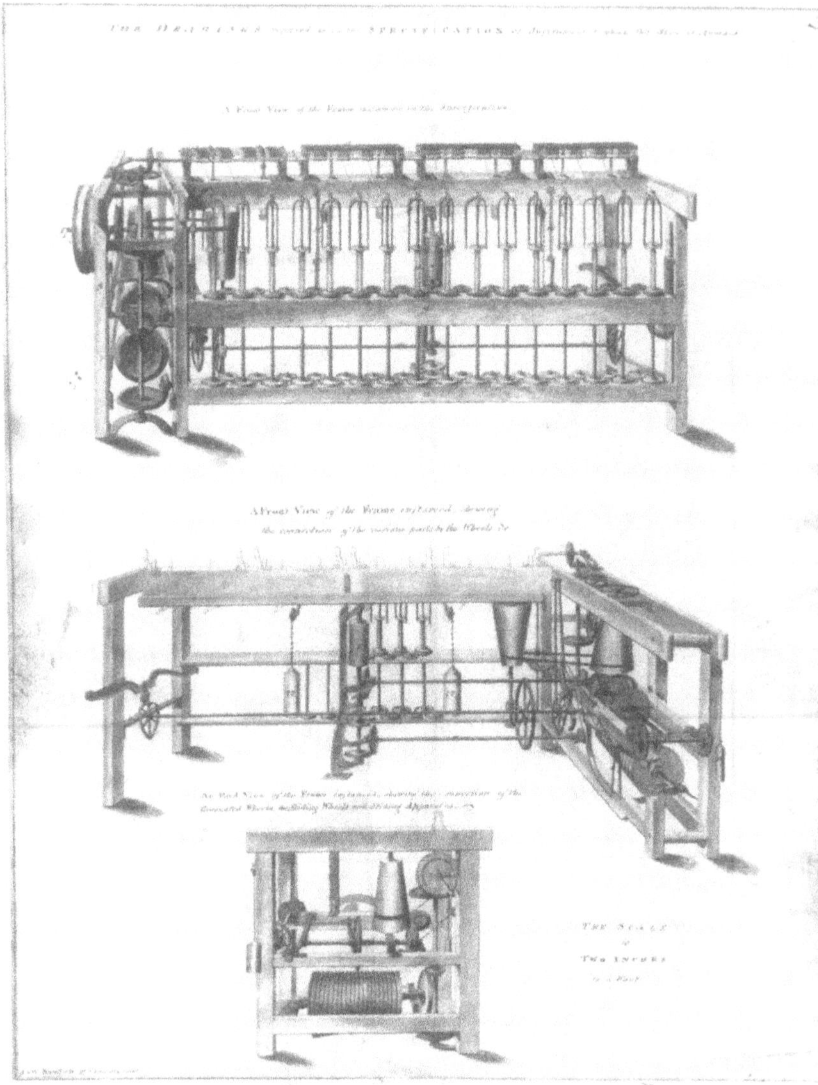

FIGURE A3.13 Joseph Raynor, English patent for the double speeder, 1 January 1813; perspective views; pen and wash on skin (76 58.2 cm). National Archives C 54/9345.

GB Patent no. 3633. Patent specification published in *RAMA* in November 1814, but not the drawing. This was "unavoidably postponed" (p.384) until the following volume but never appeared there either. Offering a promotional comment on his patent in *RAMA* 23 (November 1813) 330, Raynor noted, "Among the minor, but not unimportant, advantages of this frame may be mentioned,— the facility with which it works—the little labour it requires in tending—and the almost entire absence of waste." Fo the operating principles of this machine, see chapter 6, section 4, pp. 228–231.

Courtesy of the National Archives, London.

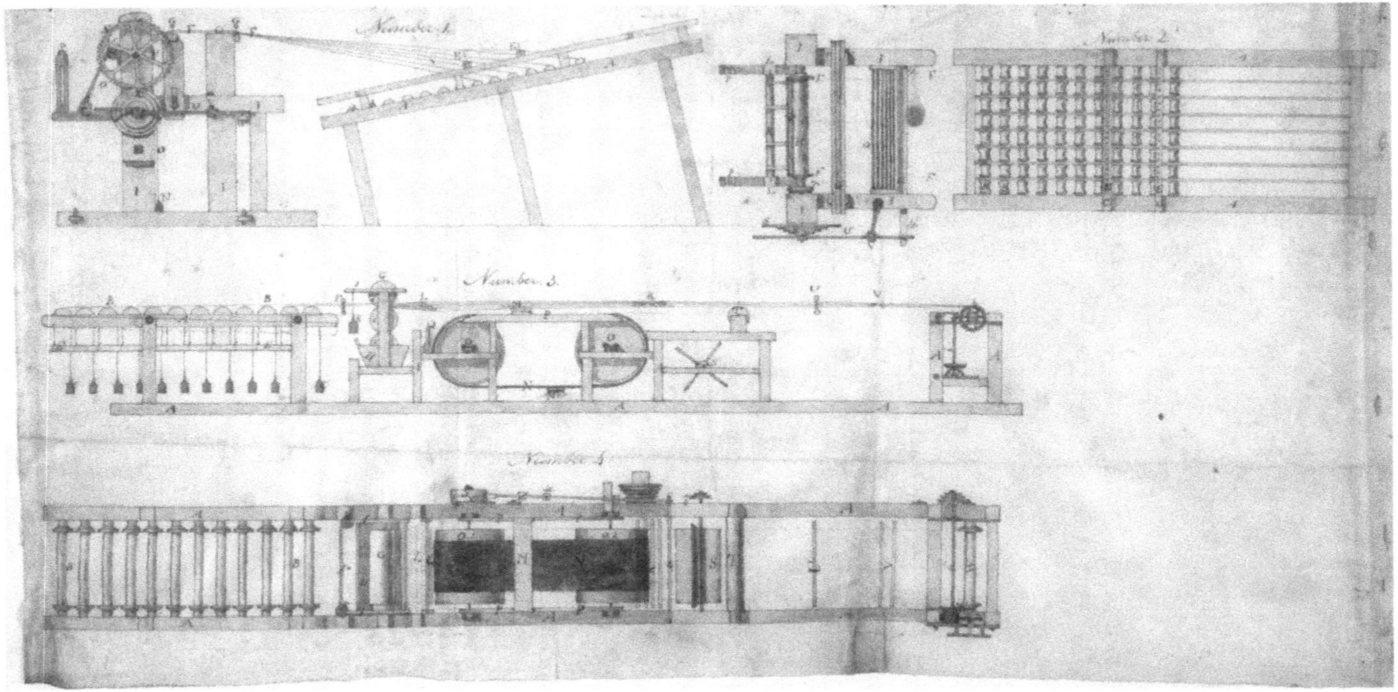

FIGURE A3.14 Thomas Johnson, English patent for warping machine and for dresser, 28 February 1803. *Top*, warping machine, side elevation at the left; plan at right. *Bottom*, dresser, side elevation above, plan below. Pen and wash on skin (36 x 71.5 cm). National Archives C 210/94. GB Patent no. 2684. Not published in *RAMA*.

Figures 1 and 2 show, respectively, the side elevation and plan of a warping frame. Figures 3 and 4, respectively, the side elevation and plan of a dresser. William Radcliffe, the Stockport cotton manufacturer, claimed in his memoirs that he invented the dresser and that he took out the two related patents (here Figs. A3.14 and A 3.15) under the name of his employee, Thomas Johnson, to conceal from foreign competitors his interest in the invention—thereby, he hoped, preventing them from learning about it. His solicitor advised him to take out secret patents but John Peel, Robert's brother, thought this too troublesome and expensive. His solicitor then suggested the ruse of using his workman's name. In addition Radcliffe went to Henry Wyatt, editor of the *Repertory of Arts*, and persuaded him not to print his patent specifications (Radcliffe, *Origin*, 25–27).

Courtesy of the National Archives, London.

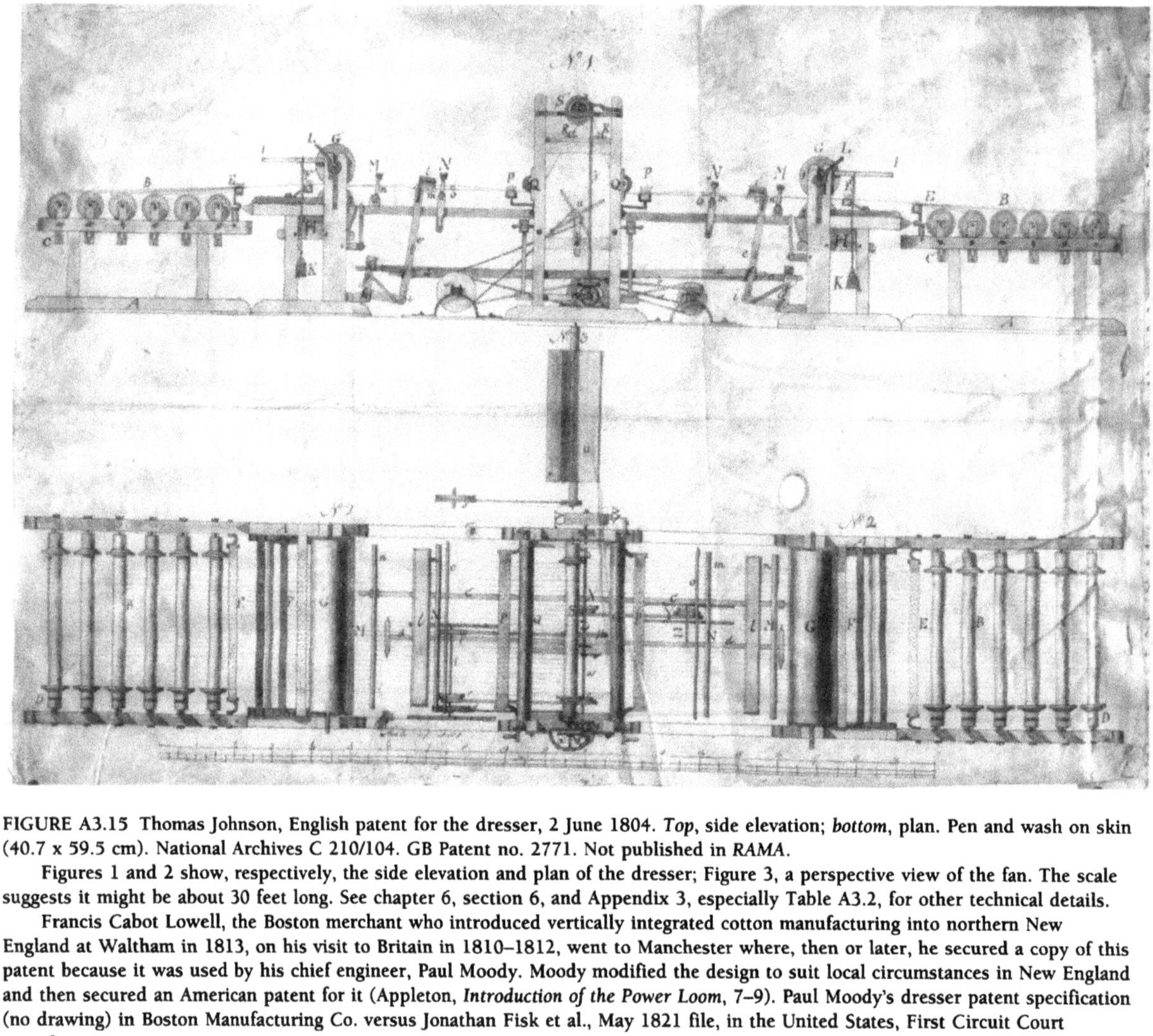

FIGURE A3.15 Thomas Johnson, English patent for the dresser, 2 June 1804. *Top*, side elevation; *bottom*, plan. Pen and wash on skin (40.7 x 59.5 cm). National Archives C 210/104. GB Patent no. 2771. Not published in *RAMA*.

Figures 1 and 2 show, respectively, the side elevation and plan of the dresser; Figure 3, a perspective view of the fan. The scale suggests it might be about 30 feet long. See chapter 6, section 6, and Appendix 3, especially Table A3.2, for other technical details.

Francis Cabot Lowell, the Boston merchant who introduced vertically integrated cotton manufacturing into northern New England at Waltham in 1813, on his visit to Britain in 1810–1812, went to Manchester where, then or later, he secured a copy of this patent because it was used by his chief engineer, Paul Moody. Moody modified the design to suit local circumstances in New England and then secured an American patent for it (Appleton, *Introduction of the Power Loom*, 7–9). Paul Moody's dresser patent specification (no drawing) in Boston Manufacturing Co. versus Jonathan Fisk et al., May 1821 file, in the United States, First Circuit Court records.

Courtesy of the National Archives, London.

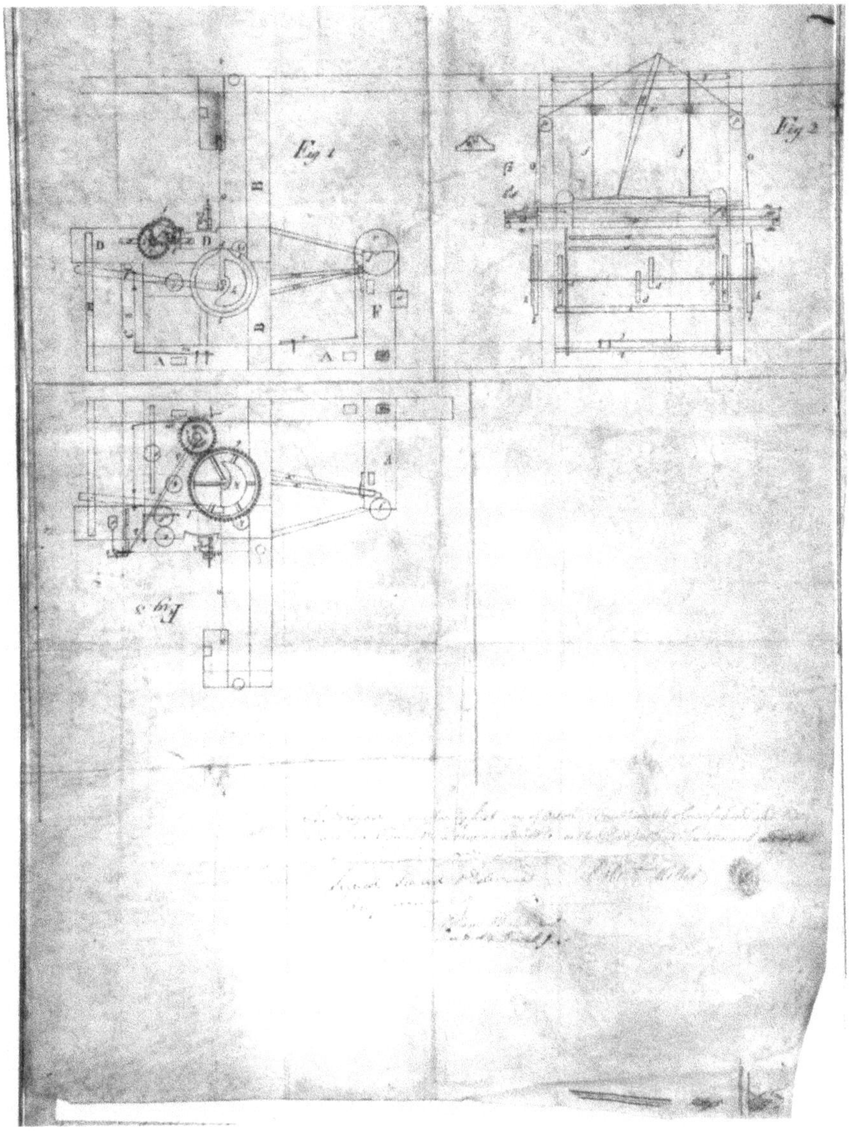

FIGURE A3.16 **Robert Miller, English patent for the power loom, 28 June 1796. Elevations of each side, on a common baseline,** ***to the left*****. Front elevation,** ***upper right*****. Pen and wash on skin (55.5 x 44 cm). National Archives C 210/53. GB Patent no. 2122. Line drawing copy published in** ***RAMA*** **in 1798.**

Figure 1 is a side elevation (warp beam to the right, cloth beam to the left). Figure 2 is a front elevation. Figure 3 (inverted but sharing the same base as Figure 1) is a side elevation of the other side of the machine (warp beam to the left, cloth beam to the right).

In the figures the cams dd control the shedding, cams kk the picking, and cams cc the beating-up motion. Unsurprisingly, since the inventor hailed from Dunbarton, in Scotland, the power loom was known as the cam loom and the Scotch loom. See chapter 6, section 7, and Appendix 4 for other technical details.

Courtesy of the National Archives, London.

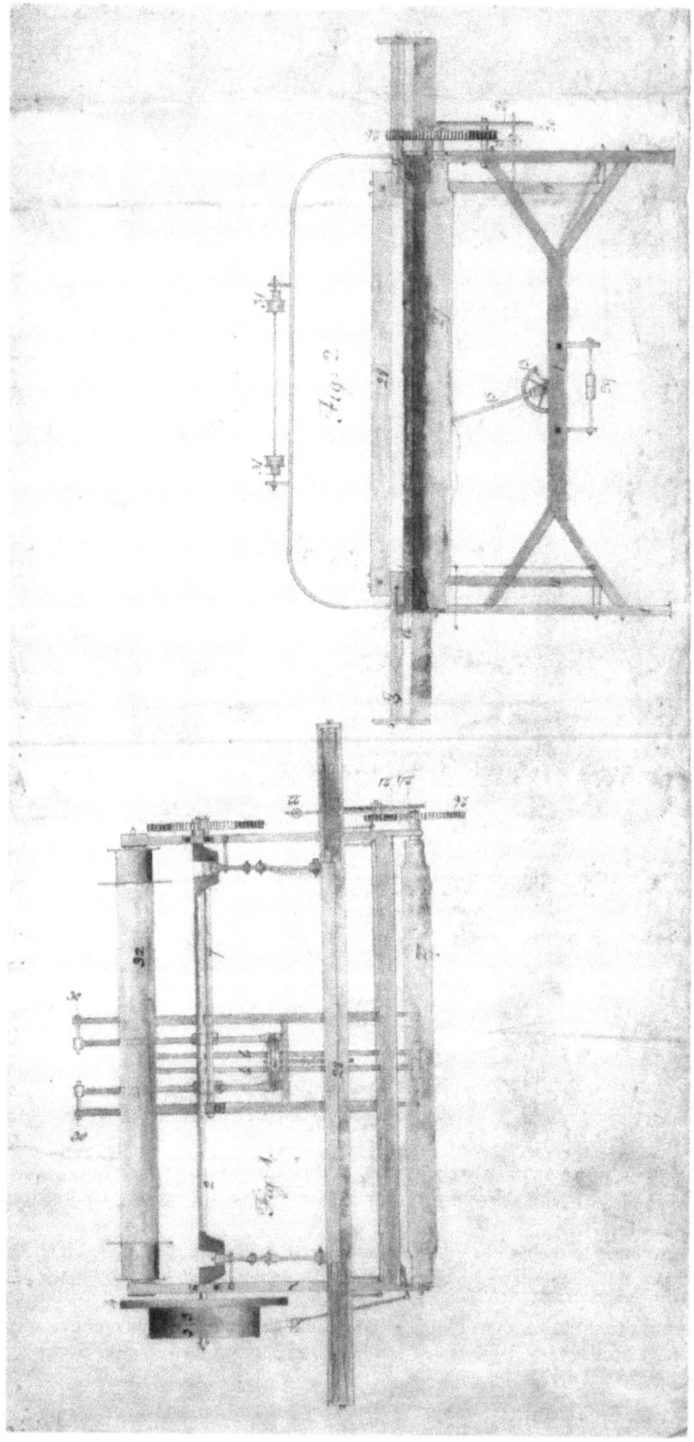

FIGURE A3.17 William Horrocks (1776–1849), English patent for the power loom, 31 July 1813. Sheet 1, Figure 1, plan. Figure 2, front elevation. Pen and wash on skin (25 x 49.5 cm). National Archives C 210/131. GB Patent no. 3728. Line drawing copy published in *RAMA* in June 1814. See chapter 6, section 7, and Appendix 4 for technical details.

Courtesy of the National Archives, London.

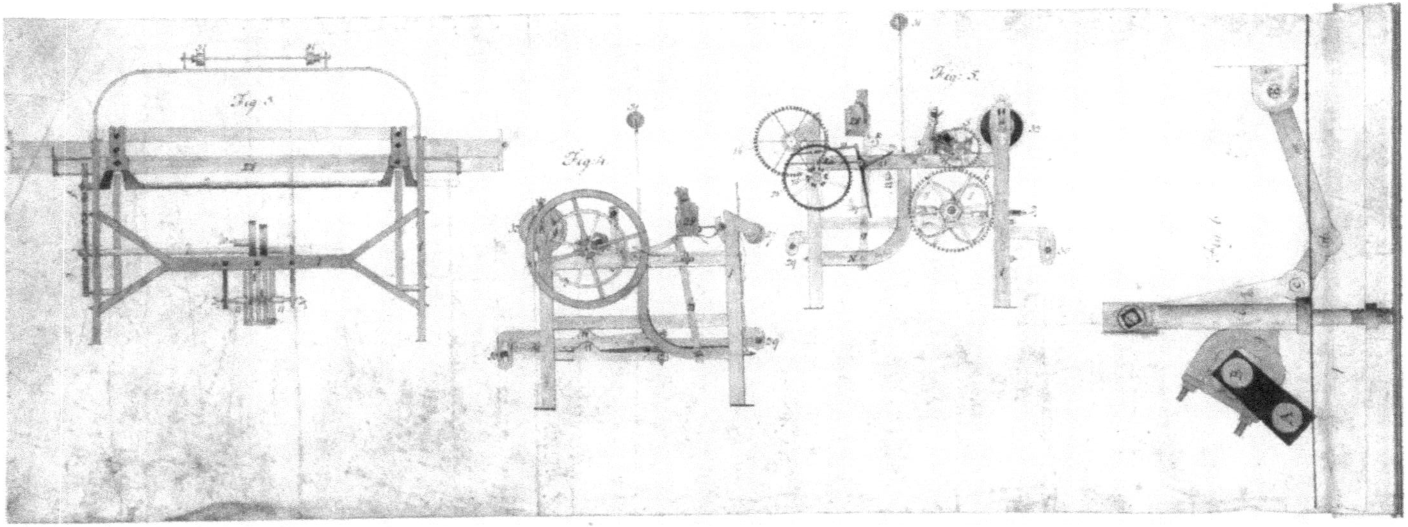

FIGURE A3.18 William Horrocks, English patent for the power loom, 31 July 1813. Sheet 2, Figure 3, rear elevation. Figure 4, side elevation. Figure 5, opposite-side elevation. Figure 6 is an enlarged side elevation of the variable batten speed motion, Horrocks's key invention, also seen in Figure 5. Pen and wash on skin (25 x 68 cm). National Archives C 210/131. GB Patent no. 3728. Line drawing copy published in *RAMA* in June 1814.

Courtesy of the National Archives, London.

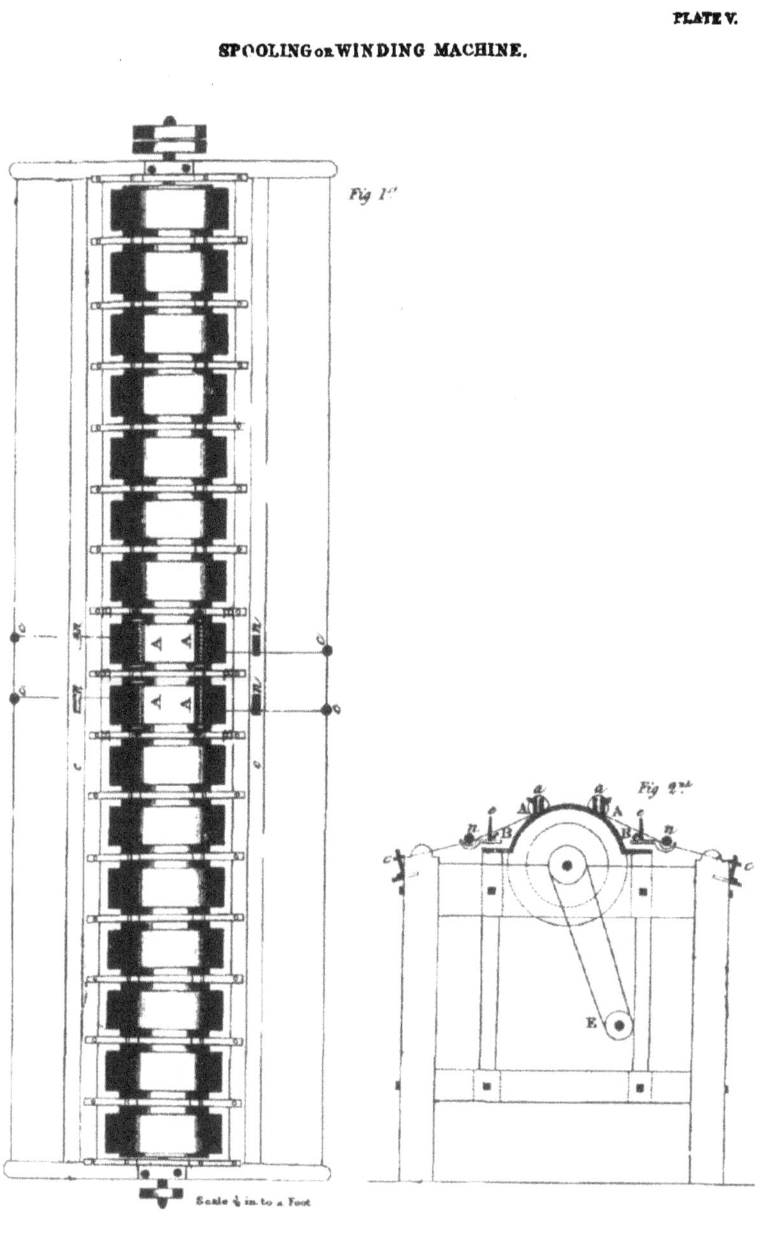

FIGURE A3.19 Spooling machine, from James Montgomery (1794–1880), *Practical Detail of the Cotton Manufacture* (1840), Plate V. See chapter 6, section 6 for technical details.

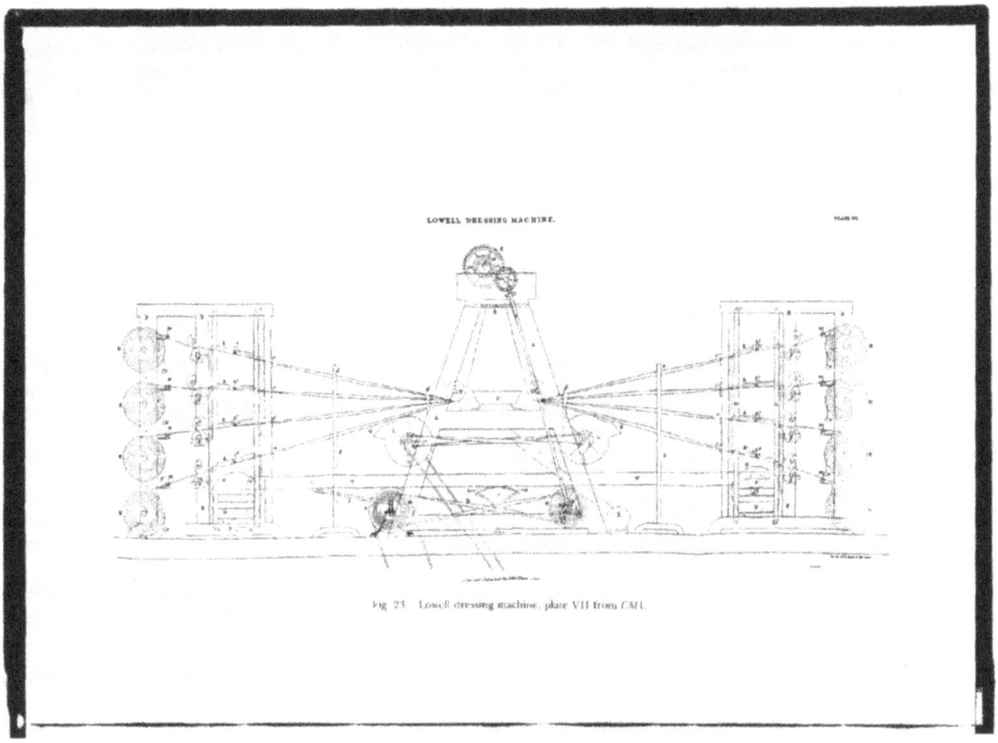

IGURE A3.20 Dresser, from James Montgomery, *Practical Detail of the Cotton Manufacture* (1840), Plate VII. See chapter 6, section
, and Appendix 3, especially Table A3.2, for other technical details.

Table A3.2 Comparisons Between British and American Dresser Designs

Elements	English dresser (Johnson's patent of 2 June 1804)	Waltham (Boston Mfg Co) dresser (Moody's patent of 17 January 1818)	Markham's dresser, as evidenced by his notes and workshop drawings, ca 1818–1825 (figure numbers in parentheses)	Lowell dresser of 1840 (Montgomery, *Practical Detail*, 91–101 and Pla VII, reproduced here a Fig. A3.20)
Dimensions	about 28.5 ft	n/a	26 ft, 1 inch (Fig. 6.21) ca. 23 ft (Fig. 6.75)	18 to 36 ft (p. 91)
Point of power input	main shaft and gearing in center of dresser	main shaft and gearing in central frame of dresser	main shaft and gearing in center (Fig. 6.75) and possibly again (Fig. 6.76); to one side of dresser (Fig. 6.77)	same as Waltham machine (p. 91)
Position of section beams	behind one another, in horizontal line	above one another, in vertical line	above one another (Fig. 6.76), behind one another, in horizontal line (Fig. 6.77)	Above, as at Waltham machine (p. 91)
Position of loom warp beam	center	center	center (Figs. 6.75, 6.76)	center (p. 91)
Variable motion device on loom warp beam (needed as warp increased beam's diameter)	strap and weighted lever brake on warp beam	strap and weighted lever brake on warp beam maybe cones if Fig. 6.3 refers to Waltham	twin-cone belt drive, as in dresser (Fig. 6.3, 6.17A, 6.17B, Fig. 6.22); and twin cones drawn in Figs. 6.28, 6.32, 6.33, 6.51B, 6.66	twin-cone belt drive to loom warp beam, and friction strap and weigh on section warp beam (pp. 91–92)
Size source	two troughs	eight troughs	four troughs (Fig. 6.17A, 6.17B)	eight troughs (p. 91)
Size applicator	brushes on cranked and parallelogram frame for reciprocating movement above warp	brushes on cranked and parallelogram frame for reciprocating movement above warp	brushes on rocking frame (Fig. 6.75); brush on small frame or belt (Fig. 6.76)	brushes on cranked and parallelogram frame for reciprocating movemen (p. 93)
Number of section beams	twelve (two sets of six)	eight (two sets of four)	four in sets of two (Fig. 6.76)	eight in two sets of fou (p. 91)
Point of merger of warps from section beams	before sizing	after sizing	before sizing (Fig. 6.77); after sizing (Figs. 6.75, 6.76)	after sizing (pp. 91–92)
Drying devices	one fan	two fans; steam pipe	two fans (Fig. 6.77)	two fans and a hot-air pipe (p. 95)
Measuring device	none	clock to measure 32-yard length of warp sized	measuring clock (Figs. 6.17A, 6.17B, 6.21)	measuring clock (p. 96)
Stop-motion	none	stop-motion linked to clock, stops machine when 32 yards are sized	none, unless linked to measuring clock without reference to extra parts (Fig. 6.17A, 6.17B)	none
Output	none given	twice that of the English dresser	none given	14 to 60 pieces of 30-y lengths per day, varying with yarn fineness and drying system (pp. 96–97)

Appendix 4

The Scotch and Waltham Power Looms and Markham's Power-Loom Designs

☆ ☆ ☆ ☆ ☆ ☆ ☆

AMERICAN POWER LOOMS BUILT IN 1820 WERE MASSIVE improvements on the machines available just eight years earlier, as the comparison with a patented power loom of 1812 shows (Figure A4.1; for a description of weaving technology see chapter 6, section 8: "Drawings: Weaving").

Here, not only is the frame wooden but nearly all the working parts appear to be wooden also. Further, the evident purpose of the loom, to weave figured goods, was far too ambitious for the technology illustrated. It demanded a complex shedding sequence. A single barrel of pegs, serving as a complex cam, activating a large number of treadles, would offer a few degrees of movement for each treadle and at full speed would not have separated the harnesses sufficiently. Neither does the picking action look very robust.

Other, unsatisfactory, attempts to introduce a British type of power loom were made during the War of 1812 in the Philadelphia area.[1] The two power looms that succeeded technically and commercially, those of Francis Lowell and William Gilmour, produced the simplest form of cloth, plain weave, which required just two harnesses (and therefore, treadles): Moving from the simple to the complex is invariably the best technological path to follow. The question that then arises is: To what extent were Markham's power-loom designs original, or based on alternative models available in New England ca. 1820?

Although Americans made a number of experiments with power looms, only two, both derived from Britain, became widespread in the United States during the period 1812–1830. The so-called Scotch loom, distinguished by cams or wipers, was patented in 1796 by Robert Miller, calico printer of Milton Printfield, Dumbarton, Scotland (see Fig. A3.16, p. 363). This loom was significantly improved by the compound lever mechanism of the variable batten speed motion patented in 1813 by William Horrocks, cotton manufacturer of Stockport (see Figs. A3.17 and A 3.18, pp. 364, 365).[2] Both loom designs were widely published. The patent specification and drawing of the Miller loom appeared in the *Repertory of Arts* in 1798 and the Horrocks loom in the same patent journal in June 1814, within a year of its English patent date. This Scotch loom was introduced to New England by William Gilmour, an immigrant Scottish machine maker, who built the first, with the help of a local weaver, in the Lyman Mills in Providence,

[1]Jeremy, "Immigrant Textile Machine Makers."

[2]For evidence that the Miller loom was transformed by Horrocks's improvements. see Duncan, "Weaving"; Baines, *Cotton Manufacture*, 234–235; Ure, *Cotton Manufacture* II, 289.

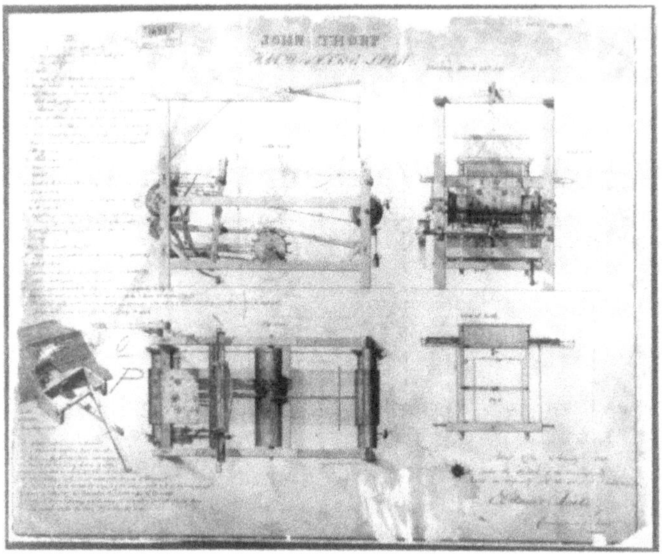

FIGURE A4.1 John Thorp (1784–1848), United States patent for water loom, 28 March 1812. Restored Patent Drawings (X-Drawings), 1693-X, Records of the Patent and Trademark Office, Record Group 241, National Archives, Washington, DC.

Rhode Island in 1816–1817.[3] So satisfactory for its purpose was the Scotch loom that Henry B. Lyman, writing in 1861, could report, "To this day, the same loom, with trivial alterations, is in use in all our mills."[4]

The Scotch or Gilmour loom had only one rival, namely, a mechanism that also derived from Scottish and English models, built under the direction of Francis Cabot Lowell in the works of the Boston Manufacturing Co. at Waltham, Massachusetts. Because no drawings of the Waltham power loom, developed in 1813–1814 and patented on 23 February 1815, seem to have survived, neither the original design nor any changes made in the succeeding ten years are known precisely today.

Job Manchester, a Rhode Island machinist, inventor, and Baptist minister, recalled first encountering the Waltham loom in 1824 and, comparing it to a loom of his own invention, remembered its distinctive features:

> I had no provision for stopping the loom, in case the shuttle failed to reach the opposite box; the motion of the lay was not positive, being arrested whenever it met a sufficient obstruction. On my first acquaintance with the Waltham loom, in the year 1824, I found that it operated on this same principle. That loom had no shuttle protection, as it is called, until the crank motion was applied to the lay, after Gilmour's plan. It was held at Waltham, as a fixed fact, that cloth should be made by an accelerated motion of the lay, whereas the crank motion, as used by Gilmour made the cloth while passing the minimum point; hence Waltham was slow to adopt it. Rhode Island operatives had, ere this, learned by experience that the secret of making good cloth, with ease, consisted quite as much in a proper elevation of the yarn back of the harnesses, so as to give a suitable slackness to the top shade

[3]For evidence that Gilmour's loom was a crank loom of the Horrocks design, see Manchester, "Recollections" quoted in the next paragraph; Bagnall, *Textile Industries*, 426, 547, 549.

[4]Lyman, "Recollections," 78.

of the web, as in the manner or force of the blow from the lay, so be that the filling was moved forward by it to the proper point. [5]

This recollection suggests at least two differences between the Waltham (and Job Manchester's) loom and the Gilmour or Scotch loom in the early 1820s.[6]

1. In the Waltham loom the batten or lay was not positively driven, implying that it was activated by pulleys and a belt, or cams and counterweights, as in the Miller loom of 1796. If the shuttle was caught in the warp shed, the batten was halted by the jammed shuttle. Evidently this would have been possible with the strong coarse yarns (no. 14) spun and then woven into heavy sheeting at Waltham.[7] With warps sufficiently strong to restrain the shuttle from smashing out of the warp shed, the remaining reciprocating energy of the batten was absorbed by belt slippage on the pulleys.

In contrast, the Gilmour loom batten was positively driven, by gears and crank. Because the force of the positively driven batten would smash a jammed shuttle out and through the warp, the loom needed a warp-protector stop-motion (what Job Manchester called "shuttle protection").

2. The second difference concerns the appearance of the cloth produced by the two looms. At Waltham, the weaving overseers relied on a fast batten speed, with no variation within the reciprocal cycle. This gave them the quality of sheeting and shirting cloth they wanted.

Rhode Island weave shed overseers, using the variable batten speed of Horrocks, believed that better results came when the "top shade of the web" ("top shed" to English loom tacklers) was slightly slacker as the batten and its reed drove the weft home into the fell of the cloth.

In relating "making the cloth while passing the minimum point" to the Gilmour loom, Job Manchester was evidently referring to cloth "cover."[8] "Cover" or "face" described the appearance of a thick surface on a cloth (a phenomenon partly related to the refraction of light), often required for a plain weave. Cover improved not only the surface of the cloth but also its salability. Cover was achieved by setting the timings of the power loom's primary movements (shedding, picking, beating up, let-off and take-up motions) to allow the batten carrying the reed to beat up the weft into the fell or leading edge of the cloth when the warp shed was widest. At this point the healds (set in a harness frame and controlling the warp yarns) of the "top shade of the web"—the upper web of alternate warp yarns—would pass through the "minimum point."

---

[5]Job Manchester, "Recollections," 67–68. The advantages of the lay (lathe, or batten) mechanism, as described by Manchester (the variable batten speed motion), were set out by its inventor, William Horrocks, in his British patent of 1813. The variable crank imparted different phases to the reciprocal movement of the lay. It moved slowly when not in the vicinity of the fell of the cloth. This meant that the shuttle could pass through the warp shed while the batten was stationary, thereby minimizing the chances of a shuttle being smashed out of the warp. Slow phases of the cycle also reduced the amount of tension required in the warp and cut down the number of warp breakages and consequent knotting repairs. Further, the slow phases in the cycle allowed the harnesses guiding the warp threads to separate more widely, hence permitting the use of larger shuttles with larger cops and less waste. A wider shed, incidentally, allowed better cover or cloth surface, the weft yarns being forced into the upper web of the warp (see later). In the shorter, high-speed phases of the batten's cycle, the lay struck the weft yarn sharply into the fell of the cloth, allowing more yarns per inch of filling and creating a stronger, more uniform cloth. See William Horrocks, GB Patent of 31 July1813 (3728); Oliver, *Weaving and Designing*, 76, 83–84; Nelson, *Practical Loom Fixing*, 41-43; Jeremy, *Transatlantic Industrial Revolution*, 62, 65, 67.

[6]I am much indebted to Mr. Ian Birtles, loom tackler (loom fixer, in American nomenclature) at the National Trust's Quarry Bank Mill, Styal, Cheshire, UK, for taking time on 12 August 2008 to discuss with me Job Manchester's account.

[7]Appleton, *Introduction of the Power Loom*, 11–12. A sample of the Boston Manufacturing Co.'s cotton shirtings and sheetings cloths, dated 1820, is in the National Archives, RG 92, Coxe and Irvine Papers, 1794–1842, box 32 A, enclosure in Benjamin C. Ward & Co (BMC's selling agents) to Callender Irvine (Commissary General). The sample was "Chain 48 [ends] to the inch; Filling 52" (noted in Jeremy, *Transatlantic Industrial Revolution*, 324, n24).

[8]Thanks to Ian Birtles for making this identification.

The minimum point can be imagined as lying on the shortest horizontal line between the warp beam (or the whip beam, positioned between warp beam and reed) and the cloth beam, in a lengthwise cross-section of the loom. The warp yarns of the top shed would be at their slackest at this minimum point. In contrast, the warp yarns of the bottom shed (the remaining alternate warp yarns) would be pulled downward, away from the minimum line, and therefore at their most tight. Beating up at this juncture would spread the tight and loose warp yarns evenly across the cloth and also push the weft yarns into the surface of the cloth, achieving good cover and preventing reed marks from showing.

Gilmour's (Horrocks's) power loom, with its variable batten speed, permitted a wider range of adjustments among the power loom's primary motions. Four different ways of achieving cover by loom tuning were familiar to power-loom mechanics at the end of the nineteenth century. One is hinted at by the phrase, "a proper elevation of the yarn back of the harnesses, so as to give a suitable slackness to the top shade of the web." This suggests that the level of the warp and cloth beams was raised so as to make the top shade slacker than the bottom shade. Another technique, recalled by Ian Birtles from his commercial experience in the 1970s, used a cord linking the reed to the lease rod of the top shade, a device also familiar to handloom weavers.[9]

Another piece of evidence about the Waltham loom is provided by the print drawing of "Power Loom with Wooden Frame, for Cotton Goods," (Fig. A4.2) consisting of a front elevation and two side elevations, published by the Lowell Machine Shop in 1848.[10] This was three years after the formation of the Lowell Machine Shop, successor to the Locks & Canals Co., which in the 1820s and 1830s built lines of machinery for the cotton-manufacturing companies at Lowell.[11] The print shows a positively driven (geared) let-off motion, releasing the warp; a cam-and-treadle shedding movement (left-hand side elevation); a cam-and-treadle picking movement (front elevation); and a crank-driven beating-up motion (right-hand side elevation). It does not appear to show the compound lever mechanism that distinguished the variable batten speed motion of the Horrocks power loom and, presumably, the Gilmour loom. In view of Henry Lyman's testimony to the longevity of a popular power-loom design, it would not be surprising if the wooden-frame loom advertised by the Lowell Machine Shop in 1848 was very similar to a much earlier Waltham loom. If this were so, it would have been a successor to the first Waltham power loom, as recalled by Job Manchester.

So what kind of power loom was Markham building, a Waltham, a Gilmour, or a different loom again? As they stand, Markham's four power-loom drawings (see Figs. 6.79–6.82, pp. 306–312) show machines with features similar to those of a Waltham loom on the one hand, and to a Gilmour/Rhode Island loom on the other. Markham's looms have a positively driven batten or lay (like the Gilmour/Rhode Island loom, unlike the Waltham one). They seem to have a warp-protector motion (like the Gilmour/Rhode Island loom, unlike the Waltham one). They do not have a Horrocks variable batten speed motion (unlike the Gilmour/Rhode Island loom, like the Waltham one).

A definite answer turns on the date of the power-loom drawings. The date on the cover of the portfolio in which the drawings were stored is 1814 (see Fig. 6.84, p. 317). Unfortunately none of the drawings is dated separately. However, David Page's testimonial of 1819 (quoted in full in chapter 1, "Isaac Markham: Working Life" section) makes it quite clear that Markham made power-loom drawings before 1819,

---

[9]See Latham, *Art of Managing the Patent Power Loom*, 22–23; Taylor, *Cotton Weaving*, 53–55; Oliver, *Weaving and Designing*, 76, 83–84, for English practice. Nelson, *Practical Loom Fixing*, 41–43; *Cyclopaedia of Textile Work* IV, 89, for American practice. Tovey, *Technique of Weaving*, 50, 73, for hand weavers' interest in cover.

[10]This 1848 drawing was reprinted by lithography in the trade catalogue *Lowell Machine Shop* (1898), 111.

[11]Gibb, *Saco-Lowell Shops*, 63–103.

FIGURE A4.2 Lowell Machine Shop, cotton power loom with wooden frame, 1848, from *Saco-Lowell trade catalogue*, 1898. Original in Broadside Collection, Division of Manuscripts, Baker Library, Harvard University.

Courtesy of the American Textile History Museum, Lowell, MA.

probably before he went to Waltham. Consequently, their design was almost certainly influenced by Gordon, who brought knowledge of power looms used in Scotland to Middlebury in 1817. Alternatively the loom drawings were made after Markham had been to Waltham, when he was settled at Middlebury as superintendent and agent of the Middlebury Manufacturing Co.

Assuming that the drawings were pre-Waltham, and Markham's loom designs were derived from Joseph Gordon, then Markham was deriving some of his ideas from looms in use around Glasgow, perhaps from Kennedy Watt & Co's mill at Johnstone and the Elderslie Mill, both in the Paisley district near Glasgow, mills cited in Markham machinery notes (see Fig. 6.11, p. 144).

Assuming a later date for the loom drawings, say after 1821, when he had accumulated experience at Waltham and might have replicated features of the Boston Manufacturing Co's loom, then it might be expected that he would have a belt-driven (i.e., nonpositive) batten motion, rather than a gear-driven one—to allow a shuttle jammed in the warp shed to halt the loom without it smashing through the warp. He would also then not need a warp-protector stop-motion, as in the Waltham loom. However, the drawing shown in Fig. 6.80 (p. 308) shows a gear-and-crank-driven batten motion (i.e., a positive one) and Fig. 6.82 (p. 312) has a mechanism that might have been a warp-protector stop-motion.

Markham's loom designs clearly rejected the Waltham model's shedding motion, with possible implications for its picking and beating up movements. Nor did he slavishly adopt the Rhode Island loom design. What was unique about the Markham loom? One feature is apparently unique because nothing like it appears in any British or American pre-1820 loom patent drawings that have survived. This is the beating-up mechanism indicated in Fig. 6.80 (p. 308).

The mechanism has two elements: a geared wheel and a rod connecting it to the batten. The eccentric motion imparted by the gear wheel is complicated by the slide, running along a quarter of the gear's circumference, allowing that end of the connecting rod a degree of freedom during rotation. What would be the effect of this mechanism on the beating-up motion?

First, the regular eccentric motion imparted to the connecting rod over one complete rotation produces four phases: fast, slow, fast, slow (a sinusoidal path). Second, the connecting rod slide in the eccentric has important implications at other end of the connecting rod. Attached to the pivoted batten, it controls the beating of the weft yarn into the fell or leading edge of the cloth. The play in the slide permits the batten to be resisted by the fell of the cloth—so although the batten compresses the weft into the newly emerging cloth it does not overcompress and crush it. Without this play, it would be difficult to ensure that the batten did not reverse before it reached the fell of the cloth or that it struck the fell of the cloth with excessive force.[12]

Markham's arrangement is a very neat mechanism mimicking the action of the handloom weaver. Did it work? Markham's employer, David Page, author of the testimonial Markham could show any prospective customer or employer, assured the world that it did and that Markham's looms were the best he knew. If they were so successful, why did they not spread and persist and be recorded by posterity? Too little is known about the technology of power-loom weaving in the United States before 1836 to provide any kind of answer to these questions.

One other feature of some significance is that the eccentric wheel, with its slide, could readily be made from cast iron. In contrast, the mechanism in Horrocks's English patent drawing of 1813 (see Figs. A3.17 and A3.18, Figure 6, pp. 364, 365) shows a compound lever mechanism that might demand machining to tolerances that Markham's machine shop could not supply. In any case, Markham's solution was a typically American one in opting for a simpler design. Although Horrocks's design permitted varied and finer adjustments, Markham's design was possibly less costly in a mechanic's time, perhaps more costly in operative time. Given the early-nineteenth-century skill shortage in the United States, presumably more acute among textile mechanics in remote communities, this made sense.

Yet with the batten drive in his looms, there was a feature of the Gilmour/Rhode Island power loom that Markham copied, namely, the warp-protector stop-motion. A geared drive to the batten required the use of a stop-motion, to halt the loom if the shuttle was trapped in the warp shed. Americans, especially those in the Waltham network, were gaining experience with labor-saving stop-motions at this time and the warp-protector motion was available from Rhode Island mechanics familiar with the Gilmour power loom.[13]

Where did Markham get his inspiration for this beating-up motion? Neither the Waltham, nor the Rhode Island power looms had quite the same eccentric motion. As for British antecedents, crank-driven batten is shown in both of Horrocks's earlier power-loom patents of 1803 and 1805.[14] However, in both of these the cranks were axial, on the main driving shaft of the loom, with connecting rods to each side of the batten, rather than relying on an eccentric gear wheel and slide design.

Neither of William Horrocks's first two power-loom patents was published in the *Repertory of Arts*. However, knowledge of crank mechanisms was hardly new to the world. Cranks were familiar in steam engines and news of their use in some of the first power looms could readily be transmitted across transatlantic mechanic networks. From axle cranks Markham's mind may have turned to eccentric wheels but the addition of a slide within the wheel would have been a novel American invention.[15] Again this judgment is based on very limited evidence.

In the Middlebury Cotton Factory machine shop, who was responsible for this novel design, Markham or Gordon? Nothing is known of Joseph Gordon's Scottish origins or how he came to be in Middlebury, Vermont. Apparently he was alive in 1859, or shortly before, because the local historian, Samuel Swift,

[12]My thanks to Dr. Richard J. Potton, of the Physics Discipline, School of Computing, Science and Engineering, University of Salford, England, for analyzing the mechanics of this unusual crank motion, 11 August 2009.

[13]Jeremy, "Innovation in American Textile Technology."

[14]William Horrocks, British patents of 20 April 1803 (2699) figure 2, and of 14 May 1805 (28 May 1805 (2848) figure 2.E.

[15]I checked the plates for entries in Rees, *Cyclopaedia* for Cotton Manufacture, Engines, Horology, Lathe, Mechanics, Steam Engine, and Woollen Manufacture but could see nothing comparable to the slide in the eccentric of Markham.

spoke to him and recorded his claims to have been the second person to have introduced power looms to New England. Swift also reported that Gordon had brought plans of these looms from Scotland.[16]

Gordon's subsequent career reinforces the view that the twenty-two-year-old Markham was soundly instructed. After about three years in Middlebury, in 1820, Gordon moved south down the Hudson valley to Schaghticoke, New York, a settlement on the Hoosick River, a tributary of the Hudson. At Hoosick Falls, three years later, he built the four-storied Caledonia Cotton Factory. Reportedly running 7,000 spindles and 154 looms, he employed fifty men and turned out 30,000 yards weekly. In 1826 a fall left him a cripple and he was forced to sell out to the Crocker, Knickerbaker, House, & Merritt Corporation (which ran the business until 1868).[17]

By 1834 Gordon was back at work and in demand not only back in Middlebury, but also in Bennington, Vermont. He wrote to a Mr. Stewart in Middlebury (possibly Ira Stewart, who had invested in the Page cotton mill) in March 1834 from Bennington:

> I am very sorry that it is not in my power to comply with your wishes in repairing to Middelberry immediately and the reasons are as follows, when I returned home from Middelbury last time I told Mr. Hinsdall that I should not leave him before the twentieth of April, and in adition to the factorys I have hade the Charge of the Mashin works fore the last 4 or 5 Months, and I am now engaged in Making an adition to this Establishment of 24 Looms with the other Mashinery nessary fore them, and at present I have about 20 Mashinists to attend to as also to attend to the making and putting up of all the gearing Shafts and Drums to drive the above Mashinery, and neither Mr Hensdall nor I knows of any one that cane be gote at present to take my place.[*sic*] [18]

Gordon returned once more to Middlebury to manage his old works for the Marshall's, which in the long run became profitable.[19] Compared to other British immigrants who brought either very limited manufacturing experience or unreliable social habits to New England in the early nineteenth century, Gordon was a highly competent textile manufacturer and machine maker.[20] Markham was not being misled in his introduction to the knowledge of the power loom then prevailing in transatlantic manufacturing districts.

Would Gordon have adhered to the technology of the Scottish/Horrocks power loom with which he was familiar in Scotland? Finding American machine-making resources rather more constrained than those in Britain, would he have been as ingenious as Markham was in his lifetime (Page's testimonial) and after his death (in his obituary)? Certainly Markham made a set of drawings that would be used in a workshop. Nothing is known of the quality of draftsmanship in any plans that Gordon took to America. In Markham's sketchbook there is a lever system impelling the batten, which in overall shape is reminiscent of the compound lever in the Horrocks loom (see Fig. 6.34 [p. 202], top drawing). Was this inspired by Gordon's information or plans, or was it an early attempt to find a substitute for the Horrocks mechanism?

Markham almost certainly played a big part in bringing this beating-up design into existence. Whether he was the only inventor seems unlikely when Gordon was there. However, as the author and possessor of the workshop drawings describing the loom, and in the light of what is known about Gordon's career

---

[16]Swift, *Middlebury*, 331 quoted in chapter 1, pp. 12 and 16.

[17]*McLane Report* II, 52; Sylvester, *History of Rensselaer Co.*, 386.

[18]Joseph Gordon letter to Mr. Stewart, March 7, 1834, Henry Sheldon Scrapbook 20 # 57, Sheldon Museum, Middlebury, Vermont. Two mills at Bennington established before 1831 made satinets (cotton warps and woollen weft): those belonging to Joel Volintine, established in 1825, and Stebbins D. Walbridge in 1829 (*McLane Report* I, 918–920.

[19]*McLane Report* I, 878; Jeremy, *Transatlantic Industrial Revolution, 169–172.*

[20]Compare those documented in Jeremy, *Transatlantic Industrial Revolution*, chapter 6.

and the alternative Waltham and Rhode Island power looms, this author concludes that the batten mechanism in the Markham drawings was new to New England and that Markham should get substantial credit, if not all of the credit, for its invention.

Appendix 5

Markham's Drawings and Methods of Early-Nineteenth-Century Machine Making

☆ ☆ ☆ ☆ ☆ ☆ ☆

SCALE DRAWINGS WERE ECONOMICAL OF TIME AND COST. They allowed experiment on the drawing board rather than costly trial-and-error changes during full-scale construction, or while model building. Specifying dimensions ahead of production allowed the work of production to be divided into multiple components and thus between several craftsmen or manufacturers. This division of labor in turn allowed increases in the scale of machines, hitherto too large for a single craftsman to tackle, and increases in the rate of production.[1]

"The effect of concentrating the geometric aspects of manufacture in a drawing is to give the designer a much greater 'perceptual span' than the craftsmen had."[2] He could think almost without a limit on his drawing board. Consequently mechanicians with design capabilities became the new mechanical engineers on which industrial technology increasingly relied in the nineteenth century.

As the earliest known workshop drawings of an America textile machine maker, Markham's drawings evidence this transition from the craftsman, reliant on developing his ideas during the construction process, to the mechanical engineer, experimenting boldly through design drawings ahead of production.

Mechanical drawing has been defined as "formal and detailed plans...with dimensions noted."[3] There were three main kinds of dimensional plans: design drawings, fixing the form of a machine and its capacity; working or shop drawings, to guide production; and final layout (as-fitted or contract or presentation) plans, showing the complete work when constructed.

What then were the characteristics of Markham's drawings and how did these compare to contemporary renderings of machine drawing in the early industrial transatlantic world? In the first place, it may be noted that he used a variety of paper sizes and obtained them, directly or indirectly, from a number of paper makers. The sizes of the papers on which he made his drawings ranged from about 25 x 16 centimeters up to the largest, 60 x 48 centimeters. Most approximated to 40 x 26 centimeters (see Table A5.2, p. 387). In inches, the measure of his day, Markham's papers ranged between about 10 x 6.5 inches to about 24 x 18 inches, not really conforming to the standard sizes in which paper was sold in Britain (one source he used). His largest paper approximated to a Royal size of drawing paper (24 x 19 inches).

---

[1] Baynes and Pugh, *Art*, 11.
[2] J. Christopher Jones quoted in Baynes and Pugh, *Art*, 11.
[3] Brown, "Design Plans."

The median size of 15 x 10 inches was a Pott, just under half the size of an Imperial sheet or a half-Imperial drawing board.[4] In contrast, Latrobe's mechanical drawings were made on relatively standardized sheets of paper, if his Navy Yard steam-engine drawings were typical: approximately 22 x 15 inches (8 of 17 drawings) or 30 x 22 inches (4 of the 17).[5]

The six makers of Markham's papers, as seen (see Table A1.1, p. 333), were both American and European, including William Balston whose Whatman handmade paper was also sold to the preeminent British artist of the day, J. M. W. Turner. This rather suggests that Markham was not particularly fastidious about the size or quality of his drawing paper, which would have been consonant with the limitations faced by the small-scale machine maker subordinate in space and organization to the needs of a factory and facing infrequent access to Boston or New York stationery suppliers—but this is partly conjecture.

Second, the vast majority of the sheets on which he made engineering drawings (34 out of 46 of machine drawings, not including two full-size cam patterns) recorded scale or dimensions. That is, they were definitely drawings intended to be used to aid construction, drawings intended for use in the workshop, not primarily to record machine forms.

Third, the drawings approximate to component and whole-layout plans, with 27 of the former and 21 of the latter. Although not "as finished" drawings, the whole machine ones might be what Markham would have regarded as such. Because there are no keys linking components to whole machines, presumably Markham relied on his familiarity and memory to make those connections.

Drawings by themselves were not sufficient to permit construction. They had to be supplemented by a variety of other kinds of information. These other sorts of information, recorded with the drawing or on it, were known as "conventions" (Table A5.1).[6] The first and most important of these was scale. For Markham, the most frequently used scale was one inch to one foot (in 23 out of 34 individual machine drawings) where a numerical scale was given; in contrast to another dozen drawings where the scale was marked rather like a ruler. However, Markham also resorted to 1/2″ to a foot (see Fig. 6.70, p. 282), 3/4″ to a foot (see Fig. 6.48, p. 236), 3/8″ to a foot (see Fig. 6.60, p. 262), and 2″ to a foot (see Fig. 6.54, p. 252). In six drawings he provided numbers of pinion (gear) teeth (see Figs. 6.67 [p. 276], 6.46 [p. 232], 6.47 [p. 234], 6.54 [p. 252], 6.55 [p. 254], 6.61 [p. 264]) and in four cases diameters of rollers (see Figs. 6.41A, 6.41B [pp. 218, 220], 6.46 [p. 232], 6.58 [p. 260]). In other words these were dimensional drawings intended to convey essential but often minimal information to aid exactness and uniformity of construction. For example, they did not name kinds of wood or iron used in the machines.

A second convention was the choice of machine views to record and place on the drawing paper. The overriding consideration was convenience, particularly ease of execution and clarity of communication.[7] As in the cotton-machinery plates in Rees's *Cyclopaedia*, Markham in most cases (66 out of 72 views of machines or parts of machines) chose front and side elevations and plans. Lining up or aligning drawings, a useful way of positioning that lent greater clarity for the reader, is seen in just eight drawings (see Figs. 6.38 [p. 212], 6.40 [p. 216], 6.41A and 6.41B [pp. 218, 220], 6.57 [p. 258]), 6.67 [p. 276], 6.70 [p. 282], and 6.72 [p. 288]).[8]

---

[4]For drawing paper sizes see *Whitaker's Almanack* (London, 1869–present).

[5]Stapleton (ed.), *Engineering Drawings of Benjamin Henry Latrobe*, drawings 64–80.

[6]Booker, *History of Engineering Drawing*, 171.

[7]Booker, *History of Engineering Drawing*, 155 makes the point about convenience.

[8]Booker, *History of Engineering Drawing*, 155.

Table A5.1 Drafting Conventions in Markham's Drawings

	Production stage	no.	Machine coverage	scale	no. scale	pinion, no. teeth	roller diams	alignment	elevations/ plan	elevations	plans	perspective	line only	line widths	shading	colors	Notes on drawing
1	carding	6.38	w	1″:1 ft				y	2e	2			y				
2	carding	6.39	w	1″:1 ft					e	1			y	y			
3	carding	6.40	w	y				y	2e	2					y	grays	
4	carding	6.41A	w	1″:1 ft			y	y	e	1			y				
5	carding	6.41B	w	1″:1 ft			y	y	e	1			y				
6	carding	6.42	w		n							y			y	grays	
7	carding	6.43	c	1″:1 ft					2e	2			y				
8	carding	6.44	w		n				p?		1		y				
9	drawing & roving	6.46	c	1″:1 ft		y	y		3e	3							y
10	drawing & roving	6.47	c	1″:1 ft		y			4e	4			y			yellow gears; brown top roller; black gears	
11	drawing & roving	6.48	c	3/4″:1 ft & 1″:1 ft				poss	e+p	1	1		y				
12	drawing & roving	6.49	w	y					e+p	1	1					grays	
13	drawing & roving	6.50	c	1″:1 ft					3e	3						yellow gears; brown top roller; black gears	

Table A5.1 *(continued)*

	Production stage	no.	Machine coverage	scale	no. scale	pinion, no. teeth	roller diams	alignment	elevations/ plan	elevations	plans	perspective	line only	line widths	shading	colors	Notes on drawing
14	drawing & roving	6.51A	c		calcs only: reverse												
15	drawing & roving	6.51B	c		n				e+2 details	1			y				
16	spinning	6.52	w	1″:1 ft					e	1			y				
17	spinning	6.53	w		n				e	1			y				
18	spinning	6.54	w	2″:1 ft		y			e	1						black gears dark-gray shafts	
19	spinning	6.55	c	1″:1 ft		y			e+p	1	1		y				
20	spinning	6.56	w	1″:1 ft					e	1			y				
21	spinning	6.57	c	1″:1 ft				y	2e+p	2	1		y				
22	spinning	6.58	w	1″:1 ft			y		e	1			y				
23	spinning	6.59	c	1″:1 ft					p		1		y				
24	spinning	6.60	c	3/8″:1 ft & 1″:1 ft					2p		2		y				
25	spinning	6.61	c		n	y			p		1					yellow gears black shafts	
26	spinning	6.62	c		n				3e(?)	3			y				
27	spinning	6.63	w	1″:1 ft					e	1					y	grays	
28	spinning	6.64	w	1″:1 ft					e	1					y	grays	

Table A5.1 *(continued)*

	Production stage	Machine no.	Machine coverage	scale	no. scale	pinion, no. teeth	roller diams	alignment	elevations/ plan	elevations	plans	perspective	line only	line widths	shading	colors	Notes on drawing
29	spinning	6.65	w		n							y			y	grays & blue	
30	spinning	6.66	w	y					e	1					y	grays dark for curvature	
31	spinning	6.67	c	1″:1 ft		y		y	2e	2						yellow gears & cam black supports	
32	spinning	6.68	c		n				e	1						grays	
33	spinning	6.69	w	1″:1 ft					e	1						black for gears shaft & support	
34	spinning	6.70	c	1/2 ″:1 ft				y	2e & 2p	2	2		y				
35	spinning	6.71	c	y					3e	3						grays	
36	warp prep'n	6.72	w	y				y	2e	2					y	grays	
37	warp prep'n	6.73	c	y					2e	2						grays	
38	warp prep'n	6.74	w		n				p? or e?	1?	1?	y?			y	grays	
39	warp prep'n	6.75	w	y					e	1					y	grays	
40	warp prep'n	6.76	c	y					e	1					y	grays	

Table A5.1 *(continued)*

	Production stage	no.	Machine coverage	scale	no. scale	pinion no. teeth	roller diams	alignment	elevations/ plan	elevations	plans	perspective	line only	line widths	shading	colors	Notes on drawing
41	warp prep'n	6.77	w	y					e	1						grays	
42	warp prep'n	6.78	w	y					e	1							
43	weaving	6.79	w	y					e	1					y	grays & black for shadows	
44	weaving	6.80	w	y					2e	2					y	grays & black for shadows	
45	weaving	6.81	w		n											grays & black for depth	
46	weaving	6.82	w		n											grays & black for depth	
47	weaving	6.83A	c	full size									y				
48	weaving	6.83B	c	full size									y				
	Totals	48	27w 21c	34 + 2 full size	11	6	4	8 + 1 poss	71	57	12	5	21		12	25	1
				1 calcs													

Note: Machine coverage: Whole (27) or Component (21).

Of the eight instances when no scale was provided, four (see Figs. 6.42 [p. 221], 6.65 [p. 273], 6.81 [p. 311], and 6.82 [p. 312]) were "perspective drawings." That is, they showed a three-dimensional-picture type of view. The term "perspective view" was used ca. 1800 of any picture type of mechanical drawing but until 1820, when the Reverend William Farish, Professor of Natural Philosophy at Cambridge University, published his *Isometrical Perspective*, few were truly isometrical drawings (one-plane orthographic projections).[9] Markham's four perspective drawings were not isometric. Nor did they conform to First Angle projection, the system pioneered in France by Gaspard Monge, close friend of Napoleon, who published his *Géométrie Descriptive* in 1795.[10] The merit of perspective drawings was that they provided an artist's view of a machine, such as it would appear when built. The problem was that perspective drawings gave few clues about building a machine—because the perspective filter could not accurately record the third dimension (see chapter 3 on Visual Knowledge).

Seven other conventions: shadows, line thickness, aligning drawings (already noted), hatching, broken lines, symbols, and color were open to Markham. The use of thick lines, to indicate edges vertical to the plane of the paper, a practice of eighteenth-century British engineering draughtsman, derived from an earlier use of shadows.[11] Thick lines appear in only one of Markham's drawings (see Fig. 6.39, p. 214) but shadows (by European standards obsolete) are used in 14 of 48 sheets. Neither hatching nor broken lines figure in Markham's drawings. There is no evidence of symbols (unless stylized flyers, as in Fig. 6.57 [p. 258], are regarded as such) or of keys to symbols.

However, in 25 of 48 sheets there is evidence of the use of color, some of which may have faded over time. Gray, which might once have been blue, frequently occurs but seems to indicate machine frames, which, generally in early-nineteenth-century America, were wooden. In four drawings (see Figs. 6.47 [p. 234], 6.50 [p. 240], 6.61 [p. 264], 6.67 [p. 276]) the use of yellow is distinctive. In three it is confined to pinions (gears) but in the fourth (Fig. 6.67, p. 276) it is used for the core of a heart-shaped cam and for the bearing of a tin roller. A late-nineteenth-century practice was to employ yellow (gamboge with a little sienna) to signify a brass component.[12] Was this the color convention that Markham was following? Brass appears in the notes but it is for reeds (see Fig. 6.17A, p. 160) or for unspecified castings (see Fig. 6.23B, p. 178) so this may possibly affirm conformity to a British practice. On the other hand, this was not Latrobe's color code: blue for cast iron and light orange for an iron wedge appear in details of cog wheels and cam ring on his plans of 1810 for the Washington Navy Yard steam engine; on other plans, light yellow-orange signified wood.[13]

The vital accompaniment to these workshop drawings were explanatory notes. Only one (Fig. 6.46, p. 232) has these written on the drawing. Before crucial operating data were widely known (from handbooks first published in the 1830s),[14] physical dimensions, as much as machine settings or economic information about costs or output rates, were closely guarded trade secrets. It is noticeable that most constructional, operating, and economic data pertaining to his textile machines (Figs. 6.1–6.14 [pp. 124–150]) were held by Markham in his notebooks, held separately from the drawings and surviving

---

[9]Booker, *History of Engineering Drawing*, 114–127.

[10] Booker, *History of Engineering Drawing*, 86–113; 155–170; Belofsky, 'Engineering Drawing'. First Angle (or quadrant) projections, based on Monge's descriptive geometry, place the front elevation in the center of the drawing but put the top (plan) view below it, the right-hand elevation on its left, the bottom plan above it, and the left-hand elevation on its right. Third Angle projection, developed in mid-nineteenth century United States, placed the top plan above the front elevation, the right-hand elevation on its right, the bottom plan below it, and the left-hand elevation on its left. This was much simpler to draw and to read.

[11]Booker, *History of Engineering Drawing*, 171.

[12]Booker, *History of Engineering Drawing*, 173.

[13]Stapleton (ed.), *Engineering Drawings of Benjamin Henry Latrobe*, Plates 4, 9, 10.

[14]Primarily Montgomery, *Carding and Spinning the Master's Assistant*. See Jeremy, *Technology and Power*; Huberman, *Escape from the Market*.

sometimes as pages. These served as the other half of his technical memory, complementing his workshop drawings. What do the notes record?

First, there are more dimensions. Although the drawings provide overall dimensions for the form of the machine, the notes supplied dimensions of parts (see Figs. 6.18–6.22 [pp. 164–174]). Of course, further information was needed if the various components of a particular machine were to work in a coordinated, synchronized, and smooth-running fashion. So, second, there are speeds of parts (e.g., Figs. 6.1–6.6 [pp. 124–134]) and relationships between moving parts (Figs. 6.1–6.6 [pp. 124–134]), mostly in individual machines. In one instance the notes provide relationships among machines in a production line, among carding, drawing, roving, and spinning (Fig. 6.4, p. 130).

Third, there is some information on the materials used to build the machines. Wood is mentioned but kinds of wood are never identified. Perhaps that sort of knowledge was carried in the memory. There are sorts of metal, for example, cast iron (Fig. 6.23B, p. 178), wrought iron (Fig. 6.16, p. 158), brass (Fig. 6.23B, p. 178), and steel (Fig. 6.23B, p. 176). There is reference to castings for a 17" gear wheel (Fig. 6.17A, p. 160), to forging iron rollers and levers (Fig. 6.23A, p. 176) to cast steel for spindles (Fig. 6.23B, p. 178), and to wire for flyers (Fig. 6.23B, p. 178). All of this presupposed sources for ferrous and nonferrous metals, either at a regional ironworks for cast iron, a more local blacksmith's shop for wrought iron, and specialist suppliers for nonferrous metals; or else a distant market with a variety of metal merchants, for example, in the Boston or New York City districts. As Page's testimonial of 1819 (see chapter 1) makes clear, Markham was "perfectly capable of doing every part of the iron and brass work (after forged)." Because Markham was capable of working with the metals needed to build these textile machines—invariably for moving parts and nearly always for the critical components of the machines—he had no reason to record notes to himself about the methods of construction he used. The same reason would apply if he relied on a carpenter to make wooden machine frames, agreeing with the carpenter on the type of wood or type of joint or fixing.

Fourth, components are named. The most complete list is that given for three spinning frames. Besides the familiar flyers, spindles, and rollers, there are stands, top saddles, top rollers, saddle bars, stirrups, levers, lever staples, spindle guides, spindle rails, flyer washers, "whurrs" (throstle roller?), shafts, gears, pulleys, flywheels, flywheel uprights, and a heart motion, that is, heart-shaped cam, and woodwork (see Figs. 6.23A [p. 176], 6.23B [p. 178]). Because, in all likelihood, only the machine frame was built of wood, almost 20 different parts had to be made in cast iron, wrought iron, brass castings, or drawn from thick wire. Castings would be made from wooden patterns or moulds carefully kept and catalogued by the machine maker. Wrought-iron parts would be shaped at the smith's forge or in a furnace at the machine shop, under the machine maker's supervision. All of this illustrated the complexity of machine building in a small firm of textile machine makers, and this list is for only one of perhaps fifteen or more different machines, to which many a mechanic would, like Markham, make his own modifications. Consequently, the small-scale machine maker needed to learn or to develop multiskill competencies.[15]

Other competencies are illustrated by the last two sorts of information found in the notes to these workshop drawings, namely, reasonably accurate estimations of the output of each machine and an even more accurate estimation of the cost of building a production-line machine. In these areas Markham's papers are most deficient. Only one estimate of output is recorded in surviving documents, namely,

[15]A point made in Hindle and Lubar, *Engines*, 181.

Figure 6.24A (p. 180), for the output of a speeder. Only one estimate of construction costs has survived, namely, Figures 6.23A (p. 176) and 6.23B (p. 178); this indicates the cost of building three spinning frames.

From the economic information in these incomplete lists of machine specifications and costs, it is possible to establish an approximate price differential between the large-scale, urban district machine maker and a small-scale, up-country rural machine maker. At $960 for three spinning frames with a total of 192 spindles, Markham's cost price of $5 a spindle was a half to two thirds the current selling price charged by the Boston Manufacturing Co.—$8 per spindle for a warp-spinning throstle, $12 per spindle for a filling-spinning throstle.[16]

One intriguing feature to emerge from a scan of the notes, the drawings, and the sketchbook is the frequent appearance of one mechanism. This was the twin-cone-and-belt drive. In the notes to the drawings it is listed in the parts to a lap winder (see Fig. 6.19, p. 166), a speeder (Fig. 6.15B, p. 156), a warping frame (Fig. 6.16, p. 158), and a dresser (Figs. 6.3, 6.17A [pp. 128, 160, 162] and 6.22 [p. 174]). In the Sketchbook, single cones or pairs of cones appear in four drawings (6.28 [p. 190], 6.31 [p. 196], 6.32 [p. 198], and 6.33 [p. 200]). It also appears in one of his spinning machines (Fig. 6.66, p. 274). The mechanical function of the twin-cone-and-belt drive was to produce variable speed and to do so smoothly. This was most useful where winding increased the diameter of the roller receiving the sheet of fibers or yarns being wound. Was Markham particularly fascinated by the potential of this mechanism? It would appear so because this was a mechanism that seems to have distinguished his dressing frame (see Table A3.2, p. 350).

Additional information in the notes to the workshop drawings comprises references to technical authorities, customers, or rivals. So there are references to (John) Smeaton's theory, (Oliver) Evans's theory, and William Waring's theory (see Fig. 6.1, p. 124); the Waltham carding machine (Fig. 6.3, p. 128); William Dunn's (or Durnn's) mule, unidentified (Fig. 6.7, p. 136); Kennedy Watt & Co's Mill at Johnstone (presumably near Paisley in Scotland) and the Elderslie Mill close by (Fig. 6.11, p. 144); Dover Cotton Mill (Fig. 6.14, p. 150); and Arthur Latham Co (Fig. 6.15B, p. 156). These references further confirm Markham's role as a member of the early-nineteenth-century transatlantic fraternity of mechanics. They also indicate how new technical knowledge was relayed, and relayed quickly over vast distances, achieving a diffusion of new technology, as in the case of the English-derived dresser of 1804, which was modified at Waltham in 1818, and built by Markham with some improvement in distant Vermont within three years or so.

So much for the evidence of Markham's techniques of mechanical drawing. How did he proceed from designing to building his machines? The physical evidence of Markham's papers offers some clues here. First, in a sketchbook (see chapter 6, section 2: Sketchbook, see also Figs. 6.26–6.37) he recorded possibilities, dreams even, for designs. Having sifted the realistic from the fanciful, in other notebooks he made careful drawings of designs with technical and commercial possibilities (for example, see Figs. 6.46–6.48 [pp. 232–236], 6.52 [p. 248], 6.53 [p. 250], 6.56 [p. 256], 6.57 [p. 258]). These notebooks may even have been used to show potential customers what he was capable of building, if his portfolio was not convenient to carry around. At a third stage, having secured an order, he moved to scale drawings, either those already in his portfolio or new ones he had to draft. From the scale drawings, machine parts were constructed. Sometimes, as in the loom cams (see Figs. 6.83A, 6.83B [pp. 314, 315]), designs could be cut out of paper and used as an outline of the machine part, either to make a wooden pattern (mould)

[16]Gibb, *Saco-Lowell Shops*, 47.

for iron casting, or else to convey the shape and dimensions of the machine part to a carpenter, blacksmith, whitesmith, and so on. Benjamin Henry Latrobe, in his machine drawings, made copies of his master drawings by pin-pricking the lines and shapes through to two or more sheets held below the master drawing. These copies were then distributed to various component makers.[17] Markham appears not to have used this technique, though over half his drawings had pinholes (Table A5.2). His pinholes were mostly in the corners. This suggests that his scale drawings served as masters. As a result Markham had to rely on more master drawings of components (e.g., see Figs. 6.59–6.62 [pp. 261–266]) than had he made copies from fewer masters.

Beyond the drawing board, little more is known about the metal-working facilities available to Markham. The Middlebury historian reported,

> At an early day, Rufus Wainwright and Jonathan Wainwright, Jun., sons of Jonathan Wainwright, Esq., of Salisbury, established themselves in the tin and iron business, on a small scale, and having enterprize and energy, they enlarged their business from time to time. Not long after the close of the war of 1812, they erected a furnance [*sic*] below the mills, built by Appleton Foot, on the site of the former forge, for casting stoves and other articles.[18]

Until 1826 their furnace was located across the creek from Page's cotton mill. This business continued until after 1859. Before that there had been a trip-hammer and forge in their location. Whether they supplied the 64 lb of steel castings for spindles and who supplied the brass castings that Markham listed (see Fig. 6.23B, p. 178) is unclear.

If the Wainwrights made Markham's castings and did some of his forging work, other wrought-iron work was done at his cotton factory, which had its own blacksmith shop. A deed from David Page to Joseph Hough conveyed all his interest in the cotton factory, blacksmith shop, and boarding houses, as well as all tools, machines, and personal property.[19] Presumably most of the wood work was also performed in the cotton-factory machine shop.

---

[17]Stapleton (ed.), *Engineering Drawings of Benjamin Henry Latrobe*, 63.
[18]Swift, *History of Middlebury*, 338.
[19]Middlebury Land Records, VIII, 310.

Table A5.2 Physical Evidence in Markham's Drawings

		Drawing no.	Size of paper (a x b cms)		Watermark (paper maker)	Pinholes in corners	Pinholes	Booklet pages
	Letter from Waltham				C Baker & Co			
	Loose sheets	6.24A & B			E Burbank			
	Sketchbook	cover (6.26)			William Hurd			
	Sketchbook	6.28			a dove			
	Sketchbook	all pages (6.30–6.35)			C Baker & Co			
	Production stage							
1	carding	6.38	26.9	22.3				y
2	carding	6.39	28.0	23.0				
3	carding	6.40	42.7	27.0	T G & Co	y		
4	carding	6.41A	44.0	26.0		y		
5	carding	6.41B	44.0	26.0		y		
6	carding	6.42	44.0	26.8		y		
7	carding	6.43	44.0	26.0				
8	carding	6.44	27.5	27.3		y		
9	drawing & roving	6.46	27.1	22.3				y
10	drawing & roving	6.47	27.1	22.2				y
11	drawing & roving	6.48	26.9	22.3				y
12	drawing & roving	6.49	42.5	27.0	T G & Co	y		
13	drawing & roving	6.50	45.5	27.7		y	y	
14	drawing & roving	6.51A	30.4	21.5				
15	drawing & roving	6.51B	30.4	21.5				
16	spinning	6.52	26.9	22.2				y
17	spinning	6.53	27.6	22.7				
18	spinning	6.54	26.9	22.3				y
19	spinning	6.55	27.1	22.2				y
20	spinning	6.56	27.6	22.7				y
21	spinning	6.57	26.9	22.2				y
22	spinning	6.58	29.0	22.8		y		
23	spinning	6.59	29.0	22.8		y		

Table A5.2 *(continued)*

		Drawing no.	Size of paper (a x b cms)		Watermark (paper maker)	Pinholes in corners	Pinholes	**Booklet pages**
24	spinning	6.60	29.0	22.8		y		
25	spinning	6.61	44.0	26.0				
26	spinning	6.62	27.5	27.3		y		
27	spinning	6.63	45.4	27.0		y		
28	spinning	6.64	45.4	27.0		y		
29	spinning	6.65	44.0	26.8		y		
30	spinning	6.66	42.6	26.5		y		
31	spinning	6.67	45.5	27.7		y		
32	spinning	6.68	42.5	26.5		y		
33	spinning	6.69	45.5	27.7		y		
34	spinning	6.70	45.5	27.7		y		
35	spinning	6.71	42.5	26.5		y		
36	warp prep'n	6.72	42.0	26.8		y		
37	warp prep'n	6.73	42.0	26.8		y		
38	warp prep'n	6.74	47.0	28.6				
39	warp prep'n	6.75	42.8	26.5		y		
40	warp prep'n	6.76	41.8	28.8		y		
41	warp prep'n	6.77	47.5	19.8	J Whatman 1816			
42	warp prep'n	6.78	42.7	27.0		y		
43	weaving	6.79	60.0	47.5			y	
44	weaving	6.80	42.5	26.0		y		
45	weaving	6.81	24.5	22.5		y		
46	weaving	6.82	25.0	21.4				
47	weaving	6.83A	20.5	16.9				
48	weaving	6.83B	20.5	16.9				
						27	2	8

NB: Paper dimensions have been rearranged here, the longer in the left-hand column.
None of the drawings have pinholes following the design lines.

Appendix 6

s a Record of One Textile Engineer's Practices ca. 1820, How Complete Is This Set of Workshop Drawings?

☆ ☆ ☆ ☆ ☆ ☆ ☆

MARKHAM'S PAPERS ARE OF ABIDING VALUE FOR THOSE INTERESTED in the development of early industrial technology and society. His workshop drawings are among the oldest American drawings to survive outside printed works, all patent drawings having been destroyed in the Patent Office fire of 1836 (after which a number of the earlier ones were reconstructed). They certainly seem to be the earliest set for a nigh-complete line of textile machinery. But how complete are they?

Compared to the papers of American Presidents and other great men, Markham's papers appear few and fragmentary. And so they are. As part of the working records of a mill mechanic, and one moving among widespread mill companies, they are unsurprisingly few. Itinerant, practical men, based in a busy workshop for a few years or months and dealing with measurements, materials, implements, and machines, relied much less on the paper record than did the mill manager and his bookkeeper. Even the far more complete archives of the large textile corporations at Waltham and Lowell have preserved no drawings of machine designs or of workshop layout for the period 1810–1825, formative years for the American textile industry.[1]

Just how complete Markham's papers are as a record of an early-19th-century textile machine maker's workshop is not easy to gauge. There are no explicit references to drawings that are missing. However, two tests might be applied. First, a complete set of mechanical drawings would comprise three types of drawing for every machine that Markham made: the rough sketch; elevations and plans for every component of every machine; and, for each machine, a final layout plan showing where components fitted in. Plainly only a fraction of these has survived, if they were ever made. Keys linking components' drawings to final layout drawings have not survived. Possibly these links were carried in Markham's head.

Second, if the drawings in these papers are compared to the drawings of textile machines in Rees's *Cyclopaedia*, a contemporary and objective measure of British textile technology (albeit some years out of date but still mostly representative) is possible.

Using the Rees measure, it is evident that nearly all the machine forms are present in Markham's papers. The big omission is of any drawing of a calico-printing machine. But this might be explained by the fact that it was not until the mid-1820s that the large New England cotton-manufacturing firms

[1]See the early chapters of Gibb, *Saco-Lowell Shops*.

turned their interest to calico printing, creating emulators and a demand for calico-printing machines.[2] When Markham's drawings are compared to Markham's notes, another major gap appears: a drawing of the warping frame that is specified in Figure 6.16 (p. 158). Nor is there any drawing relating to the waterwheel specified in Figure 6.1 (p. 124). More scale drawings of the looms might have existed, showing detail of the shedding and picking motions. Last, it is noticeable that Markham built a limited line of woolen-manufacturing equipment, if his designs spanned the whole of his range in the early 1820s. Woolen-specific equipment consisted of his carding machine and his variants on the spinning jenny. Both sorts were appropriate for the semi-domestic manufacturers of northern New England and upstate New York, where they were widely scattered. His range did not include the gigmills (cloth-raising machines) and cloth-cropping machines that had been used for several centuries in the West of England to produce high-quality cloth, and which were being copied by ambitious and large-scale American manufacturers in southern New England, like Zachariah Allen.[3]

Markham's notes on the drawings seem to be more incomplete than the drawings themselves. Absent, but very desirable, would be a thorough specification, citing parts, materials, speeds, costs, outputs, selling prices, for each and every machine. So also would be methods of construction, an inventory of workshop tools and equipment, and a premises layout, as well as wages lists for all employees. But all this, given the nature of the work and the hazards threatening documentary survival, is a less than realistic wish-list, particularly because such a totally comprehensive record is not even available for the best documented machine shop of this time, that of the Boston Manufacturing Co.

---

[2] Jeremy, *Transatlantic Industrial Revolution*, 104–117.

[3] Cole, *American Wool Manufacture*, I, 175–243; Jeremy, *Transatlantic Industrial Revolution*, 118–137, 218–224.

Bibliography

☆ ☆ ☆ ☆ ☆ ☆ ☆

ABBREVIATIONS USED IN THE BIBLIOGRAPHY

ANB	*American National Biography*
Co.	Company
DAB	*Dictionary of American Biography*
DNB	*Dictionary of National Biography*
ODNB	*Oxford Dictionary of National Biography*
OED	*Oxford English Dictionary*
ODQ	*Oxford Dictionary of Quotations*
RAMA	*Repertory of Arts, Manufactures, and Agriculture*
TRISEDI	*Transactions of the Rhode Island Society for the Encouragement of Domestic Industry*

SOURCES

Archives

American Textile History Museum, Lowell, Massachusetts, microfiche: Bagnall, William R. "Sketches of Manufacturing Establishments in New York City, and of Textile Establishments in the Eastern States." Edited by. Victor S. Clark (typescript, 4 vols., 1908). Cited here as Bagnall, *Textile Industries*, Volume II.

Baker Library, Harvard University, Cambridge, Massachusetts, Boston Manufacturing Co. Papers.

Baker Library, Harvard University, Cambridge, Massachusetts, Wendell Collection.

Connecticut State Library, Hartford, Connecticut, Scholfield Papers.

Middlebury, Vermont, Town Clerk, Middlebury Land Records.

Rhode Island Historical Society, Providence, Rhode Island, Zachariah Allen Papers.

Sheldon Museum, Middlebury, Vermont:

Battell, Philip. "Minute Book of Conversations," 1848, Swift Family Papers,
Center Library (Middlebury, VT), Records in Records of the Middlebury Congregational Church.
Hagar Family Papers.
Houghton, John, Account book.
Judd and Harris Family Papers.
Letter Collection.
Markham Family Papers.
Markham Family Papers Collection File.
Sheldon, Henry L. Papers.
United States, First Circuit Court records, Federal Record Centre, Waltham, Massachusetts.
United States, National Archives, RG 92, Coxe and Irvine Papers, 1794–1842.
United States, National Archives, Census of Manufactures, 1820, film M-279.

Newspapers and Journals

Edinburgh Review

Franklin Journal

Middlebury Mercury, Middlebury, Vermont

Middlebury Register, Middlebury, Vermont

National Standard, Middlebury, Vermont

North American Review

Pennsylvania Magazine

Philosophical Magazine and Journal

Repertory of Arts, Manufactures, and Agriculture

Transactions of the Rhode Island Society for the Encouragement of Domestic Industry.

Vermont Mirror

Books, Articles, Theses

Académie Royale des Sciences. *Descriptions des Arts et Métiers*. 25 vols. Paris, 1761–1788.

Alder, Ken. *Engineering the Revolution: Arms and Enlightenment in France, 1763–1815*. Princeton: Princeton University Press, 1997.

Alder, Ken. "Making Things the Same: Representation, Tolerance and the End of the *Ancien Regime* in France." *Social Studies of Science* 28 (1998).

Ambrosini, Veronique and Bowman, Cliff. "Tacit Knowledge: Some Suggestions for Operationalization." *Journal of Management Studies* 38 (2001).

American National Biography. Edited by John A. Garraty and Mark C. Carnes. 24 vols. New York: Oxford University Press, 1999.

Appleton, Nathan. *The Introduction of the Power Loom and Origin of Lowell*. Lowell: B. H. Pennhallow, 1858.

Artz, Frederick B. *The Development of Technical Education in France, 1500–1850*. Cambridge, MA: MIT Press, 1966.

Aspin, Christopher and Chapman, Stanley D. *James Hargreaves and the Spinning Jenny*. Helmshore, Lancashire: Helmshore Local History Society, 1964.

Aspin, Christopher, "James Thomson," *ODNB*.

Bagnall, William R. *The Textile Industries of the United States Volume 1: 1639–1810*. 1893; reprint, New York: Augustus M. Kelley, 1971.

Bailey, William. *One Hundred and Six Copper Plates of Mechanical Machines and Implements of Husbandry: Approved and Adopted by the Society for the Encouragement of Arts, Manufactures, and Commerce, and Contained in Their Repository*. London: B. White, 1782.

Baines, Edward Jr. *History of the Cotton Manufacture in Great Britain*. London: H. Fisher, R. Fisher, and P. Jackson, 1835; reprint, London: Frank Cass, 1966.

Bakewell, Robert. "Wool." 1818. In Abraham Rees, ed. *The Cyclopaedia; or, Universal Dictionary of Arts, Sciences, and Literature*. 39 vols. text, 6 vols. plates, London: Longman et al., 1802–1820; 41 vols. text, 6 vols. plates, Philadelphia: Samuel F. Bradford et al., 1806–1820.

Bakewell, Robert. "Woolen Manufacture." 1818. In Abraham Rees, ed. *The Cyclopaedia; or, Universal Dictionary of Arts, Sciences, and Literature*. 39 vols. text, 6 vols. plates, London: Longman et al., 1802–1820; 41 vols. text, 6 vols. plates, Philadelphia: Samuel F. Bradford et al., 1806–1820.

Banks, John. *A Treatise on Mills*. Kendal: W. Richardson & W. Pennington, 1795.

Banks, John. *On the Power of Machines*. Kendal: W. Pennington, 1803.

Barber, Thomas W. *The Engineer's Sketchbook of Mechanical Movements, Devices, Appliances, Contrivances, and Details Employed in the Design and Construction of Machinery for Every Purpose*. London: Spon, 1918.

Barnhill, Georgia B. "The Dissemination of Technology through Book and Periodical Illustrations." *The Chronicle of the Early American Industries Association* 61 (2008).

Barrow, John D. *Cosmic Imagery: Key Images in the History of Science*. London: Bodley Head, 2008.

Bathe, Greville and Dorothy. *Oliver Evans: A Chronicle of Early American Engineering*. Philadelphia: Historical Society of Pennsylvania, 1935; reprint, New York: Arno Press, 1972.

Baynes, Ken and Pugh, Francis. *The Art of the Engineer*. Guildford: Lutterworth, 1981.

Belofsky, Harold. "Engineering Drawing—A Universal Language in Two Dialects." *Technology and Culture* 32 (1991).

Ben-Atar, Doran S. *Trade Secrets: Intellectual Piracy and the Origins of American Industrial Power*. New Haven: Yale University Press, 2004.

Berg, Maxine and Bruland, Christine, eds. *Technological Revolutions In Europe: Historical Perspectives*. Cheltenham: Edward Elgar, 1998.

Blunt, Charles. *An Essay on Mechanical Drawing*. London: R. Ackermann, 1811.

Boden, Margaret A. *Mind as Machine. A History of Cognitive Science*. 2 vols. Oxford: Clarendon Press, 2006.

Booker, Peter J. *A History of Engineering Drawing*. London: North Gate Publishing, 1979.

Bower, Peter. *Turner's Papers: A Study of the Manufacture, Selection and Use of His Drawing Papers, 1787–1820*. London: Tate Gallery, 1990.

Bower, Peter. *Turner's Later Papers: A Study of the Manufacture, Selection and Use of His Drawing Papers, 1820–1851*. London: Tate Gallery, 1999.

Bramah, Joseph. *A Letter to the Rt Hon Sir James Eyre, Lord Chief Justice of the Common Pleas* (1797). Quoted in MacLeod, *Inventing*.

Brown, John K. "Design Plans, Working Drawings, National Styles: Engineering Practice in Great Britain and the United States, 1775–1945." *Technology and Culture* 41 (2000).

Brown, Richard D. *Knowledge Is Power: The Diffusion of Information in Early America, 1700–1865*. New York: Oxford University Press, 1989.

Brunton, Robert. *Compendium of Mechanics; or Text Book for Engineers, Mill-wrights, Machine-makers, Founders, Smiths &c.* Glasgow: John Niven Jr., 1824.

Buchanan, Robertson. *Practical Essays on Mill Work and Other Machinery, Mechanical and Descriptive.* Edinburgh: J. Taylor, 1814.

Buchanan, Robertson. *A Practical Treatise on Propelling Vessels by Steam.* London: R. Ackermann, 1816.

Buchanan, Robertson. *Practical Essays on Mill Work and Other Machinery, Mechanical and Descriptive.* 2nd ed. Corrected, with notes and additional articles by Thomas Tredgold. London: for J. Taylor, 1823.

Burnett, John. *A History of the Cost of Living.* Harmondsworth: Penguin, 1969.

Candee, Richard, "Architecture and Corporate Planning in the Early Waltham System." In R. Weible (ed.), *Essays from the Lowell Conference in Industrial History, 1982 and 1983.* North Andover, MA: Museum of American Textile History, 1985.

Cantrell, John A. and Cookson, Gillian, eds. *Henry Maudslay and the Pioneers of the Machine Age.* Stroud: Tempus Publishing, 2002.

Catling, Harold. *The Spinning Mule.* Newton Abbot: David & Charles, 1970.

Chambers, Ephraim. *Cyclopaedia; or, an Universal Dictionary of Arts and Sciences.* 4 vols. London: J. F. and C. Rivington, 1779–1786.

Child, Hamilton (comp.). *Gazetteer and Business Directory of Windsor County, Vermont for 1883-84.* Syracuse, New York: Hamilton Child, Journal Office, 1884.

Chipman, Daniel. *An Essay on the Law of Contracts for the Payment of Specific Articles.* Middlebury, Vermont, 1822.

Claxton, Timothy, *Memoir of a Mechanic: Being a Sketch of the Life of Timothy Claxton Written by Himself.* Boston: G. W. Light, 1839.

Clayton, Tim. *The English Print, 1688–1802.* New Haven: Yale University Press, 1999.

Clayton, Tim. "The Engraving and Publication of Prints of Joseph Wright's Paintings." In J. Egerton, *Wright of Derby.* London: Tate Gallery, 1990.

Cochran, Thomas C. *Railroad Leaders, 1845–1890: The Business Mind in Action.* Cambridge, MA: Harvard University Press, 1953.

Cole, Arthur H. *The American Wool Manufacture* 2 vols. Cambridge, MA: Harvard University Press, 1926; reprint, New York: Harper, 1969.

Conrad, James L. Jr. "'Drive that Branch': Samuel Slater, the Power Loom, and America's Textile History." *Technology & Culture* 36 (1995).

Cooke, Jacob E. *Tench Coxe and the Early Republic.* Chapel Hill, NC: University of North Carolina Press, 1978.

Cooper, Carolyn C. *Shaping Invention: Thomas Blanchard's Machinery and Patent Management in Nineteenth Century America.* New York: Columbia University Press, 1991.

Cossons, Neil, *Rees's Manufacturing Industry: A Selection from "The Cyclopaedia."* 6 vols., Newton Abbot: David & Charles, 1972.

Cross, Whitney R. *The Burned-Over District: The Social and Intellectual History of Enthusiastic Religion in Western New York, 1800–1850.* Utica, NY: Cornell University Press, 1950

Crouzet, Francois. *The First Industrialists: The Problem of Origins.* Cambridge, UK: Cambridge University Press, 1985.

Crozet, Claudius. *A Treatise on Descriptive Geometry.* New York: A. T. Goodrich, 1821.

Cyclopaedia of Textile Work. 7 vols., Chicago: American Technical Society, 1914. Vol. I, *Cotton Fibre and Cotton Spinning.* Vol. IV, *Warp Preparation, Weaving.*

Cyert, R. M., and March, J.G. *A Behavioural Theory of the Firm.* Englewood Cliffs, NJ: Prentice Hall, 1963.

Dalzell Robert F. Jr., *Enterprising Elite: The Boston Associates and the World They Made*. Cambridge, MA: Harvard University Press, 1987.

Davis, Joseph S. *Essays in the Earlier History of American Corporations*. 2 vols. Cambridge, MA: Harvard University Press, 1917.

Diamond, Sigmund. *The Reputation of the American Businessman*. Cambridge, MA: Harvard University Press, 1956.

Dickinson, Henry W. "James White and His 'New Century of Inventions.'" *Newcomen Society Transactions* 27 (1949–1951).

Dickinson, Henry W. and Jenkins, Rhys O. *James Watt and the Steam Engine*. Oxford: Clarendon Press, 1927.

Dictionary of American Biography. Edited by Allen Johnson and Dumas Malone. 22 vols. New York: Charles Scribner's Sons, 1928–44.

Dictionary of National Biography. Edited by Leslie Stephen and Sidney Lee. 63 vols. London: Oxford University Press, 1885–1933.

Diderot, Denis et al., eds. *Encyclopédie ou Dictionnaire Raisonné des Arts et des Métiers*. 35 vols. Paris, 1751–1780.

D'Oench, Ellen G. *'Copper into Gold': Prints by John Raphael Smith, 1751–1812*. New Haven: Yale University Press, 1999.

Donne, Benjamin. *An Essay on Mechanical Geometry, Chiefly Explanatory of a Set of Schemes and Models*. Bristol: the author, 1796.

Dublin, Thomas. *Women at Work: The Transformation of Work and Community at Lowell, Massachusetts, 1826–1860*. New York: Columbia University Press, 1979.

Duffy, John J., Hand, Samuel B., and Orth Ralph H., eds. *The Vermont Encyclopedia*. Hanover, NH: University Press of New England, 2003.

Duncan, John. *Practical and Descriptive Essays on the Art of Weaving*. Glasgow: James and Andrew Duncan, 1808.

Duncan, John. "Weaving." 1818. In Abraham Rees, ed. *The Cyclopaedia; or, Universal Dictionary of Arts, Sciences, and Literature*. 39 vols. text, 6 vols. plates, London: Longman et al., 1802–1820; 41 vols. text, 6 vols. plates, Philadelphia: Samuel F. Bradford et al., 1806–1820.

Dyer, Joseph Chesseborough. "Notes on the Origin of Several Mechanical Inventions and Their Subsequent Application to Different Purposes." *Memoirs of the Manchester Literary and Philosophical Society* 23 (1868), 134–149.

Easterby-Smith, Mark and Lyles, Marjorie A., eds. *The Blackwell Handbook of Organisational Learning and Knowledge Management*. Oxford: Blackwell Publishing, 2003.

Edgerton, Samuel Y., Jr. *The Renaissance Discovery of Linear Perspective*. New York: Basic Books, 1975.

Edgeworth, Maria and Richard L. *Practical Education*. 2 vols. London: J. Johnson, 1798.

Edwards, Michael M. *The Growth of the British Cotton Trade, 1780 to 1815*. Manchester: Manchester University Press, 1967.

Egerton, Judy. *Wright of Derby*. London: Tate Gallery, 1990.

Eisenhardt, Kathleen M. and Santos, Filipe M. "Knowledge-Based View: A New Theory of Strategy?" In A. Pettigrew, H. Thomas, and R. Whittington, eds. *Handbook of Strategy and Management*. London: Sage Publications, 2002.

Emerson, William. *The Doctrine of Combinations, Permutations, and Compositions of Quantities, Clearly and Succinctly Demonstrated*. London: J. Nourse, 1770.

Emerson, William. *The Principles of Mechanics*. 5th ed. London: G. G. & J. Robinson, 1800.

Encyclopaedia Judaica. 17 vols. Jerusalem: Encyclopaedia Judaica, 1971–1972.

English, Walter. *The Textile Industry: An Account of the Early Inventions of Spinning, Weaving, and Knitting Machines*. London: Longman, 1969.

Erickson, Charlotte. *Invisible Immigrants: The Adaptation of English and Scottish Immigrants in Nineteenth-Century America*. Coral Gables: University of Miami Press, 1972.

Evans, Oliver. *The Young Mill-wright and Miller's Guide*. Philadelphia: the author, 1795; reprint, Foreword by Eugene S. Ferguson. Wallingford, Pennsylvania: The Oliver Evans Press, 1990.

Evans, Oliver. *The Abortion of the Young Steam Engineer's Guide*. Philadelphia: Fry and Kammerer, 1805; reprint, Foreword by Eugene S. Ferguson. Wallingford, Pennsylvania: The Oliver Evans Press, 1990.

Farey, John Jr. "Machinery." 1812 in Abraham Rees, ed. *The Cyclopaedia; or, Universal Dictionary of Arts, Sciences, and Literature*. 39 vols. text, 6 vols. plates, London: Longman et al., 1802–1820; 41 vols. text, 6 vols. plates, Philadelphia: Samuel F. Bradford et al., 1806–1820.

Farish, William. "On Isometrical Perspective." *Transactions of the Cambridge Philosophical Society* 1 (1822). This paper was read to the CPS on 21 February and 6 March 1820. Reprinted as *On Isometrical Perspective*. Cambridge: Cambridge University Press, 1822.

Farnie, Douglas A. "Samuel Crompton." In *ODNB*.

Farnie, Douglas A. and Jeremy, David J., eds. *The Fibre that Changed the World: The Cotton Industry in International Perspective, 1600–1990s*. Oxford: Oxford University Press for Pasold Research Fund, 2004.

Farnsworth, Harold and Rodgers, Robert. *New Haven in Vermont, 1761–1983*. New Haven, VT: Town of New Haven, 1984.

Fenwick, Thomas. *Essays on Practical Mechanics*. 3rd ed. Durham: George Andrews, 1822.

Ferguson, Eugene S., ed. *Early Engineering Reminiscences (1815–40) of George Escol Sellers*. Washington, DC: Smithsonian Institution, 1965.

Ferguson, Eugene S. *Bibliography of the History of Technology*. Cambridge, MA: MIT Press, 1968.

Ferguson, Eugene S. *Oliver Evans: Inventive Genius of the American Industrial Revolution*. Greenville, DE: Hagley Museum, 1980.

Ferguson, Eugene S. *Engineering and the Mind's Eye*. Cambridge, MA: MIT Press, 1992.

Ferguson, Eugene S. and Baer, Christopher. *Little Machines: Patent Models in the Nineteenth Century*. Greenville, DE: Hagley Museum, 1979.

Fitton, Robert S. *The Arkwrights: Spinners of Fortune*. Manchester: Manchester University Press, 1989.

Folsom, Michael Brewster and Lubar, Steven D., eds. *The Philosophy of Manufactures: Early Debates over Industrialisation in the United States*. Cambridge, MA: MIT Press, 1982.

Ford, John. "Rudolph Ackermann." *ODNB*.

Forman, Sydney. *West Point: A History of the United States Military Academy*. New York: Columbia University Press, 1950.

French, Gilbert J. *Life and Times of Samuel Crompton*. Manchester: T. Dinham & Co., 1860; reprint, Bath: Adams & Dart, 1970.

Gallatin, (Abraham Alfonse) Albert. *Report of the Secretary of the Treasury on the Subject of Public Roads and Canals*. Washington, DC: R. C. Weightman, 1808.

Gibb, George S. *The Saco-Lowell Shops: Textile Machinery Building in New England, 1813–1949*. Harvard University Press, Cambridge, MA: 1950.

Gnudi, Martha Teach and Ferguson, Eugene S., eds. *The Various and Ingenious Machines of Agostino Ramelli: A Classic Sixteenth Century Illustrated Treatise on Technology*. New York: Dover, 1987.

Gombrich, Ernst H. *Art and Illusion: A Study in the Psychology of Pictorial Representation*. London: Phaidon Press 1960.

Gonzales, Lois Bayer. "Female Millworkers and the Mechanization of Textile Production: The Boston Manufacturing Company of Waltham, Massachusetts, 1813 to 1822." PhD dissertation, University of New Mexico, 1995.

Goodhue, Josiah F. *History of the Town of Shoreham Vermont*. Middlebury, VT: A. H. Copeland, 1861.

Gravell, Thomas L. and Miller, George. *American Watermarks, 1690–1835*. 2nd ed. Newcastle, DE: Oak Knoll Press, 2002.

Gravell, Thomas L. and Miller, George. *A Catalogue of Foreign Watermarks Found on Paper Used in America, 1700–1835*. New York: Garland, 1983.

Gray, Andrew. *A Treatise on Spinning Machinery: Illustrated with Plans of Different Machines Made Use of in That Art*. Edinburgh: Archibald Constable & Co., 1819.

Gregory, Olinthus. *A Treatise of Mechanics, Theoretical, Practical, and Descriptive*. 2 vols. London: George Kearsley, 1806.

Guyatt, Mary, "Wilson Lowry." *ODNB*.

Hall, Frederick. *Statistical Account of the Town of Middlebury, in the State of Vermont. Part First*. Boston: S. Phelps, 1821.

Hamilton, Alexander. [United States, Department of the Treasury, Report of the Secretary.] *Report on the Subject of Manufactures*. Philadelphia: Childs & Swain, 1791.

Hargrove, John. *The Weavers Draft Book and Clothier's Assistant*. Baltimore: I. Hagerty, 1792.

Hargrove, John. *The Weavers Draft Book and Clothier's Assistant*. With new intro. by Rita Adrosko. Worcester, MA: American Antiquarian Society, 1979.

Harris, John R. *Industrial Espionage and Technology Transfer: Britain and France in the Eighteenth Century*. Aldershot: Ashgate, 1998.

Harte, Negley B. "On Rees's *Cyclopaedia* as a Source for the History of the Textile Industries in the Early Nineteenth Century." *Textile History* 5 (1974).

Hemenway, Abby Maria et al., eds. *The Vermont Historical Gazetteer: A Magazine, Embracing a History of Each Town, Civil, Ecclesiastical, Biographical and Military*. 2 vols. Burlington, VT: Miss A. M. Hemenway, 1868–1891.

Hills, Richard L. *Power in the Industrial Revolution*. Manchester: Manchester University Press, 1970.

Hills, Richard L. *James Watt. Vol. I. His Time in Scotland*. Ashbourne: Landmark, 2002.

Hills, Richard L. *The Life and Inventions of Richard Roberts, 1789-1864*. Ashbourne: Landmark Publishing, 2002.

Hills, Richard L. "Richard Roberts." In J.A.Cantrell and G. Cookson, eds. *Henry Maudslay and the Pioneers of the Machine Age*. Stroud: Tempus Publishing, 2002.

Hindle, Brooke. *Emulation and Invention*. New York University Press, 1981.

Hindle, Brooke and Lubar, Steven. *Engines of Change: The American Industrial Revolution, 1790–1860*. Washington, DC: Smithsonian Institution Press, 1986.

Hoke, Donald. *Ingenious Yankees: The Rise of the American System of Manufactures in the Private Sector*. New York: Columbia University Press, 1990.

Honour, Hugh and Fleming, John. *A World History of Art*. 7th ed. London: Laurence King Publishing, 2005.

Hounshell, David. *From the American System to Mass Production, 1800–1932: The Development of Manufacturing Technology in the United States*. Baltimore: Johns Hopkins University Press, 1984.

Howe, Daniel Walker. *What Hath God Wrought: The Transformation of America, 1815–1848*. Oxford: Oxford University Press, 2007.

Huberman, Michael. *Escape from the Market: Negotiating Work in Lancashire*. Cambridge, UK: Cambridge University Press, 1996.

Hunter, Louis C. *A History of Industrial Power in the United States, 1780–1930. Vol. I. Water Power in the Century of the Steam Engine*. Hagley, DE: Hagley Museum, 1979.

Hunter, Louis C. *A History of Industrial Power in the United States, 1780–1930. Vol. II. Steam Power*. Hagley, DE: Hagley Museum, 1985.

Imison, John. *The School of Arts; or, An Introduction to Useful Knowledge, Being a Compilation of Real Experiments and Improvements, in Several Pleasing Branches of Science*. London: the author, 1785.

Imison, John. *The School of Arts; or, An Introduction to Useful Knowledge, Being a Compilation of Real Experiments and Improvements, in Several Pleasing Branches of Science*. London: J. Murray and S. Highley, 1796.

Imison John. *A Treatise on the Mechanical Powers*. 2nd ed. London: H. Murray, 1794.

Ivins, William M. Jr. *Prints and Visual Communication*. London: Routledge & Kegan Paul, 1953.

Jackson, Benjamin Dayton. *An Attempt to Ascertain the Actual Dates of Publication of the Various Parts of the Rees Cyclopaedia*. London: Pewtress & Co, 1895.

Jamieson, D. R. "Introduction." In B. Woodcroft, *Alphabetical Index of Patentees of Invention*. 1854; reprint, London: Evelyn, Adams & Mackay, 1969.

Jefferson, Thomas. *Notes on the State of Virginia* (1785). In M.B. Folsom and S.D. Lubar, eds. *The Philosophy of Manufactures: Early Debates over Industrialisation in the United States*. Cambridge, MA: MIT Press, 1982.

Jeremy, David J., ed. *Henry Wansey and His American Journal, 1794*. Philadelphia: American Philosophical Society *Memoirs* vol. 82, 1970.

Jeremy, David J. "British and American Yarn Count Systems: An Historical Analysis." *Business History Review* 45 (1971). Reprinted in Jeremy, *Artisans, Entrepreneurs, and Machines: Essays on the Early Anglo-American Textile Industries, 1770–1840s*. Aldershot: Ashgate, 1998.

Jeremy, David J. "Innovation in American Textile Technology during the Early Nineteenth Century." *Technology and Culture* 14 (1973). Reprinted in Jeremy, *Artisans, Entrepreneurs, and Machines: Essays on the Early Anglo-American Textile Industries, 1770–1840s*. Aldershot: Ashgate, 1998.

Jeremy, David J. "British Textile Technology Transmission to the United States: The Philadelphia Region Experience, 1770–1820." *Business History Review* 47 (1973). Reprinted in Jeremy, *Artisans, Entrepreneurs, and Machines: Essays on the Early Anglo-American Textile Industries, 1770–1840s*. Aldershot: Ashgate, 1998.

Jeremy, David J. "Damming the Flood: British Government Efforts to Check the Outflow of Technicians and Machinery, 1780-1843." *Business History Review* 51 (1977). Reprinted in Jeremy, *Artisans, Entrepreneurs, and Machines: Essays on the Early Anglo-American Textile Industries, 1770–1840s*. Aldershot: Ashgate, 1998.

Jeremy, David J., *Transatlantic Industrial Revolution: The Diffusion of Textile Technologies between Britain and America, 1790–1830s*. Cambridge, MA: MIT Press, 1981.

Jeremy, David J. "Technological Diffusion: The Case of the Differential Gear." *Industrial Archaeology Review* 5 (1981). Reprinted in Jeremy, *Artisans, Entrepreneurs, and Machines: Essays on the Early Anglo-American Textile Industries, 1770–1840s*. Aldershot: Ashgate, 1998.

Jeremy, David J. "Immigrant Textile Machine Makers along the Brandywine, 1810-1820." *Textile History* 13 (1982). Reprinted in Jeremy, *Artisans, Entrepreneurs, and Machines: Essays on the Early Anglo-American Textile Industries, 1770–1840s*. Aldershot: Ashgate, 1998.

Jeremy, David J. *Technology and Power in the Early American Cotton Industry: James Montgomery, the Second Edition of His "Cotton Manufacture" (1840), and the 'Justitia' Controversy about Relative Power Costs*. Philadelphia: American Philosophical Society *Memoirs* vol. 189, 1990.

Jeremy, David J. "British and American Entrepreneurial Values in the Early Nineteenth Century: A Parting of the Ways?" In Robert A. Burchell (ed.), *The End of Anglo America: Historical Essays in the Study*

of Cultural Divergence. Manchester: Manchester University Press, 1991. Reprinted in Jeremy, *Artisans, Entrepreneurs, and Machines: Essays on the Early Anglo-American Textile Industries, 1770–1840s*. Aldershot: Ashgate, 1998.
Jeremy, David J. *Artisans, Entrepreneurs, and Machines: Essays on the Early Anglo-American Textile Industries, 1770–1840s*. Aldershot: Ashgate, 1998.
Jewish Encyclopaedia. A Descriptive Record of the History, Religion, Literature and Customs of the Jewish People from the Earliest times to the Present Day. 12 vols. New York: Funk & Wagnalls, 1925.
Jones, Caroline A. and Galison, Peter, eds. *Picturing Science Producing Art*. New York: Routledge, 1998.
Kemp, Martin. *The Science of Art: Optical Themes in Western Art from Brunelleschi to Seurat*. New Haven: Yale University Press, 1990.
Kemp, Martin. *Seen/Unseen: Art, Science, and Intuition from Leonardo to the Hubble Telescope*. Oxford: Oxford University Press, 2006.
Khan, B. Zorina. *The Democratization of Invention: Patents and Copyrights in American Economic Development, 1790–1920*. Cambridge, UK: Cambridge University Press, 2005.
Klingender, Francis D. *Art and the Industrial Revolution*. 1947. Rev ed. Arthur Elton (ed.), New York: Augustus M. Kelly, 1968.
Kubovy, Michael. *The Psychology of Perspective and Renaissance Art*. Cambridge University Press, 1986.
Lakwete, Angela. *Inventing the Cotton Gin: Machine and Myth in Antebellum America*. Baltimore: Johns Hopkins University Press, 2003.
Latham, James. *The Art of Managing the Patent Power Loom*. Bury: John Trimble, 1855.
Latour, Bruno. "Drawing Things Together." In Michael Lynch and Steve Woolgar, eds., *Representation in Scientific Practice*. Cambridge, MA: MIT Press, 1990.
Leavitt, Thomas W (ed.). *The Hollingworth Letters: Technical Change in the Textile Industry, 1826–1837*. Cambridge, MA: MIT Press, 1969.
Lee, Clive H. *A Cotton Enterprise, 1795–1840. A History Of M'Connel & Kennedy, Fine Cotton Spinners*. Manchester: Manchester University Press, 1972.
Leupold, Jacob. *Theatrum Machinarum*. 8 vols. Leipzig: author et al., 1774.
Lloyd-Jones, Roger and Lewis, M. J. "The Size of Firms in the Cotton Industry: Manchester 1815–41." *Economic History Review* 33 (1980).
Lloyd-Jones, Roger and Lewis, M. J. *Manchester and the Age of the Factory: The Business Structure of Cottonopolis in the Industrial Revolution*. London: Croom Helm, 1988.
Lozier, John W. *Taunton and Mason: Cotton Machinery and Locomotive Manufacture in Taunton, Massachusetts, 1811–1861*. New York: Garland, 1986.
Lowell Machine Shop. Lowell, Massachusetts: Lowell Machine Shop trade catalogue, 1898.
Lyman, Henry B. "Recollections." *TRISEDI* 1861.
MacLeod, Christine. *Inventing the Industrial Revolution: The English Patent System, 1660–1800*. Cambridge, UK: Cambridge University Press, 1988.
MacLeod, Christine. "James Watt, Heroic Invention and the Idea of the Industrial Revolution." In Maxine Berg and Christine Bruland, eds. *Technological Revolutions In Europe: Historical Perspectives*. Cheltenham: Edward Elgar, 1998.
MacLeod, Christine. *Heroes of Invention: Technology, Liberalism and British Identity, 1750–1914*. Cambridge, UK: Cambridge University Press, 2007.
McCorison, Marcus A., "John Holbrook." In John J. Duffy, Samuel B. Hand and Ralph H. Orth, eds. *The Vermont Encyclopedia*. Hanover, NH: University Press of New England, 2003.
McConnell, Anita. "William Farish." *ODNB*.
McLane Report: US Congress, Documents Relative to the Manufactures of the United States. Executive Document 308, 1st Session, 22nd Congress. 2 vols. Washington DC: Duff Green, 1833.

Manchester, Job, "Recollections," *TRISEDI* 1864.
Mann, Julia de L. *The Cloth Industry in the West of England from 1642 to1880*. Oxford: Clarendon Press, 1971.
Mann, Julia de L. "The Cardmaking Machine." *Textile History* 7 (1976).
Martin, Thomas. *The Circle of Mechanical Arts*. London: Richard Rees, 1813.
Meyer, David R. *Networked Machinists: High-Technology Industries in Antebellum America*. Baltimore: Johns Hopkins University Press, 2006.
Mitchell, W.J.Thomas, ed. *The Language of Images*. University of Chicago Press, 1980.
Moir, Esther. *The Discovery of Britain. The English Tourists, 1540–1840*. London: Routledge, 1964.
Mokyr, Joel. *The Gifts of Athena: Historical Origins of the Knowledge Economy*. Princeton: Princeton University Press, 2002.
Monge, Gaspard. *Géométrie Descriptive*. Paris: Baudouin, 1798.
Montgomery, James. *The Carding and Spinning Master's Assistant: or the Theory of Cotton Spinning*. Glasgow: John Niven Jr., 1832.
Montgomery, James. *A Practical Detail of the Cotton Manufacture of the United States of America*. Glasgow: John Niven Jr., 1840.
Morrison, Elting E. *From Know-How to Nowhere*. New York: Basic Books, 1974.
Moss, Michael S. "Robertson Buchanan." In *ODNB*.
Musson, Albert E. and Robinson, Eric. *Science and Technology in the Industrial Revolution*. Manchester: Manchester University Press, 1969.
Nelson, John R., Jr. *Liberty and Property: Political Economy and Policymaking in the New Nation, 1789–1812*. Baltimore: Johns Hopkins University Press, 1987.
Nelson, Thomas. *Practical Loom Fixing*. Charlotte, NC: Ray Printing Co., 1917.
Nicholson, John. *The Operative Mechanic and British Machinist*. 2 vols. 1st editions, London: 1825; Philadelphia: 1826.
North, Douglass C. *Institutions, Institutional Change and Economic Performance*. Cambridge, UK: Cambridge University Press, 1990.
O'Brien, Patrick. "The Micro Foundations of Macro Invention: The Case of the Reverend Edmund Cartwright." *Textile History* 28 (1997).
Oliver, Thomas. *Weaving and Designing*. 2nd ed. Galashiels: Craighead Brothers, 1911.
Oxford Dictionary of National Biography. Colin Matthew and Brian Harrison, eds. 60 vols. Oxford: Oxford University Press, 2004.
Oxford Dictionary of Quotations. 4th ed. Angela Partington (ed.). Oxford: Oxford University Press, 1992.
Oxford English Dictionary. 2nd ed. J.A. Simpson and E.S.C.Weiner, eds. Oxford: Clarendon Press, 1989.
Parliamentary Papers 1829 (332). Vol.III. "Report from the Select Committee [on]...the Law and Practice Relative to the Granting of Patents for Invention."
Parliamentary Papers 1851 (486). Vol. XVIII. "Report and Minutes of Evidence Taken before the Select Committee of the House of Lords [on] ... the Bills for the Amendment of the Law Touching Letters Patent for Inventions."
Pettigrew, Andrew, Howard Thomas and Richard Whittington, eds. *Handbook of Strategy and Management*. London: Sage, 2002.
Pinker, Steven. *How the Mind Works*. London: Penguin Books, 1998.
Polanyi, Michael. *The Tacit Dimension*. 1966; reprint, Gloucester, MA: Peter Smith.
Porter, Michael E. *Competitive Advantage: Creating and Sustaining Superior Performance*. New York: Free Press, 1985.

Potash, P. Jeffrey. *Vermont's Burned Over District: Patterns of Community Development and Religious Activity, 1761–1850*. Brooklyn, NY: Carlson Publishing Co., 1991.

Prude, Jonathan. *The Coming of Industrial Order: Town and Factory Life in Rural Massachusetts, 1810–1860*. Cambridge, UK: Cambridge University Press, 1983.

Pursell, Carroll W. Jr. *Early Stationary Steam Engines in America: A Study in the Migration of a Technology*. Washington, DC: Smithsonian Institution, 1969.

Radcliffe, William. *Origin of the New System of Manufacture Commonly Called Power-Weaving*. Stockport: James Lomax, 1828.

Raum, Hans. "Joshua Bates." In John J. Duffy, Samuel B. Hand, and Ralph H. Orth, eds. *The Vermont Encyclopedia*. Hanover, NH: University Press of New England, 2003.

Rees, Abraham, ed. *The Cyclopaedia; or, Universal Dictionary of Arts, Sciences, and Literature*. 39 vols. text, 6 vols. plates, London: Longman et al., 1802-20; 41 vols. text, 6 vols. plates, Philadelphia: Samuel F. Bradford et al., 1806–1820.

Ribeiro, Aileen. *A Visual History of Costume: The Eighteenth Century*. London: Batsford, 1983.

Rink, Evald. *Technical Americana: A Checklist of Technical Publications Printed before 1831*. New York: Kraus, 1981.

Roland de la Platière, Jean Marie. *L'Art du Fabricant de Velours de Coton*. Paris, 1780.

Rolt, Lionel T.C. *Tools for the Job*. Rev. ed. London: Her Majesty's Stationery Office, 1985.

Rosenberg, Nathan and Vincenti, Walter. *The Britannia Bridge: The Generation and Diffusion of Technological Knowledge*. Cambridge, MA: MIT Press, 1978.

Rosenbloom, Joshua L., "Path Dependence and the Origins of Cotton Textile Manufacturing in New England." In Douglas A. Farnie and David J. Jeremy, eds. *The Fibre that Changed the World: The Cotton Industry in International Perspective, 1600–1990s*. Oxford: Oxford University Press for Pasold Research Fund, 2004.

Schumpeter, Joseph A. *The Theory of Economic Development: An Inquiry into Profits, Capital, Credit, Interest, and The Business Cycle*. Cambridge, MA: Harvard University Press, 1934; reprint. Cambridge, MA: Harvard University Press, 1961.

Selleck, James E. *Bicentennial History of Union Lodge # 2 F & AM*. Middlebury, VT: Union Lodge Bicentennial Committee, 1993.

Sherman, Michael, Sessions, Gene, and Potash, P. Jeffrey. *Freedom and Unity: A History of Vermont*. Barre, VT: Vermont Historical Society, 2004.

Sinclair, Bruce. *Philadelphia's Philosopher Mechanics: A History of the Franklin Institute, 1824-1865*. Baltimore: Johns Hopkins University Press, 1974.

Singleton, John. "The Lancashire Cotton Industry, the Royal Navy, and the British Empire, c.1700–1960." In Douglas A. Farnie and David J. Jeremy, eds. *The Fibre that Changed the World: The Cotton Industry in International Perspective, 1600–1990s*. Oxford: Oxford University Press for Pasold Research Fund, 2004.

Smeaton, John. *Experimental Enquiry Concerning the Natural Powers of Wind and Water to Turn Mills and Other Machines*. 2nd ed. London: L. & J. Taylor, 1796.

Smiles, Samuel. *Self Help*. London: Murray, 1859.

Smiles, Samuel, ed. *James Nasmyth. An Autobiography*. London: John Murray, 1883.

Smith, Adam. *An Inquiry into the Nature and Causes of the Wealth of Nations*. 1776. Edited by Edwin Cannan. New York: Modern Library, 1937.

Smith, Denis, ed. *Perceptions of Great Engineers: Fact and Fantasy*. London: Science Museum, 1994.

Smith, Henry Perry, ed. *History of Addison County, Vermont*. Syracuse, NY: D. Mason & Co, 1886.

Smith, James. *The Mechanic, or Compendium of Practical Inventions*. 2 vols. Liverpool: Henry Fisher, 1816.

Smith, Merritt Roe. *Harpers Ferry Armory and the New Technology: The Challenge of Change*. Ithaca, New York: Cornell University Press, 1977.

Snyder, Joel. "Picturing Vision." In W.J. Thomas Mitchell, ed. *The Language of Images*. University of Chicago Press, 1980.

Stapleton, Darwin H., ed. *The Engineering Drawings of Benjamin Henry Latrobe*. New Haven: Yale University Press, 1980.

Stapleton, Darwin H., *The Transfer of Early Industrial Technologies to America*. Philadelphia: American Philosophical Society, *Memoirs*, vol. 177, 1987.

Stevens, Edward W. *The Grammar of the Machine: Technical Literacy and Early Industrial Expansion in the United States*. New Haven: Yale University Press, 1995.

Sunderland, David. *Social Capital, Trust and the Industrial Revolution, 1780–1880*. London: Routledge, 2007.

Swift, Samuel. *History of the Town of Middlebury in the County of Addison, Vermont: To Which Is Prefixed a Statistical and Historical Account of the County*. Middlebury, VT: A. H. Copeland, 1859.

Sylvester, Nathaniel Bartlett. *History of Rensselaer County, New York*. Philadelphia: Everts & Peck, 1880.

Tann, Jennifer. *The Development of the Factory*. London: Cornmarket, 1970.

Taussig, Frank W. *The Tariff History of the United States*. 8th ed. New York: G. P. Putnam's Sons, 1931.

Taylor, George Rogers. *The Transportation Revolution, 1815–1860*. New York: Holt, Rinehart & Winston, 1951.

Taylor, John T. *Cotton Weaving and Designing*. London: Longmans, Green & Co, 1893.

The Marble Border of Western New England: Its Geology and Marble Development in the Present Century. Middlebury, VT: Middlebury Historical Society, 1885.

Thomson, James. "Cotton" (1808). In Abraham Rees, ed. *The Cyclopaedia; or, Universal Dictionary of Arts, Sciences, and Literature*. 39 vols. text, 6 vols. plates, London: Longman et al., 1802–20; 41 vols. text, 6 vols. plates, Philadelphia: Samuel F. Bradford et al., 1806–1820.

Thomson, James. "Cotton Manufacture" (1808). In Abraham Rees, ed. *The Cyclopaedia; or, Universal Dictionary of Arts, Sciences, and Literature*. 39 vols. text, 6 vols. plates, London: Longman et al., 1802–20; 41 vols. text, 6 vols. plates, Philadelphia: Samuel F. Bradford et al., 1806–1820.

Thomson, James. "Manufacture of Cotton" (1812). In Abraham Rees, ed. *The Cyclopaedia; or, Universal Dictionary of Arts, Sciences, and Literature*. 39 vols. text, 6 vols. plates, London: Longman et al., 1802–20; 41 vols. text, 6 vols. plates, Philadelphia: Samuel F. Bradford et al., 1806–1820.

Thomson, James K.J. "Transferring the Spinning Jenny to Barcelona: An Apprenticeship in the Technology of the Industrial Revolution." *Textile History*, 34 (2003).

Thomson, Ross. *Structures of Change in the Mechanical Age: Technological Innovation in the United States, 1790–1865*. Baltimore: John's Hopkins University Press, 2009.

Torrens, Hugh. "Jonathan Hornblower (1753–1815): An Historical Analysis." In Denis Smith, ed.

Torrens, Hugh. "Robert Bakewell." *ODNB*.

Tovey, John. *The Technique of Weaving*. London: B. T. Batsford, 1965; new edition 1975.

Tsoukas, Haridimos. "Do We Really Understand Tacit Knowledge?" In M. Easterby-Smith and Marjorie A. Lyles, eds. *The Blackwell Handbook of Organisational Learning and Knowledge Management*. Oxford: Blackwell Publishing, 2003.

Tucker, Barbara M. *Samuel Slater and the Origins of the American Textile Industry, 1790–1860*. Ithaca, New York: Cornell University Press, 1984.

Tully, Christopher. "Spinning Jenny." *Pennsylvania Magazine; or American Monthly Museum* 1 (1775).

Tyler, Alice Felt. *Freedom's Ferment: Phases of American Social History from the Colonial Period to the Outbreak of the Civil War*. Minneapolis: University of Minnesota Press, 1944.

United States. Census Office. *Digest of Accounts of Manufacturing Establishments in the United States.* Washington DC: Gales and Seaton, 1823.

United States, Department of Commerce, Patent Office. *The Story of the United States Patent Office.* Washington DC: Government Printing Office, 1965.

United States, Patent Office. *A List of Patents Granted by the United States from April 10, 1790, to December 31, 1836.* Washington DC: Patent Office, 1872.

Ure, Andrew. *The Cotton Manufacture of Great Britain.* 2 vols. London: Charles Knight, 1836.

Ure, Andrew. *Dictionary of Arts, Manufactures, and Mines.* London: Longman, 1839.

Usher, Abbott Payson. *A History of Mechanical Inventions.* New York: McGraw-Hill Book Co., 1929; reprint, Cambridge, MA: Harvard University Press, 1954.

Van Wijk, Raymond; Van Den Bosch, Frans A.J.; and Volberda, Henk W. "Knowledge and Networks." In M. Easterby-Smith and Marjorie A. Lyles, eds. *The Blackwell Handbook of Organisational Learning and Knowledge Management.* Oxford: Blackwell Publishing, 2003.

Vermont. *Acts and Laws Passed by the Legislature of the State of Vermont: At their Session at Woodstock, on the Second Thursday of October, One Thousand Eight Hundred and Seven.* Randolph, VT: Sereno Wright, State, 1807.

Vermont. *Acts and Laws Passed by the Legislature of the State of Vermont: At their Session at Montpelier, on the Second Thursday of October, One Thousand Eight Hundred and Eight.* Bennington, VT: Anthony Haswell, 1808.

Vermont. *Acts and Laws Passed by the Legislature of the State of Vermont: At their Session at Montpelier, on the Second Thursday of October, One Thousand Eight Hundred and Ten.* Danville, VT: Ebenezer Eaton, 1810.

Vincenti, Walter G. *What Engineers Know and How They Know It: Analytical Studies from Aeronautical History.* Baltimore: Johns Hopkins University Press, 1990.

Wadsworth, Alfred P. and Mann, Julia de Lacy. *The Cotton Trade and Industrial Lancashire, 1600-1780.* Manchester: Manchester University Press, 1931.

Wallace, Anthony F. C. and Jeremy, David J. "William Pollard and the Arkwright Patents." *William & Mary Quarterly* 3rd ser. 34 (1977).

Wallace, Anthony F. C. *Rockdale: The Growth of an American Village in the Early Industrial Revolution.* New York: Alfred A. Knopf, 1978.

Wallis, Jane. *Joseph Wright of Derby.* Derby: City of Derby Museum and Art Gallery, 1997.

Ware, Caroline F. *The Early New England Cotton Manufacture: A Study in Industrial Beginnings.* Boston: Houghton Mifflin, 1931; reprint. New York: Russell & Russell, 1966.

Weeks, John M. *History of Salisbury, Vermont.* Middlebury, VT: A. H. Copeland, 1860.

Weible, Robert, ed. *Essays from the Lowell Conference in Industrial History, 1982 and 1983.* North Andover, MA: Museum of American Textile History, 1985.

White, George. *A Practical Treatise on Weaving, by Hand and Power Looms.* London: Simpkin, Marshall & Co., n.d. but 1846.

White, George S. *Memoir of Samuel Slater, Father of American Manufactures. Connected with a History of the Rise and Progress of the Cotton Manufacture in England and America with Remarks on the Moral Influence of Factories in the United States.* Philadelphia: 46, Carpenter Street, 1836.

White, James. *Memoir on a New System of Cog or Toothed Wheels.* Manchester: Executors of S. Russell, 1819.

White, James. *A New Century of Inventions, Being Designs and Descriptions of One Hundred Machines Relating to Arts, Manufactures and Domestic Life.* Manchester: Leech & Cheetham, 1822.

White, John H. *Cincinnati Locomotive Builders, 1845–1868.* Washington, DC: Smithsonian Institution Press, 1965.

Wood, George H. *The History of Wages in the Cotton Trade during the Past Hundred Years*. London: Sherratt & Hughes, 1910.

Woodcroft, Bennet. *Alphabetical Index of Patentees of Invention*. London: Patent Office, 1854; reprint, London: Evelyn, Adams & Mackay, 1969.

Woolrich, A. P. "John Farey, Jr., Technical Author and Draughtsman: His Contribution to Rees's *Cyclopaedia*." *Industrial Archaeology Review* 20 (1998).

Woolrich, A.P. "Abraham Rees." *ODNB*.

Woolrich, A.P. "John Farey." *ODNB*.

Woolrich, A.P. "London Engineering Industry." In John Cantrell and Gillian Cookson, eds. *Henry Maudslay and the Pioneers of the Machine Age*. Stroud: Tempus Publishing, 2002.

York, Neil Longley. *Mechanical Metamorphosis: Technological Change in Revolutionary America*. Westport, CT: Greenwood Press, 1985.

Index

☆ ☆ ☆ ☆ ☆ ☆ ☆

W

Y

FIGURE 1.3 View of Middlebury, by George Heriot (1766–1844), ca. 1810, watercolor ($6^{1}/2$ x 9 inches, or 16.5 x 23 cm). Painted by a Canadian artist, this painting shows the buildings around the falls during Markham's youth. The large building on the right with four chimneys may be the Markham family's home and tavern.

Courtesy of the Sheldon Museum.

FIGURE 2.1 View of Boston Manufacturing Co. mills at Waltham, Massachusetts, by Isaac Ebenezer Markham, ca. 1819–1821, watercolor sketch (26 x 42.5 cm). This is the earliest known view of the Boston Manufacturing Co. mills at Waltham.

The view is taken from the north side of the mills, probably from near a line of single-storey structures that are visible in an 1830 painting viewing the mills from the south side, across the Charles River.

The building with the cupola (*to the right of the picture*) is the first mill; built 1813–1814, it measured 90 feet long by 45 feet wide and was four storeys above the basement. The second mill (*to the left of the picture*), built 1816–1818, measured 150 feet long by 40 feet wide. Its external but adjoining stair tower was a novel feature that was repeated on the south side of the mill. The 3,584 spindles in this mill became the unit of measurement of one mill power in northern New England. See Candee, "Architecture and Corporate Power in the Early Waltham System"; Jeremy, *Transatlantic Industrial Revolution*, 97.

Courtesy of the Sheldon Museum.

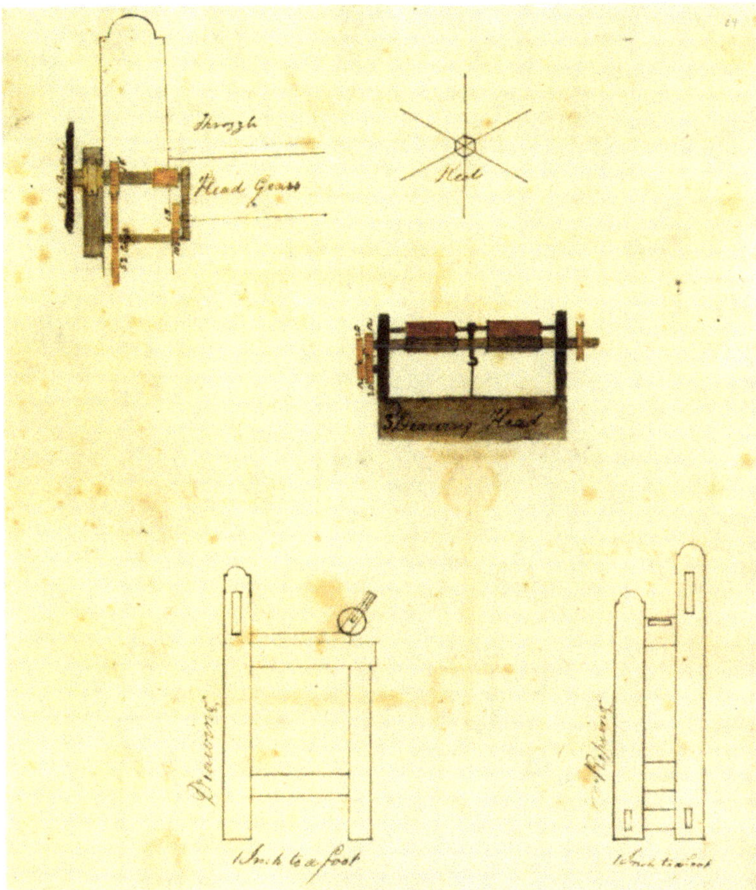

FIGURE 6.47 Drawing head (front and side elevations) for drawing and roving machines and for throstle spinning; framing for drawing and roving machines. Scale drawings, elevations.

Catalogue number: IEM 2.7

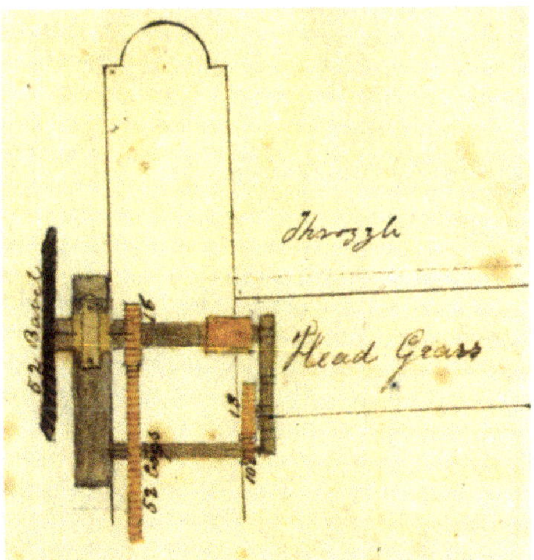

Detail of drawing head for throstle spinning.

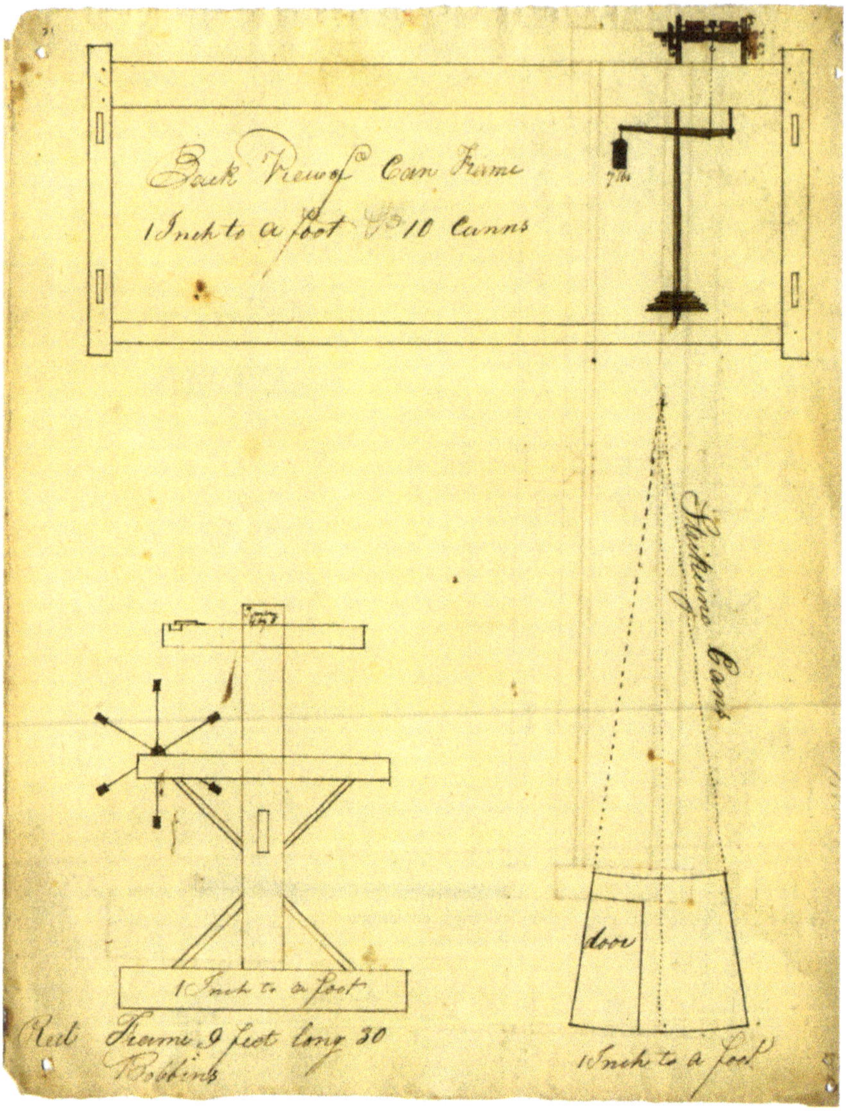

FIGURE 6.50 Roving-can frame, components, and reeling frame. Scale drawings, elevations.

Catalogue number: IEM 4.2

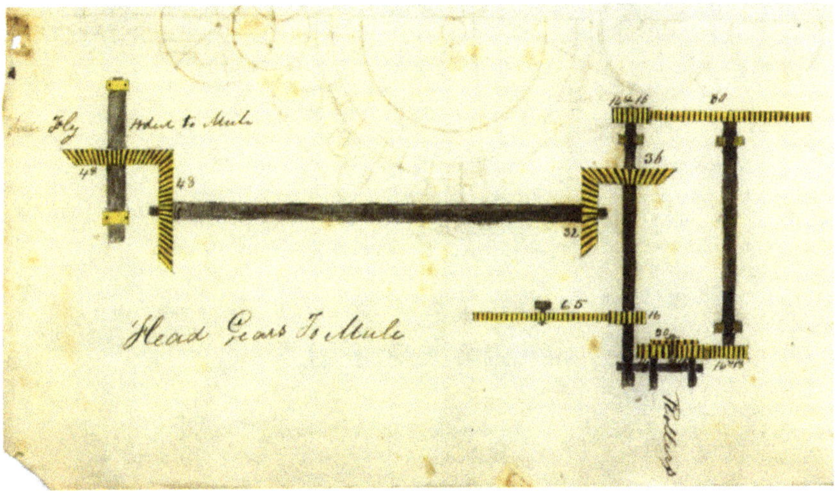

FIGURE 6.61 **"Head Gears to Mule," plan.**

Catalogue number: IEM 8.3

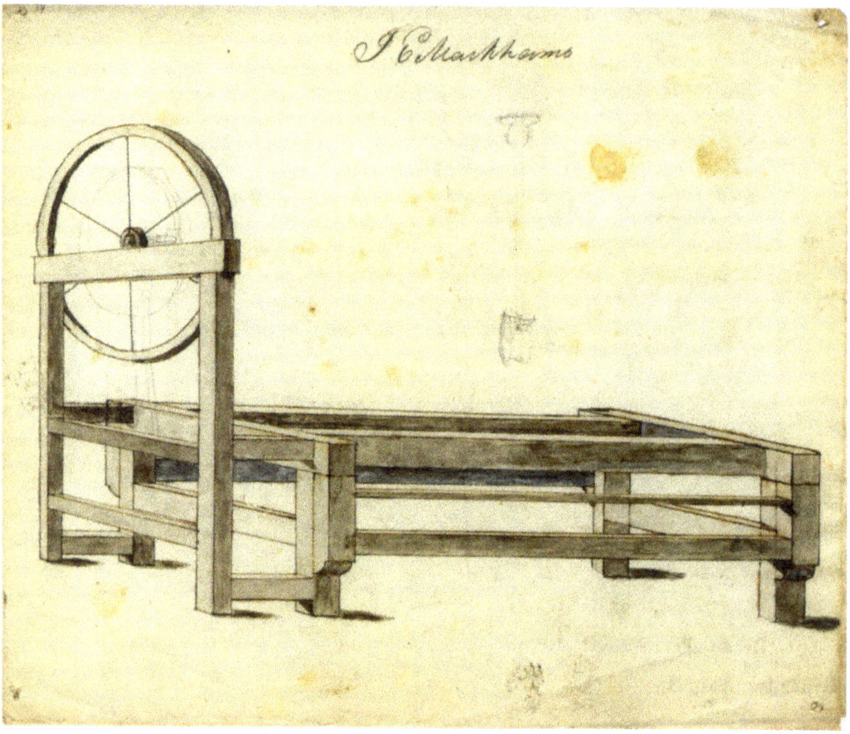

FIGURE 6.65 "I E Markhams" Spinning machine, headstock, and driving wheel. Possibly the "fly wheel spinning" machine seen at Figure 6.69. Perspective sketch.

Catalogue number: IEM 10.1

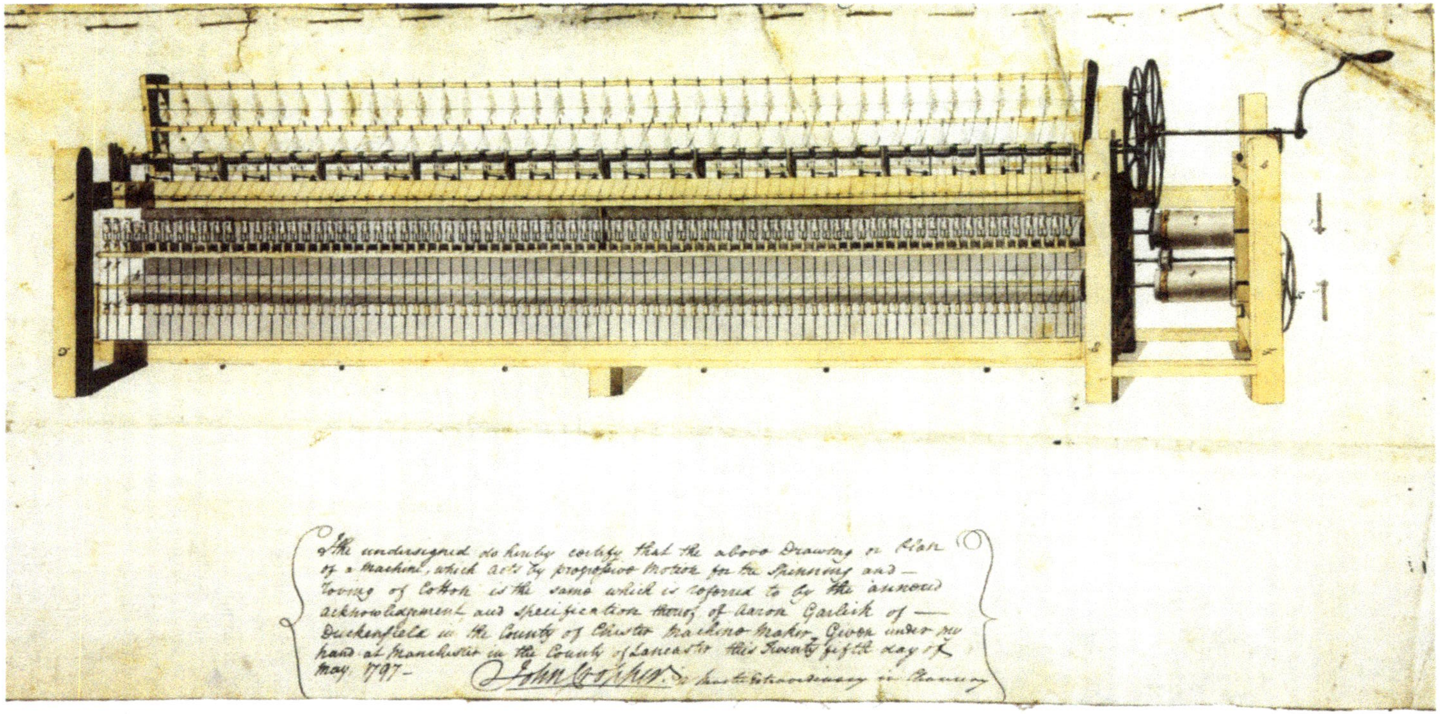

FIGURE A3.12 Aaron Garlick, English patent for the double speeder, 7 February 1797; perspective view; pen and wash on linen (?) (33.5 x 67.5 cm). National Archives C 210/54. GB Patent no. 2166. Not published in *RAMA*. See chapter 6, section 4, pp. 228–231, for technical details.

Courtesy of the National Archives, London.

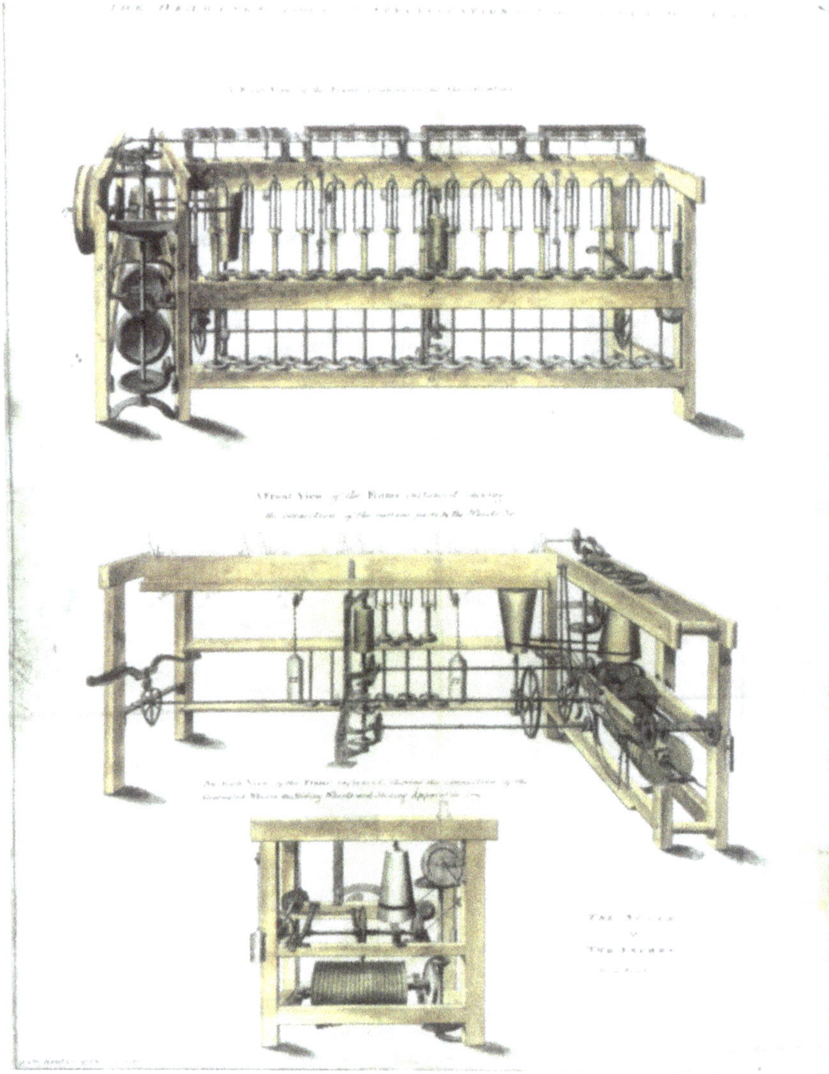

FIGURE A3.13 Joseph Raynor, English patent for the double speeder, 1 January 1813; perspective views; pen and wash on skin (76 x 58.2 cm). National Archives C 54/9345.

GB Patent no. 3633. Patent specification published in *RAMA* in November 1814, but not the drawing. This was "unavoidably postponed" (p.384) until the following volume but never appeared there either. Offering a promotional comment on his patent in *RAMA* 23 (November 1813) 330, Raynor noted, "Among the minor, but not unimportant, advantages of this frame may be mentioned,— the facility with which it works—the little labour it requires in tending—and the almost entire absence of waste." For the operating principles of this machine, see chapter 6, section 4, pp. 228–231.

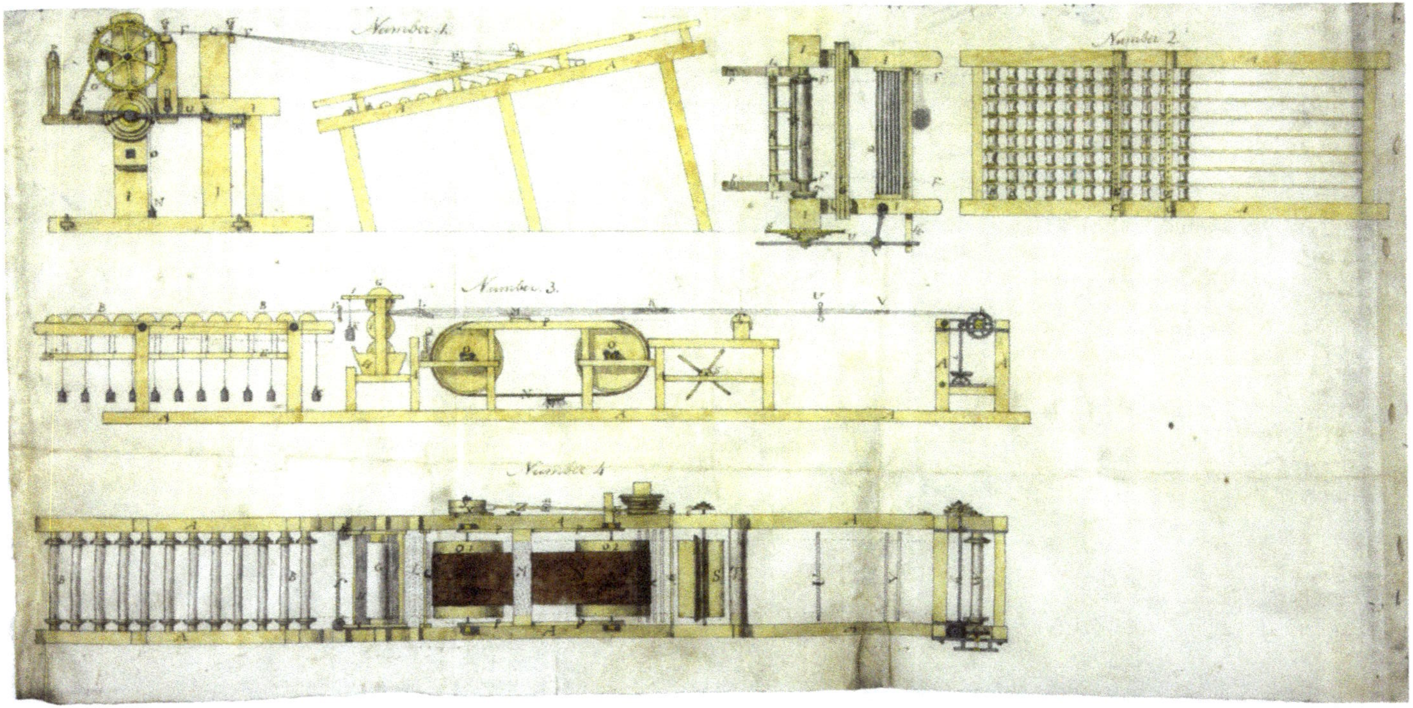

FIGURE A3.14 Thomas Johnson, English patent for warping machine and for dresser, 28 February 1803. *Top*, warping machine, side elevation at the left; plan at right. *Bottom*, dresser, side elevation above, plan below. Pen and wash on skin (36 x 71.5 cm). National Archives C 210/94. GB Patent no. 2684. Not published in *RAMA*.

Figures 1 and 2 show, respectively, the side elevation and plan of a warping frame. Figures 3 and 4, respectively, the side elevation and plan of a dresser. William Radcliffe, the Stockport cotton manufacturer, claimed in his memoirs that he invented the dresser and that he took out the two related patents (here Figs. A3.14 and A 3.15) under the name of his employee, Thomas Johnson, to conceal from foreign competitors his interest in the invention—thereby, he hoped, preventing them from learning about it. His solicitor advised him to take out secret patents but John Peel, Robert's brother, thought this too troublesome and expensive. His solicitor then suggested the ruse of using his workman's name. In addition Radcliffe went to Henry Wyatt, editor of the *Repertory of Arts*, and persuaded him not to print his patent specifications (Radcliffe, *Origin*, 25–27).

Courtesy of the National Archives, London.

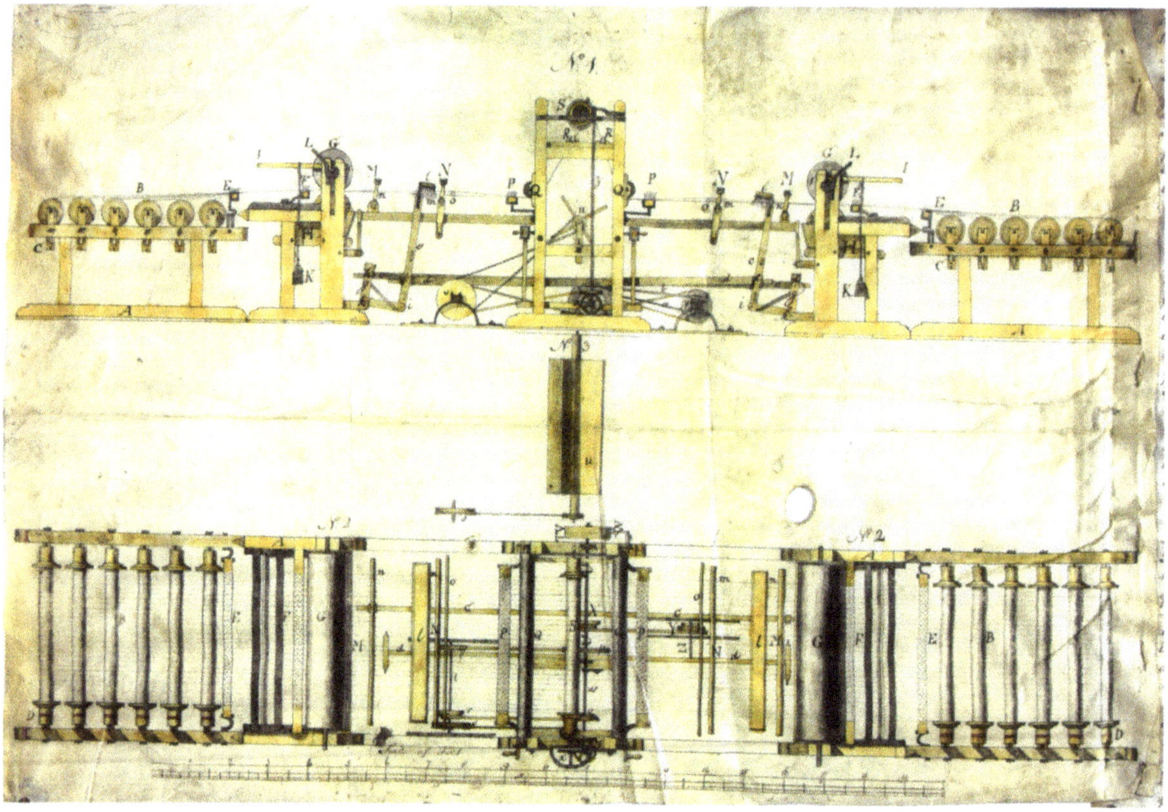

FIGURE A3.15 Thomas Johnson, English patent for the dresser, 2 June 1804. *Top*, side elevation; *bottom*, plan. Pen and wash on skin (40.7 x 59.5 cm). National Archives C 210/104. GB Patent no. 2771. Not published in *RAMA*.

Figures 1 and 2 show, respectively, the side elevation and plan of the dresser; Figure 3, a perspective view of the fan. The scale suggests it might be about 30 feet long. See chapter 6, section 6, and Appendix 3, especially Table A3.2, for other technical details.

Francis Cabot Lowell, the Boston merchant who introduced vertically integrated cotton manufacturing into northern New England at Waltham in 1813, on his visit to Britain in 1810–1812, went to Manchester where, then or later, he secured a copy of this patent because it was used by his chief engineer, Paul Moody. Moody modified the design to suit local circumstances in New England and then secured an American patent for it (Appleton, *Introduction of the Power Loom*, 7–9). Paul Moody's dresser patent specification (no drawing) in Boston Manufacturing Co. versus Jonathan Fisk et al., May 1821 file, in the United States, First Circuit Court records.

Courtesy of the National Archives, London.

FIGURE A3.18 William Horrocks, English patent for the power loom, 31 July 1813. Sheet 2, Figure 3, rear elevation. Figure 4, side elevation. Figure 5, opposite-side elevation. Figure 6 is an enlarged side elevation of the variable batten speed motion, Horrocks's key invention, also seen in Figure 5. Pen and wash on skin (25 x 68 cm). National Archives C 210/131. GB Patent no. 3728. Line drawing copy published in *RAMA* in June 1814.

www.ingramcontent.com/pod-product-compliance
Lightning Source LLC
La Vergne TN
LVHW081936200926
848484LV00009B/2445